Distributed Algorithms and Frameworks of Mobile Networks

Distributed Algorithms and Frameworks of Mobile Networks

Edited by **Grant Harper**

Willford Press
New York

Published by Willford Press,
118-35 Queens Blvd., Suite 400,
Forest Hills, NY 11375, USA
www.willfordpress.com

Distributed Algorithms and Frameworks of Mobile Networks
Edited by Grant Harper

International Standard Book Number: 978-1-68285-170-8 (Hardback)

Printed in China.

Contents

Preface

Distributed algorithms and frameworks are applied to solve various standard problems in telecommunications. This book is a compilation of chapters that discuss some of the vital concepts of this field like complexity analysis of algorithms for mobile environments, security and privacy of mobile networks, algorithms and modeling for wireless networks, architectures and protocols of ad-hoc and sensor networks, etc. It consists of contributions made by international experts, which makes this book a resource guide for students and researchers alike.

This book is a result of research of several months to collate the most relevant data in the field.

When I was approached with the idea of this book and the proposal to edit it, I was overwhelmed. It gave me an opportunity to reach out to all those who share a common interest with me in this field. I had 3 main parameters for editing this text:

1. Accuracy – The data and information provided in this book should be up-to-date and valuable to the readers.
2. Structure – The data must be presented in a structured format for easy understanding and better grasping of the readers.
3. Universal Approach – This book not only targets students but also experts and innovators in the field, thus my aim was to present topics which are of use to all.

Thus, it took me a couple of months to finish the editing of this book.

I would like to make a special mention of my publisher who considered me worthy of this opportunity and also supported me throughout the editing process. I would also like to thank the editing team at the back-end who extended their help whenever required.

Editor

1

FINE SYNCHRONIZATION THROUGH UWB TH-PPM IMPULSE RADIOS

Moez Hizem[1] and Ridha Bouallegue[2]

[1]6'Tel Research Unit, Higher School of Communications of Tunis, Sup'Com, Tunisia
moezhizem@yahoo.fr
[2]Sup'Com, Tunisia
ridhabouallegue@gmail.com

ABSTRACT

In this paper, a novel fine timing algorithm has been tested and developed to synchronize Ultra-Wideband (UWB) signals with pulse position modulation (PPM). By applying this algorithm, we evaluate timing algorithms in both data-aided (DA) and non-data-aided (NDA) modes. Based on correlation operations, our algorithm remains operational in practical UWB settings. The proposed timing scheme consists of two complementary floors or steps. The first floor consists on a coarse synchronization which is founded on the recently proposed acquisition scheme based on dirty templates (TDT). In the second floor, we investigate a new fine synchronization algorithm which gives an improved estimate of timing offset. Simulations confirm performance improvement of our timing synchronization compared to the original TDT algorithm in terms of mean square error.

KEYWORDS

Time-Hopping (TH); pulse position modulation (PPM); estimation; timing acquisition; synchronization; performance; ultra-wideband (UWB)

1. INTRODUCTION

UWB radios have attracted increasing importance in short-range high-speed wireless communications due to their potential to offer high user capacity with low-complexity and low-power transceivers [1]-[3]. Most of these benefits initiate from the distinctive characteristics inherent to UWB wireless transmissions [3]. However, to harness these benefits, one of the most critical challenges is to obtain an accurate timing synchronization and more specifically timing offset estimation. The complexity of which is accentuated in UWB, compared to narrowband systems, owing to the fact that information bearing waveforms are impulse-like and have low amplitude. In addition, the difficulty of timing UWB signals is furthermore induced by the dense multipath channel that remains unknown at the synchronization step. These reasons give explanation why synchronization has obtained so much importance in UWB research [4]-[7].

Typically, pulse position modulation (PPM), pulse amplitude modulation or on/off keying (OOK) is employed. PPM modulation transmits pulses with constant amplitude and encodes the information according to the position of the pulse, while PAM and OOK use the amplitude for this purpose. Moreover, PPM is regularly implemented to reduce transceiver complexity in UWB systems. But unlike pulse amplitude modulation (PAM) applied in the context of UWB systems, the difficulty of accurate synchronization is accentuated in PPM UWB systems owing to the fact that information is transmitted by the shifts of the pulse positions.

In the last years, numerous timing algorithms have been studied for UWB impulse radios under various operating environments (see e.g., [4], [5], [7]-[16]). In [10], a blind synchronization algorithm that takes advantage of the shift invariance structure in the frequency domain is proposed. An accurate signal processing model for a Transmit-reference UWB (TR-UWB) system is given in [11]. The model considers the channel correlation coefficients that can be estimated blindly. In [12], the authors proposed a code-assisted blind synchronization (CABS) algorithm which relies on the discriminative nature of both the time hopping code and a well-designed polarity code. Timing with dirty templates (TDT), which is the starting point of this paper, was introduced in [15] for rapid synchronization and was developed in [16] for PPM-UWB signals with direct sequence (DS) and/or time hopping (TH) spreading. This technique is based on correlating adjacent symbol-long segments of the received waveform. Except [16], all these timing algorithms are developed for PAM-UWB signals. Since their operations greatly rely on zero-mean property of PAM, these presented timing algorithms are not appropriate to PPM-UWB signals.

In this paper, we develop a fine synchronization algorithm for UWB TH-PPM signals. To improve the synchronization performance of the original TDT developed in [16], our contribution will be to implement a new fine synchronization step and insert it after the coarse one, which is original TDT. The principle of our fine synchronization algorithm is to make a fine search to find the exact moment of pulse beginning (fine estimation of timing offset). This is realized by correlating two consecutive symbol-long segments of the received waveform but this time in an interval that corresponds to the number of frames included in one data symbol. Simulation results show that this new synchronizer using TDT can realize a lower mean square error (MSE) than the original TDT in both non-data-aided (NDA) and data-aided (DA) modes.

The rest of this paper is organized as follows. In Section 2, we outline the PPM-UWB system model. Section 3 describes first the TDT algorithm in a single-user links and upper bounds on the mean square error of TDT estimators in both NDA and DA modes are derived. Then, we defined and developed our fine synchronization algorithm. In Section 4, simulations are yielded to corroborate our analysis. Conclusions are given in Section 5.

2. System Model for Single-User Links

In UWB impulse radios, each information symbol is transmitted over a T_s period that consists of N_f frames [1]. During each frame of duration T_f, a data-modulated ultra-short pulse p(t) with duration $T_p \ll T_f$ is transmitted from the antenna source. With PPM modulation (see e.g., [17], [18]), the transmitted signal in single-user links is described by the following model:

$$v(t) = \sqrt{\varepsilon}\sum_{k=0}^{+\infty}\sum_{i=0}^{N_f-1} p(t - iT_f - c_{th}(i)T_c - kT_s - d_i\delta) \tag{1}$$

where ε is the energy per pulse, $d_i \in (0,1)$ represents the i-th information bit transmitted, and δ is the time shift associated with binary PPM. User separation is realized with pseudo-random TH-codes $c_{th}(i)$, which time-shift the pulse positions at multiples of the chip duration T_c [1]. In this paper, we focus on a single user link and treat multi-user interference (MUI) as noise.

The transmitted signal propagates through the multipath channel with impulse response:

$$g(t) = \sum_{l=0}^{L-1} \alpha_l \delta(t - \tau_l) \tag{2}$$

where $\{\alpha_l\}_{l=0}^{L-1}$ and $\{\tau_l\}_{l=0}^{L-1}$ are amplitudes and delays of the L multipath elements, respectively. The channel is assumed quasi-static and among $\{\tau_l\}_{l=0}^{L-1}$, τ_0 represents the propagation delay of the channel.

Then, the received waveform is given by:

$$r(t) = \sqrt{\varepsilon}\sum_{l=0}^{L}\alpha_l\sum_{k=0}^{+\infty}p_T(t - kT_s - \tau_{l,0} - \tau_0 - d_i\delta) + \eta(t) \quad (3)$$

where $\tau_{l,0}$ is arbitrary reference at the receiver representing the delay relative to the arrival moment of the first pulse, $\eta(t)$ is the additive noise and $p_T(t)$ denotes the received symbol waveform as:

$$p_T(t) = \sum_{i=0}^{N_f-1} p(t - iT_f - c_{th}(i)T_c) * g(t + \tau_0) \quad (4)$$

where * indicates the convolution operation. We define the timing offset as $\Delta\tau \coloneqq \tau_{l,0} - \tau_0$. Let us suppose that $\Delta\tau$ is in the range of $[0, T_s)$ and we will show in the rest of this paper that this assumption will not affect the timing synchronization. Let $p_R(t)$ the overall received symbol-long waveform defined as follows:

$$p_R(t) = \sum_{l=0}^{L}\alpha_l p_T(t - \tau_{l,0}) \quad (5)$$

Using (5), the received waveform in (3) becomes:

$$r(t) = \sqrt{\varepsilon}\sum_{k=0}^{+\infty} p_R(t - kT_s - \tau_0 - d_i\delta) + \eta(t) \quad (6)$$

In the next section, we will develop a low-complexity fine synchronization approach using TDT synchronizer in order to find the desired timing offset. It will be evaluated in both non-data-aided (NDA) and data-aided (DA) modes, without acquaintance of the transmitted sequence and the multipath channel.

3. Fine Synchronization Approach for UWB TH-PPM Impulse Radios

As mentioned previously, our proposed timing scheme consists of two complementary floors or steps. The first is based on a coarse (or blind) synchronization that is timing with dirty templates (TDT) developed in [16]. The second consists of a new fine synchronization algorithm which further improves the timing offset found in the first floor. We will first give an outline of the TDT approach for UWB TH-PPM impulse radios to better understand the overall timing synchronization suggested in this paper.

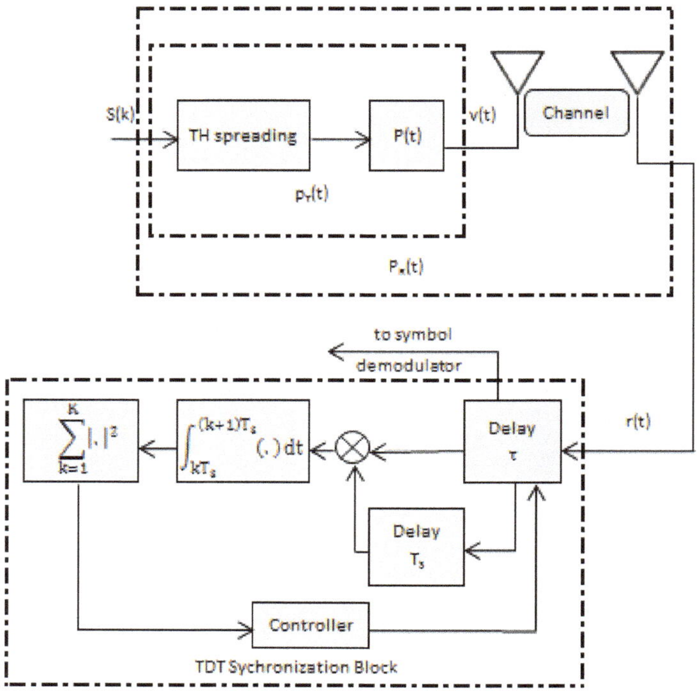

Figure 1. Description of our model with first stage synchronization

3.1. TDT for UWB TH-PPM Impulse Radios

The basic idea behind TDT is to find the maximum of square correlation between pairs of successive symbol-long segments. The description of our system model with first stage synchronization (TDT) is illustrated in Fig.1. First, we will start by analyzing $\tilde{\tau}_0$ which represents estimate offset of τ_0 with deriving upper bounds on their mean square error (MSE) in both non-data-aided (NDA) and data-aided (DA) modes.

For UWB TH-PPM systems, a correlation between the two adjacent symbol-long segments $r_k(t) = r(t + kT_s)$ and $r_{k+1}(t) = r(t + (k+1)T_s)$ is achieved [16]. Let $x(k;\tau)$ the value of this correlation $\forall k \in [1, +\infty)$ and $\tau \in [0, T_s)$,

$$x(k;\tau) := \int_0^{T_s} r_{k+1}(t;\tau)\,\tilde{r}_k(t;\tau)dt$$

$$\tilde{r}_k(t;\tau) := r_k(t+\delta;\tau) - r_k(t-\delta;\tau) \tag{7}$$

By applying the Cauchy-Schwartz inequality and exploiting the statistical properties of the signal and noise [16], the mean square of the samples in (7) is given by:

$$E_{s,\zeta}\{x^2(k;\tau)\} \approx \frac{1}{2}\left(\varepsilon_R^2 - 3\varepsilon_A(\tilde{\tau}_0)\varepsilon_B(\tilde{\tau}_0) + 2\sigma_\zeta^2\right) \tag{8}$$

where $\varepsilon_A(\tau) := \varepsilon \int_{T_s-\tau}^{T_s} p_R^2(t)dt$, $\varepsilon_B(\tau) := \varepsilon \int_0^{T_s-\tau} p_R^2(t)dt$, and σ_ζ is the power of $\zeta(k;\tau)$ corresponding to the superposition of three noise terms [16] and can be approximated as an additive white Gaussian noise (AWGN) with zero mean. We notice that $\varepsilon_B(\tilde{\tau}_0) + \varepsilon_A(\tilde{\tau}_0) = \varepsilon\int_0^{T_s} p_R^2(t)dt := \varepsilon_R$ for $\tilde{\tau}_0 \in [0, T_s)$, where ε_R represents the constant energy corresponding to the unknown aggregate template at the receiver.

Similarly to PAM signals in [15], the term $E_{s,\zeta}\{x^2(k;\tau)\}$ reached its unique maximum at $\tilde{\tau}_0 = 0$. In the practice, the mean square of $x^2(k;\tau)$ is estimated from the average of different values $x^2(k;\tau)$ for k ranging from 0 to M – 1 obtained during an observation interval of duration MT_s. In what follows, we summarize the TDT algorithm for UWB TH-PPM systems in its NDA form and then in its DA form.

3.1.1. Non-Data-Aided Model for UWB TH-PPM System

For the synchronization mode NDA, the synchronization algorithm is defined as follows:

$$\hat{\tau}_{0,nda} = \arg\max_{\tau\in[0,T_s]} x_{nda}(M;\tau)$$

$$x_{nda}(M;\tau) = \frac{1}{M}\sum_{m=0}^{M-1}\left(\int_0^{T_s} r_k(t;\tau)\tilde{r}_{k-1}(t;\tau)dt\right)^2 \tag{9}$$

The estimator $\hat{\tau}_{0,\ nda}$ in (9) can be verified to be m.s.s. consistent by deriving the mean and variance of the function $x_{nda}(M;\tau)$ [16].

3.1.2. Data-Aided Model for UWB TH-PPM System

The blind (NDA) algorithm is principally attractive for its spectral efficiency. Nevertheless, the number of samples M required for reliable estimation can be reduced noticeably if a data-aided (DA) approach is pursued. For this reason, the training sequence $\{s_k\}$ should be considered such that no successive symbols are the same. Therefore, the training sequence for DA TDT is considered to comprise a repeated pattern (for example (1,0, 1,0)); that is:

$$s(k) = \{k+1\}_2 \tag{10}$$

With this pattern, it can be easily verified that the mean square in (8) becomes:

$$E_{s,\zeta}\{x^2(k;\tau)\} = \varepsilon_R^2 - 4\varepsilon_A(\tilde{\tau}_0)\varepsilon_B(\tilde{\tau}_0) + \sigma_\zeta^2 \tag{11}$$

With the NDA approach, it is necessary to take expectation with respect to s_k in order to remove the unknown symbol effects; while the DA mode, this is not needed. Hence, the sample mean $M^{-1}\sum_{m=0}^{M-1} x^2(k;\tau)$converges faster to its expected value in (11). This pattern is particularly attractive, since it permits a very rapid acquisition which is a major benefit of the DA mode. Data-aided TDT for UWB TH-PPM signals can be accomplished even when TH codes are present and the multipath channel is unknown [16], using:

$$\hat{\tau}_{0,da} = \arg\max_{\tau\in[0,T_s]} x_{da}(M;\tau)$$

$$x_{da}(M;\tau) = \left(\frac{1}{M}\sum_{m=0}^{M-1}\int_0^{T_s} r_k(t;\tau)\tilde{r}_{k-1}(t;\tau)dt\right)^2 \tag{12}$$

The estimator in (12) is essentially the same as (9), except training symbols used in (12). However, theses training symbols are essential in improving the estimation performance. This is approved by the simulation results in Section 4.

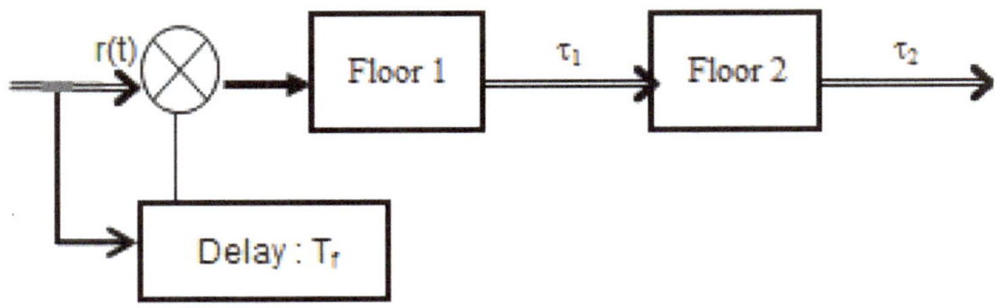

Figure 2. Block diagram of our synchronization scheme

3.2. Proposed Fine Synchronization Algorithm

In this part, we propose a low-complexity novel fine synchronization approach using the TDT synchronizer in order to have a more accurate estimate of the exact time synchronization [19]. As mentioned previously, the proposed synchronization algorithm consists of two complementary floors (or steps). The first floor consists of a coarse synchronization, which is none other than the TDT approach presented in the previous part. In the second stage which will be studied, we develop a fine synchronization algorithm that will give a better estimate of the time delay. The block diagram of our synchronization scheme is shown in Fig 2.

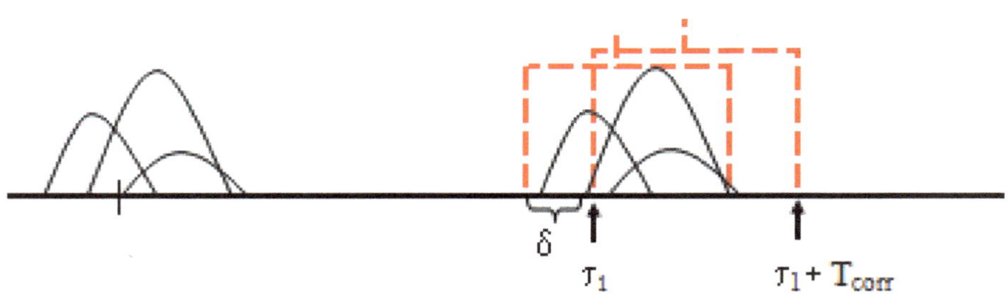

Figure 3. Principle of second synchronization floor

In what follows, we present the second floor or step of our synchronization approach in UWB TH-PPM systems. This second floor realizes a fine estimation of the frame beginning, after a coarse research in the first (TDT approach). The concept which is based this floor is really straightforward. The idea is to scan the interval $[\tau_1 - T_{corr}, \tau_1 + T_{corr}]$ with a step noted δ by making integration between the received signal and its replica shifted by T_f on a window of width T_{corr}. τ_1 being the estimate delay removed after the first synchronization floor and the width integration window value's T_{corr} will be given in Section 5. This principle is illustrated in Fig.3. We can write the integration window output for the nth step nδ as follows:

$$Z_n = \sum_{k=0}^{K-1} \left| \int_{\tau_1+n\delta}^{\tau_1+n\delta+T_s} r(t-kT_s)r(t-(k+1)T_s)dt \right| \quad (13)$$

where $n = -N + 1..0..N - 1$, $N = \lfloor T_{corr}/\delta \rfloor$ and K is the number of frames considered for improving the decision taken at the first floor. The value of n which maximizes Z_n provides the exact moment of pulse beginning that we note $\tau_2 = \tau_1 + n_{opt}\delta$. Thus, the fine synchronization is performed. Finally, note that this approach will be applied in both non-data-aided (NDA) and data-aided (DA) modes and will be applicable for any UWB transmission system. We will see later (Section 4) in what mode this approach gives better result compared to those given by the original approach TDT.

4. Simulation Results

In this section, we will evaluate the performance of our proposed fine synchronization approach in UWB TH-PPM impulse radio systems. The UWB pulse is the second derivative of the Gaussian function with unit energy and duration $T_p \approx 0.8$ns. Simulations are achieved in the IEEE 802.15.3a channel model CM1 [20]. The sampling frequency is $f_c = 50$ GHz. Each symbol contains $N_f = 32$ frames each with duration $T_f = 35$ ns. We used a random TH code uniformly distributed over $[0, N_c - 1]$, with $N_c = 35$ and $T_c = 1.0$ ns. The width integration window value's T_{corr} is 4 ns. The time shift associated with binary PPM is $\delta = 1$ns. Then, the performance of our synchronization approach is tested for various values of M.

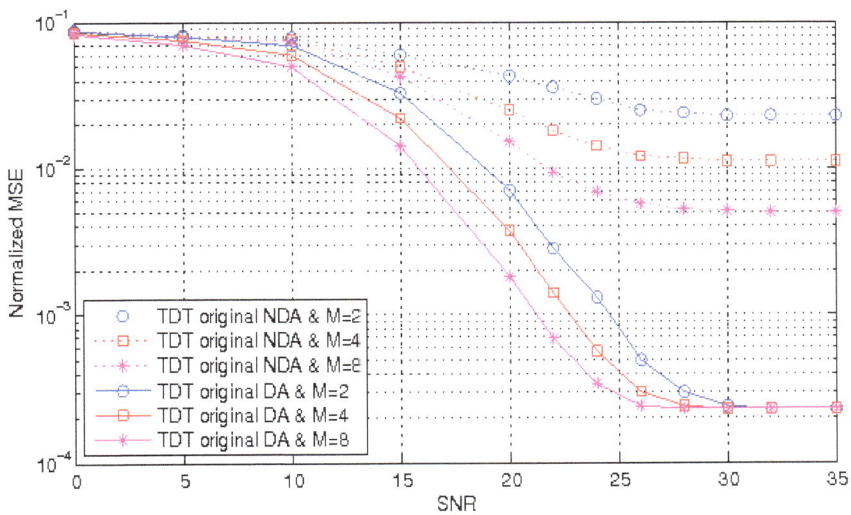

Figure 4. Normalized MSE of original TDT for UWB TH-PPM systems in both NDA and DA modes

In Fig. 4, we first test the mean square error (MSE) of NDA and DA TDT algorithms for UWB TH-PPM systems. The MSE is normalized by the square of the symbol duration T_S^2, and plotted versus signal-to-noise rate (SNR). From the simulation results obtained from Fig.4, we note that increasing the duration of the observation interval M leads to improved performance for both NDA and DA modes. We also note that the use of training sequences (DA mode) leads to improved performance compared to the NDA mode.

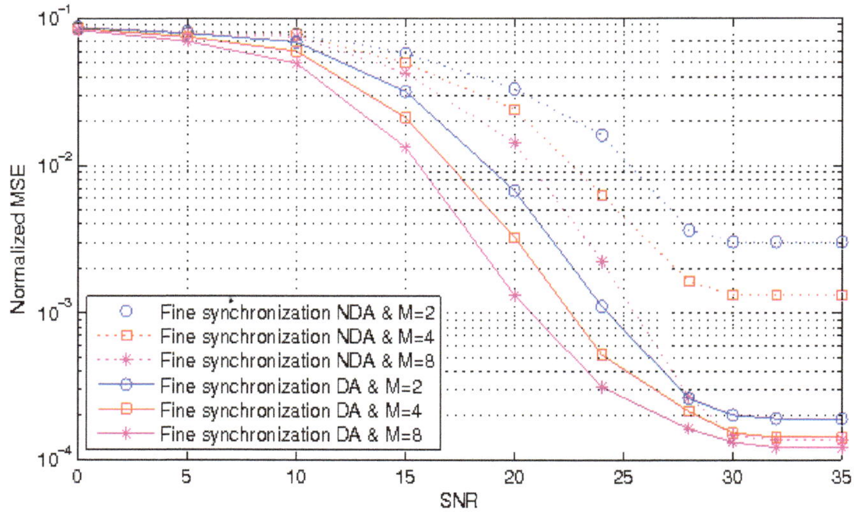

Figure 5. Normalized MSE of our fine synchronization approach for UWB TH-PPM systems in both NDA and DA modes

In Figs. 5-7, we evaluate and compare by simulation the performance of our proposed fine synchronizer with the original NDA and DA TDT algorithms in [16]. For purposes of these simulations, we kept the same channel model and the same TH code parameters as those used previously in Fig. 4. In Fig. 5, we compare the performances of the new fine synchronization approach in both NDA and DA modes.

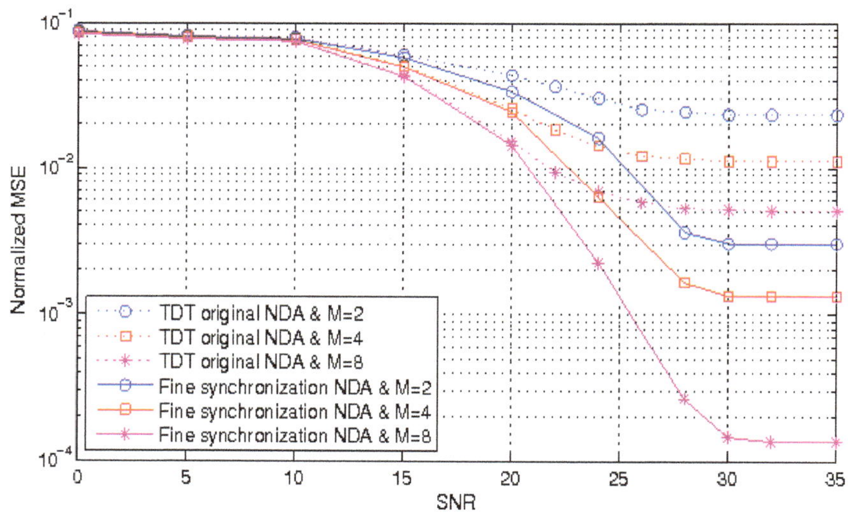

Figure 6. Performances comparison between original TDT and our fine synchronization approaches for UWB TH-PPM systems in NDA mode

In Figs. 6-7, we compare the performances of both original TDT and fine synchronization approach proposed in both NDA and DA modes for different values of M. In comparison with the original TDT approach, we note that the new approach outperforms the NDA mode and offers a slight improvement in DA mode. Even without any training symbol sequence, our synchronizer can greatly outperform the original NDA TDT especially when M is small. This performance improvement is enabled with fine synchronization approach introduced in second floor which can further improve the timing offset found in first floor (coarse synchronization approach : TDT).

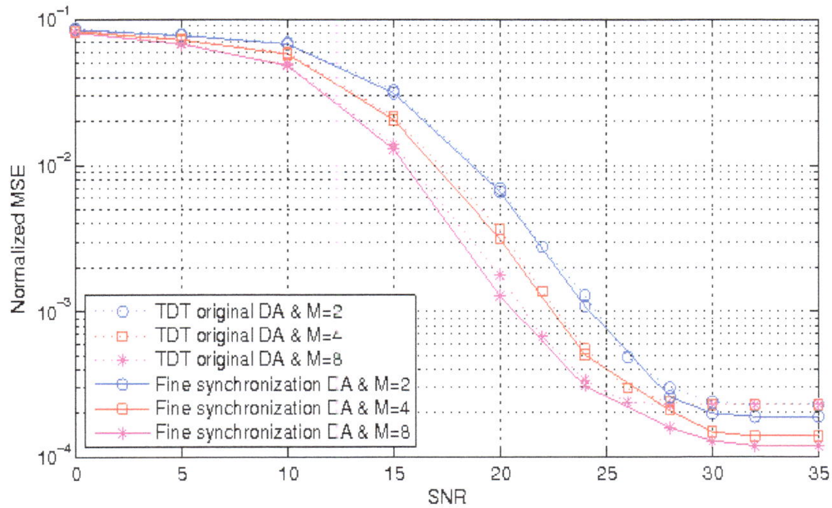

Figure 7. Performances comparison between original TDT and our fine synchronization approaches for UWB TH-PPM systems in DA mode

5. CONCLUSIONS

In this paper, we propose a novel fine synchronization scheme using TDT algorithm for UWB TH-PPM impulse radio systems in single-user links. With the fine synchronization algorithm introduced in second floor after TDT (first floor), we can achieve a fine estimation of the frame beginning. The simulation results show that even without training symbols, our new synchronizer can enable a better performance than the original TDT in NDA mode especially when M is small and offers a slight improvement in DA mode.

REFERENCES

[1] M. Z. Win and R. A. Scholtz, "Ultra wide bandwidth time-hopping spread-spectrum impulse radio for wireless multiple access communications," *IEEE Transactions of Communications*, vol. 48, no. 4, pp. 679–691, April 2000.

[2] R. C. Qiu, H. Liu and X. Shen, "Ultra-wideband for multiple access communications," *IEEE Transactions of Communications*, vol. 48, no. 4, pp. 679–691, April 2000.

[3] L. Yang and G. B. Giannakis, "Ultra-wideband communications: an idea whose time has come," *IEEE Communications Magazine*, vol. 43, no. 2, pp. 80–87, Feb. 2005.

[4] R. Fleming, C. Kushner, G. Roberts and U. Nandiwada, "Rapid acquisition for ultra-wideband localizers," in *Proceedings of Conference Ultra-Wideband System Technologies*, Baltimore, MD, May 20-23, 2002, pp. 245–250.

[5] E. A. Homier and R. A. Scholtz, "Rapid acquisition for ultra-wideband signals in the dense multipath channel," in *Proceedings of Conference Ultra-Wideband System Technologies*, Baltimore, MD, May 20-23, 2002, pp. 105–110.

[6] Z. Tian and G. B. Giannakis, "Data-aided ML timing acquisition in ultra-wideband radios," in *Proceedings of Conference Ultra-Wideband System Technologies*, Reston, VA, Nov. 16-19, 2003, pp. 245–250.

[7] L. Yang, Z. Tian and G. B. Giannakis, "Non-data aided timing acquisition of ultra-wideband transmissions using cyclostationarity," in *Proceedings of International Conference Acoustic, Speech, Signal Process.*, Hong Kong, China, Apr. 6-10, 2003, pp. 121–124.

[8] J. Y. Lee and R. A. Scholtz, "Ranging in a dense multipath environment using an UWB radio link," *IEEE Journal on Selected Areas in Communications*, vol. 20, no. 9, pp. 1677-1683, December 2002.

[9] S. Aedudodla, S. Vijayakumaran, and T. F. Wong, "Ultra-wideband signal acquisition with hybrid ds-th spreading," *IEEE Transactions on Wireless Communications*, Sep. 2006, pp. 2504-2515.

[10] R. Djapic, G. Leus, A. J. Van der Veen and A. Trindade, "Blind synchronization in asynchronous UWB networks based on the transmit-reference scheme," *EURASIP Journal on Wireless Communications and Networking*, vol. 2006, no. 2, pp. 65–75, Apr. 2006.

[11] Q. H. Dang, A. Trindade and , A. J. Van der Veen, "Signal model and receiver algorithms for a Transmit-Reference Ultra-Wideband Communication system," *IEEE Journal on Selected Areas in Communications*, vol. 24, no. 4, pp. 773–779, Apr. 2006.

[12] Y. Ying, M. Ghogho and A. Swami, "Code-Assisted synchronization for UWB-IR systems: algorithms and analysis," *IEEE Transactions on Signal Processing*, vol. 56, no. 10, pp. 5169–5180, Oct. 2008.

[13] X. Luo and B. Giannakis, "Low-complexity blind synchronization and demodulation for (ultra-) wideband multi-user ad hoc access," *IEEE Transactions on Wireless Communications*, vol. 5, no. 7, pp. 1930-1941, July 2006.

[14] L. Yang and G. B. Giannakis, "Low-complexit training for rapid timing synchronization in ultra-wideband communications," in *Proceedings of Global Telecommunications Conference*, San Francisco, CA, pp. 769-773, Dec. 2003.

[15] L. Yang and G. B. Giannakis, "Timing UWB signals using dirty templates," *IEEE Transactions on Communications*, Vol. 53, no. 11, pp. 1952-1963, Nov. 2005.

[16] L. Yang, "Timing PPM-UWB signals in ad hoc multi-access," *IEEE Journal on Selected Areas in Communications*, vol. 24, no. 4, pp. 794-800, Apr. 2006.

[17] G. Durisi, and S. Benedetto, "Performance evaluation of TH-PPM UWB systems in the presence of multiuser interference," *IEEE Communication Letters*, vol. 5, pp. 224-226, May 2003.

[18] M. Di Renzo, F. Graziosi, and F. Santuci, "A framework for performance analysis for TH-UWB communications," in *Proceedings of IEEE International Conference on Ultra-Wideband (ICUWB)*, Zurich, Switzerland, Sep. 5-8, 2005, pp. 559-564.

[19] M. Hizem and R. Bouallegue, "A framework for performance analysis for TH-UWB communications," in *Proceedings of International Information and Telecommunication Technologies Symposium (I2TS)*, Rio de Janeiro City, Rio de Janeiro State, Brazil, Dec. 11-13, 2010.

[20] J. R. Foerster, "Channel Modeling Sub-committee Report Final," *IEEE P802.15-02/368r5-SG3a, IEEE P802.15 Working Group for WPAN*, November 2002.

2

Speech Enhancement Algorithm to Reduce the Effect of Background Noise in Mobile Phones

Premananda B S [1] and Dr. Uma B V [2]

[1]Department of Telecommunication Engineering,
R.V. College of Engineering, Bangalore, India.

premanandabs@rvce.edu.in

[2]Department of Electronics and Communication Engineering,
R.V. College of Engineering, Bangalore, India.

umabv@rvce.edu.in

Abstract

In mobile phones, perceived quality of speech signal deteriorates significantly in the presence of background noise since near-end/surrounding noise also arrives at the near-end listener's ears. The quality of the received signal varies widely depending upon signal strength and unavoidable background noise in the user environment. There is a need to improve the quality of received speech signal in noisy conditions by developing the speech enhancement algorithms. This paper focuses on the impact of the various background noises on signal degradation and mechanisms to mitigate the noise impact for improved speech signal perception. Gain adjustment process with simple time domain and frequency domain approach using psychoacoustic has been adapted to improve the quality and intelligibility of the speech signal in the noisy environments by automatically enhancing the speech signals when the noise dominates. Time domain approach is less complex and frequency domain using psychoacoustic is more efficient in overcoming the degradation of speech signals in presence of near-end or background noise.

Keywords

Absolute Threshold of Hearing, Background noise, Degradation, Gain, Psychoacoustic, Speech enhancement.

1. Introduction

Mobile phones are essential component of day to day life and have become the best friend of a person. Day by day they are becoming very affordable making the communication infrastructure as the backbone of daily life. Background noise is a natural part of a conversation. A high background acoustic noise level is annoying to the listener side. Listener fatigue (the ears get tired) and difficult to understand each other. Considerable attention has been paid over the past decade for the enhancement of speech degraded by additive background noise [6]. Listening to speech or audio signals becomes more difficult as the background noise level dominates. The near-end listener perceives a combination of the clean far-end (downlink) speech and the background noise from the near-end and thus experiences an increased listening effort. As the noise signal cannot be influenced, a reasonable approach is to manipulate the clean far-end speech signal depending on the local background noise. Hence, there is a strong need to improve the quality of the speech signal in noisy conditions by developing speech enhancement algorithms to minimize the effect of background noise. In the gain adjustment process the quality of the speech signal in noisy environment is improved by automatically adjusting the output level when the background noise exceeds the noise masking threshold.

The traditional noise cancellation algorithms cannot be used as the near-end noise directly arrives at the near-end listener's ears. The approaches proposed for far-end noise cancellation discussed in literature [1], [2], [3], [4] are not suitable. In the subtractive-type approach proposed by Virag [1], a psychoacoustic model was used to guide the derivation of the spectral subtractive parameters. Jeon Yu-young et al. discusses about the compensation for partial masking effect [5]. Several approaches to mitigate the background (near-end) noise using speech enhancement are discussed by Bastian et al. [6], [7], [8], [9] and Jong Won Shin et al. [10], [11], [12]. Speech intelligibility improvement in presence of near-end noise and loud-speaker output power constraint are discussed in [6], [7]. Perceptual speech reinforcement based on partial specific loudness is discussed in [10], [11] focuses on mitigating near end noise at the speaker not at the receiver. Psychoacoustic model in frequency domain approach using wavelet-threshold multitaper magnitude spectra as alternate to Fast Fourier Transform (FFT) was incorporated by Yi Hu [13]. The speech enhancement methods today are more exploiting the auditory masking properties of human [3]. Many approaches discussed above have not fully exploited the psychoacoustic approach for speech enhancement in presence of near-end noise.

The paper is organized as follows: In section 2, proposed time domain approach is described with implementation details and results. Section 3, describes the proposed frequency domain approach using psychoacoustic with the implementation details and results. Comparison results of both the approaches are discussed in Section 4, and the conclusions are given in Section 5.

2. Proposed Time Domain Speech Enhancement Method

This paper proposes a method to overcome the relative degradation of the speech signal in the noisy environment. In order to retain the signal quality in the noisy environment different algorithms have been proposed and implemented in [6-9], but the complexity involved in determining the gain seems to be unresolved. In this paper a concept which is simple to analyze and implement in real time is presented. A multiplier is required to enhance the speech signals degraded in the presence of background noise. As the speech and noise signals are dynamic it is difficult to find a constant multiplication factor (MF) for the incoming signal. Thus the signal strength and noise present in the environment is to be analyzed. The speech signal is to be amplified by applying a dynamically varying gain depending upon the different parameters of speech and noise signal. The external volume control of the devices can't be used for this purpose as it is painful.

Block diagram of the proposed time domain approach is shown in the figure 1. Energy of the downlink speech signal and near-end noise signal is computed. By comparing energy of both the speech and noise signals, gain/MF is computed for enhancing the speech signal. Gain obtained is multiplied with speech (element by element) signal to get enhanced speech signal.

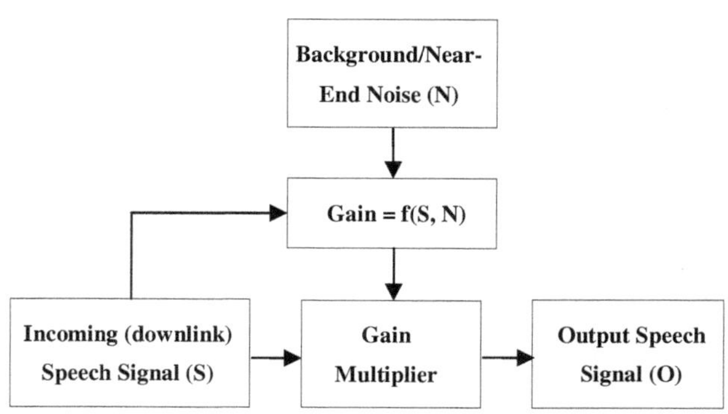

Fig. 1: Proposed time domain approach.

The sound energy is proportional to the square of the amplitude. Energy, E_{dB} in decibels is calculated using,

$$E_{dB} = 10 * \log_{10} \sum(X^2)/N \quad (1)$$

where X is the amplitude of the speech signal and N, number of samples.

The Gain is to be adaptive for being effective in most of the scenarios. It can be modelled as,

$$\text{Gain} = \sum_{i=1}^{p} c_i x_i \quad (2)$$

where $x_1 \ldots x_i \ldots x_p$ represent different parameters and $c_1 \ldots c_i \ldots c_p$ represent different coefficients to be derived corresponding to one or more of the above mentioned parameters.

The challenge lies in finding out the correct coefficients for adapting equation (2). Hit and trial methods or human expertise can be utilized for setting these values, works for limited categories of speech signals and noises.

2.1. Deriving the Gain

The correct gain/MF for a pair of speech and noise sample (selected parameters) is user specific. System provides a mechanism for user to declare whether system generated MF (auto gain) is acceptable or not. Such an assumption is easy to realize for most of the devices as volume control can be used for the purpose. In figure 1, gain block calculates the amplification factor/MF of speech signal in the presence of varying background noise which can dominate the received speech signal as well. Different parameters of signal and noise (energy in the present case) can be considered. In order to avoid the perceptual artefacts like clicks & pops due to signal saturation/ overflows and signal bursts due to sudden gain changes, optimal gain computed must be characterized by slow and configurable response time for the gain variations.

The Gain is derived using the equation,

$$\text{Gain} = A + \text{maximum}(B, (C - D)) \cdot E \quad (3)$$

where A, B, C, D, and E are experimental constants. Value of A (default gain) is set to 1 so that Gain =1 when no enhancement for the speech signal is required, B (default enhancement) is set to 0. (C – D) is extent of noise over speech signal, (if negative no amplification is required), E (compensation factor) is used to control the gain (<1).

When signal energy (SE) is sufficiently greater than noise energy (NE) then gain of 1 is used so that no amplification is required, input signal buffer value will be copied to output signal buffer. When SE is approximately equal to NE then gain is selected such that enhanced speech signal is increased nearly by 1 dB. When SE is less than NE then gain is calculated using eq. 3. Present frame gain/MF is compared with previous frame MF if the difference is more, then current frame MF is adjusted to avoid sudden change in large gain to avoid click and pop noise. If the hearing and perception choice of user varies with time, the proposed mechanism is capable of capturing such alterations. If the amplification value exceeds the maximum loudness [7] of the loud speaker [9] then end capping can be performed depending on minimum and maximum values computed using,

Case-1: for +ve amplified value (AV)

$$AV = \text{minimum}(AV, 2^{15} - 1) \quad (4)$$

Case-2: for -ve amplified value (AV)

$$AV = \text{maximum}(AV, -2^{15}) \quad (5)$$

Signal buffer value can be overridden with new amplified value and output to output speech signal buffer. Most mobile phone applications limit the overall power of the loudspeaker signal to a constant maximum power instead of the power of the original signal.

2.2. Experimental Results

Algorithm has been verified for different speech signals with varying background (tapering, hammer, parabolic type ...) noise signals. To **illustrate** a speech signal of 16-bit PCM with sampling rate of 8 kHz with duration of 5 seconds is considered. The speech signal is captured using an audio editor tool Gold Wave and saved in .wav format. The captured speech signal has 40000 (8000*5) samples, total samples are divided into frame size of 1024 each, resulting in 39 frames. Varying background noise (Fig. 2(b)) for 5 seconds, sampled at 8 kHz is captured using an audio editor tool Gold Wave and saved in .wav format for experimenting and gain/MF are calculated. Microphone in the mobile phones can be used to capture the background noise signals. Experiment is performed by considering the difference in energy of background noise (dB) and speech signal (dB). Figure 2 show the results obtained for varying background noise. The speech signal in figure 2(a) corresponds to input (downlink) speech signal, figure 2(b) represents noise signal whose amplitude is continuously decreasing, and figure 2(c) shows the amplified speech signal in presence of noise where the gain is varied from 1 to 7.

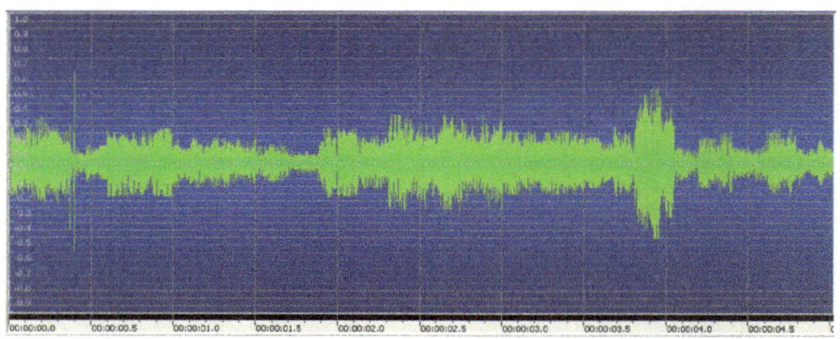

Fig. 2(a): Original down linked speech signal.

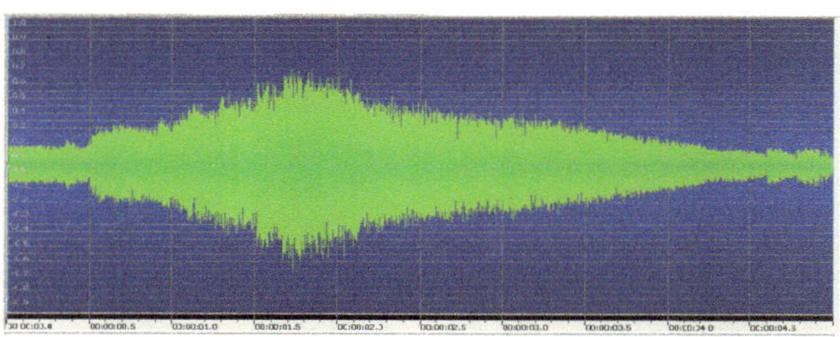

Fig. 2(b): Varying background noise signal.

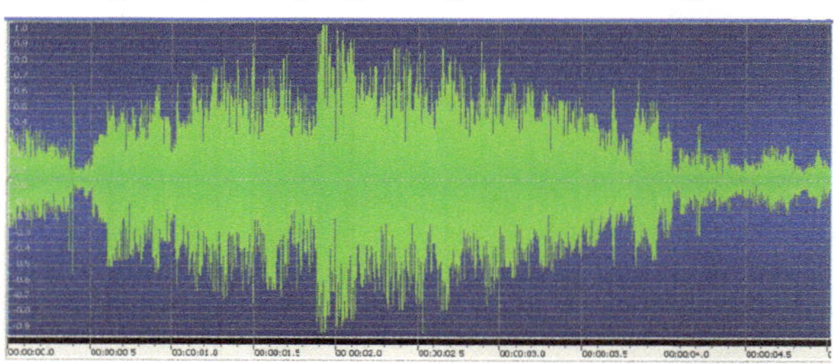

Fig. 2(c): Enhanced speech signal using time domain approach.

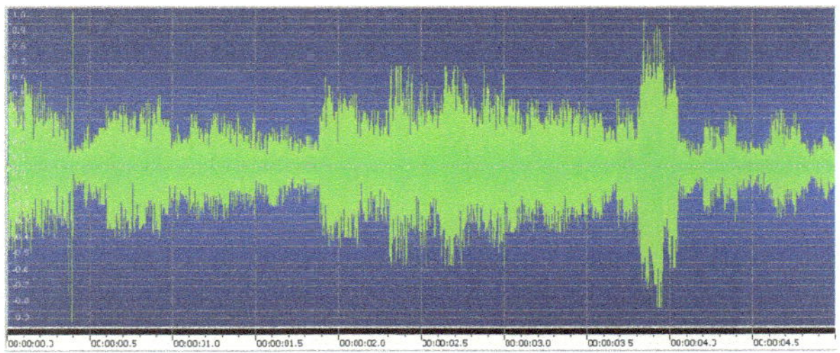

Fig. 2(d): Enhanced speech signal using frequency domain approach.

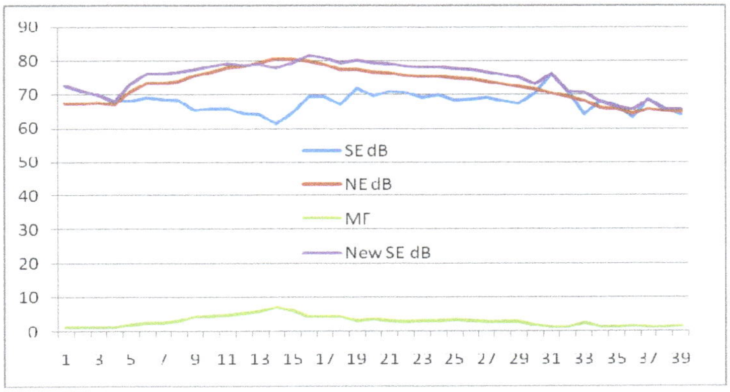

Fig. 3: Variations of speech, noise, enhanced speech signal and gain/MF using time domain approach.

Figure 2(c) reflects the variation in the amplification of speech signal based on changing noise signals. Speech signal can be further enhanced by increasing the value of gain control, E in equation 3 as desired. The obtained results are plotted in figure 3 to verify how dynamically gain/MF is varied in accordance with speech and noise signals. Number of frames is indicated in x-axis and y-axis indicates energy of the speech, noise & enhanced signal and variation of gain w.r.t. speech and noise signal. For a gain of ≈ 7, when the noise signal is at maximum, the speech signal energy is enhanced by $\approx$ 17 dB. Speech signal has enhanced to a maximum energy of $\approx$ 82 dB.

3. Proposed Frequency Domain Speech Enhancement Method

3.1 Psychoacoustics

Psychoacoustics is the study of sound perception (including speech and music) [1], [7]. The range of human hearing is about 20 Hz to 20 kHz and frequency range of the speech (voice) is typically from 500 Hz to 4 kHz. The dynamic range of sound (ratio of the maximum sound amplitude to the quietest sound that can be heard), is of the order of about 120 dB [14]. Equal-loudness contours indicate the sound pressure level (dB), over the range of audible frequencies, which are perceived as being of equal loudness. Equal-loudness contours were first measured by Fletcher and Munson using pure tones reproduced via headphones, and the data collected are called Fletcher-Munson curves.

Speech enhancement methods today are more often exploiting the human auditory masking properties [3]. Yi Hu et al. proposes speech enhancement in the frequency domain as a constrained minimization problem and includes the masking thresholds as the constraints. In [13], psychoacoustic model is integrated in the derived spectral weighting function, paper further investigate the importance of using good (low variance) spectrum estimators in speech enhancement. The human listener will not perceive any noise distortion as long as the power spectrum density of the distortion lies below the masking threshold (the masking thresholds can be obtained by performing critical band analysis of the speech signal [13]). As an alternative to time domain approach (section 2) is to amplify the frequency components of the signal so that the noise level in each critical band becomes lower than the masking threshold. In most of the above speech enhancement methods, the incorporation of auditory masking was done heuristically.

A. Absolute Threshold of Hearing

ATH is minimum sound level of a pure tone that an average ear with normal hearing can hear with no other sound present, this is also known as the auditory threshold. It is the lowest of the equal-loudness contours. If one has a low absolute threshold, it means that he is able to detect small amounts of stimulation, and thus is more sensitive. If one has a high absolute threshold, then he requires more stimulation and thus is less sensitive [14]. ATH varies with the frequency of the sound, which is shown in figure 4.

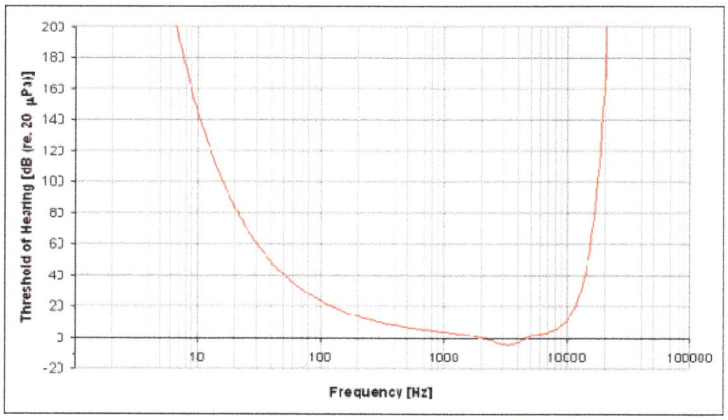

Fig. 4: Threshold of hearing.

The threshold of hearing is given by,

$$T_q(f) = 3.64\left(\frac{f}{1000}\right)^{-0.8} - 6.5e^{-0.6\left(\frac{f}{1000}-3.3\right)^2} + 10^{-3}\left(\frac{f}{1000}\right)^4 \tag{6}$$

where

$T_q(f)$ =Threshold of Hearing (dB) and f = frequency (Hz)

The threshold of hearing is generally reported as the RMS sound pressure of 20 µPa. It is the sound a young human with undamaged hearing can detect at 1000 Hz. The threshold of hearing is frequency dependent and it has been shown that the ear's sensitivity is best at frequencies between 1 kHz to 5 kHz [14].

B. Masking

The masking phenomenon occurs because any loud sound will distort the ATH, making quieter, otherwise perceptible sounds inaudible [14]. The masking effect stands for the phenomenon that a certain weak signal called a maskee cannot be heard, i.e., 'masked' in the presence of a strong signal called a masker in a nearby time or frequency region [12]. A weaker sound is masked if it is made inaudible in the presence of a louder sound. The stronger signal that masks the weaker one is called masker and the one that is masked is the maskee. The masking threshold will depend upon the sound pressure level (SPL), the frequency of the masker and the characteristics of the masker and the maskee, such as whether the masker or maskee is a tone or noise. The masking characteristics are measured by **signal to mask ratio (SMR)** and **mask to noise ratio**

(MNR). SMR at a given frequency is expressed as the difference (in dB) between the sound pressure level of the masker and the masking threshold at that frequency. MNR at a given frequency is expressed as the difference (in dB) between the masking threshold at that frequency and the noise level. To make the noise inaudible, its level should be below the masking threshold i.e. the MNR should be positive [3].

If there is more than one masker, then each masker contributes to its own masking threshold and global masking threshold is computed that describes just noticeable distortion as a function of frequency.

Simultaneous Masking: masking between two concurrent sounds often observed when the sounds share a frequency band.

Temporal Masking: characteristic of the auditory system where sounds are hidden due to maskers which have just disappeared, or even after maskers which are about to appear.

Loudness: Sound loudness is a subjective term describing the strength of the ear's perception of a sound, also referred to as volume or strength. The units used to measure loudness are the, Sone (loudness N) and Phon (loudness level L).

3.2. Deriving the Weighted Curve using Psychoacoustic

In the measurement of loudness, weighted curve is used to emphasize frequencies around 3 to 5 kHz where the human ear is most sensitive, while attenuating very low and high frequencies to which the ear is insensitive [14]. Deriving a weighted curve is considered as one of the most important factors which facilitates in measuring noise loudness. Table I shows the derived threshold of hearing (dBSPL) values using equation 6, for frequencies ranging from 20 Hz to 20 kHz (frequency range for human hearing). In table I, the column I represent 64 different frequency values ranging from 20 Hz to 20 KHz, and column 2 represents threshold of hearing computed (dB) for the respective frequencies.

The linear curve (figure 5) is derived for 64 vectors, with a fixed linear frequency interval of 312.5 Hz along the X axis, using the equation 6. The threshold of the hearing was found to be -4.983187589 dB at 3320.3125 Hz, and 160.3313427 dB at 20 kHz which shows, our ear is most sensitive to the frequency around 3.3 to 3.5 kHz. It is observed that maximum audibility lies in the range of 1 to 5 kHz; any signal with least loudness can be heard in this range.

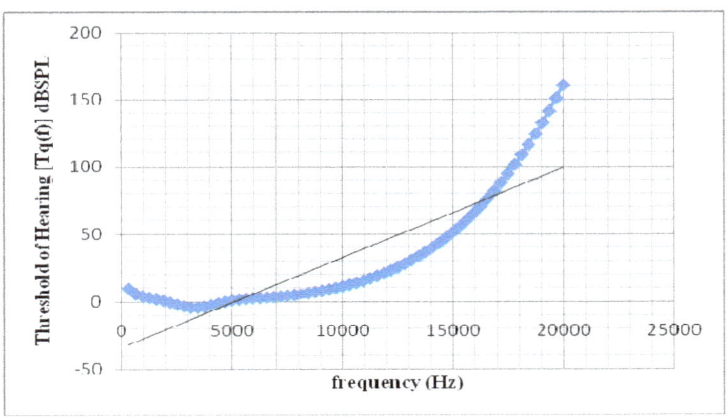

Fig. 5: Threshold of Human Hearing.

After deriving 64 Vectors for a frequency range of 20 Hz to 20 kHz, its weighted curve has to be derived, for this we have normalized 64 vectors to lie in between 0 and 1 (figure 6). To normalize these values we use equation 7 to set maximum value 160.3313427 dB (threshold) of 20 kHz to be 1 and minimum value -4.931530782 dB of 3.4375 KHz to 0.

$$X'(n) = \frac{X(n) - \min(score)}{\max(score)} \tag{7}$$

where,
X'(n) = Normalized value, *X*(n) = Value to be normalized, min (score) = minimum value, max (score) = maximum value.

The weighed curve is the inverse of Threshold of hearing curve shown in figure 6, normalized between 0 and 1. We take 1-X'(n) (represented as B(n) in table I), to derive the weighed curve shown in figure 7.

Table 3.1: F = frequency (Hz) [column I], Tq(f) = Threshold of Hearing (dB) [column II], X'(n) = Normalized Values [column III], B(n) = 1-X'(n) [column IV].

No.	I Freq. (Hz)	II Tq(f)	III X'(n)	IV B(n)	No.	I Freq. (Hz)	II Tq(f)	III X'(n)	IV B(n)
1	312.5	9.199721317	0.085507724	0.914492276	33	10312.5	11.87269673	0.101681806	0.898318194
2	625	5.212853678	0.061383324	0.938616676	34	10625	13.29388247	0.110281353	0.889718647
3	937.5	3.605332751	0.05165627	0.948343731	35	10937.5	14.84806612	0.119685665	0.880314335
4	1250	2.525140233	0.045120061	0.954879939	36	11250	16.54309066	0.1299422	0.870057821
5	1562.5	1.490733319	0.038860901	0.961139099	37	11562.5	18.38703346	0.141099835	0.858900165
6	1875	0.291649746	0.031605287	0.968394713	38	11875	20.38820559	0.15320886	0.846791141
7	2187.5	-1.12430887	0.02303737	0.976962631	39	12187.5	22.55515119	0.166320973	0.833679027
8	2500	-2.63945460	0.013869275	0.986130725	40	12500	24.8966472	0.180489284	0.819510716
9	2812.5	-3.98204419	0.005745311	0.994254689	41	12812.5	27.42170193	0.195768306	0.804231694
10	3125	-4.82336305	0.000654519	0.999345481	42	13125	30.13955663	0.212213951	0.787786049
11	3437.5	-4.93153078	5.37434E-18	1	43	13437.5	33.05968325	0.229883538	0.770116462
12	3750	-4.29421157	0.003856397	0.996143603	44	13750	36.19178507	0.248835779	0.751164221
13	4062.5	-3.12744206	0.010916479	0.989083521	45	14062.5	39.54579634	0.269130786	0.730869214
14	4375	-1.76521239	0.019159284	0.980840716	46	14375	43.13188198	0.290830068	0.709169932
15	4687.5	-0.50721883	0.02677136	0.97322864	47	14687.5	46.96043748	0.313996527	0.686003473
16	5000	0.481694717	0.032755243	0.967244757	48	15000	51.04208865	0.338694458	0.661305542
17	5312.5	1.181223029	0.036988064	0.963011936	49	15312.5	55.38769156	0.364989553	0.635010447
18	5625	1.66154186	0.039894457	0.960105543	50	15625	60.00833233	0.392948893	0.607051107
19	5937.5	2.018207613	0.04205263	0.95794737	51	15937.5	64.91532708	0.42264095	0.577359051
20	6250	2.331012374	0.0439454	0.95605461	52	16250	70.12022178	0.45413559	0.545864411
21	6562.5	2.651831043	0.045886663	0.954113337	53	16562.5	75.63479221	0.487504067	0.512495933
22	6875	3.009545465	0.048051181	0.951948819	54	16875	81.47104384	0.522819026	0.477180974
23	7187.5	3.419366485	0.050530994	0.949469006	55	17187.5	87.64121176	0.560154502	0.439845498
24	7500	3.89009146	0.053379335	0.946620665	56	17500	94.15776064	0.599585916	0.400414084
25	7812.5	4.42811903	0.056634921	0.943365079	57	17812.5	101.0333847	0.641190082	0.358809918
26	8125	5.039207051	0.060332594	0.939667406	58	18125	108.2810075	0.6850452	0.31495482
27	8437.5	5.729107249	0.064507156	0.935492844	59	18437.5	115.9137821	0.731230858	0.268769142
28	8750	6.503757032	0.069194536	0.930805464	60	18750	123.9450909	0.779828034	0.220171966
29	9062.5	7.369327338	0.074432072	0.925567928	61	19062.5	132.3885457	0.83091909	0.16908092
30	9375	8.33223132	0.080258571	0.919741429	62	19375	141.2579876	0.88458778	0.11541222

31	9687.5	9.39912406	0.086714303	0.913285697	63	19687.5	150.5674867	0.940919241	0.059080759
32	10000	10.57690112	0.093840991	0.906159009	64	20000	160.3313427	1	1.98947E-10

3.3 Deriving the Gain

Deriving the gain in frequency domain approach is more critical than time domain approach. Background noise is captured and sampled at a rate of 8 kHz. The total samples are divided into frame size of 1024 each, resulting in 39 frames as explained in section 2.2. Each frame of 1024 samples are sub-divided into 16 sub-frames for processing, each sub-frame has 64 samples. Flow representation of the approach is shown in figure 8.

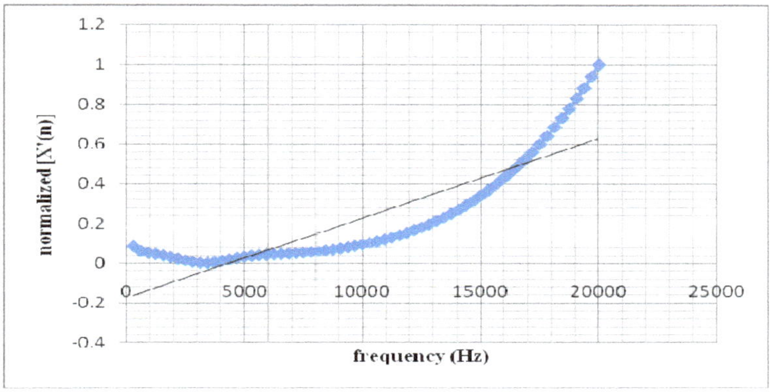

Fig. 6: Normalized vectors V_s frequency (Hz).

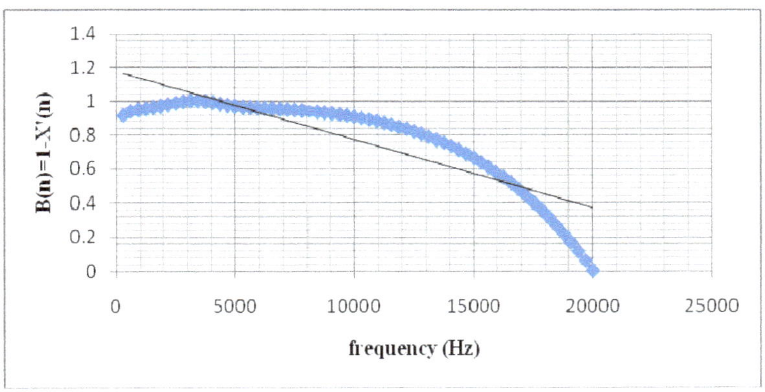

Fig. 7: Weighted Curve.

Obtain the magnitude of each sample by computing 64 point FFT of captured noise signal and compute the loudness level of varying background noise. The magnitude of frequency samples obtained by computing FFT is multiplied by 64 point weighed curve on sample basis and noise energy or noise loudness is calculated using the equation,

$$\text{Loudness} = (1/N)\sum_{i=0}^{N-1} 10(\log_{10}(R(i))^2) \tag{8}$$

where

$$R(i) = \sum_{n=0}^{N-1} (M(n) * B(n)) \tag{9}$$

M(n) = Magnitude of FFT samples

B(n) = 1-X'(n), weighted N samples derived from figure 7

Compute the magnitude of down-linked speech samples (with Fs = 8 kHz) using 64 point FFT for 16 sub-frames and calculate the energy (frame wise) of speech samples using equation 1. The energy of the speech and noise signals computed (figure 8) is compared and gain is derived for enhancing the degraded speech signal as explain in section 2. For deriving the gain equation 3 is used. Present frame gain is compared with previous frame gain if the difference is more, then current frame gain is adjusted to avoid sudden change in the output levels to avoid click & pop noise which fatigue the ears. Updated gain is multiplied with speech (element by element) signal in time or frequency domain (multiplying by constant is a linear function) to get enhanced speech signal. If the enhanced/amplified value of speech signal exceeds the maximum loudness of the loud speaker then end capping can be performed depending on minimum and maximum values computed using equations 4 and 5 as explained in section 2.1.

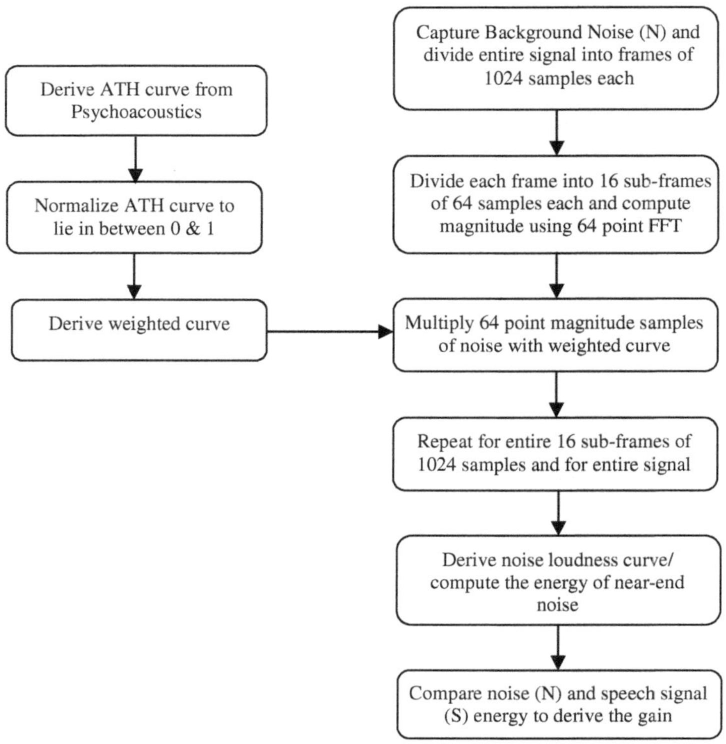

Fig. 8: Flow diagram of frequency domain approach using psychoacoustic.

3.4. Experimental Results

To **illustrate** a speech signal (Fig. 2(a)) and varying noise signal (Fig. 2(b)) of 16-bit PCM with sampling rate of 8 kHz with duration of 5 seconds is captured using an audio editor tool, Gold Wave and saved in .wav format. The captured speech signal has 40000 samples; total samples are further divided into frame size of 1024 each, resulting in 39 frames. Figure 2(d) reflects the variation in the amplification of speech signal based on changing noise signals. Speech signal can be further enhanced by increasing the gain control, E in equation 3 as desired.

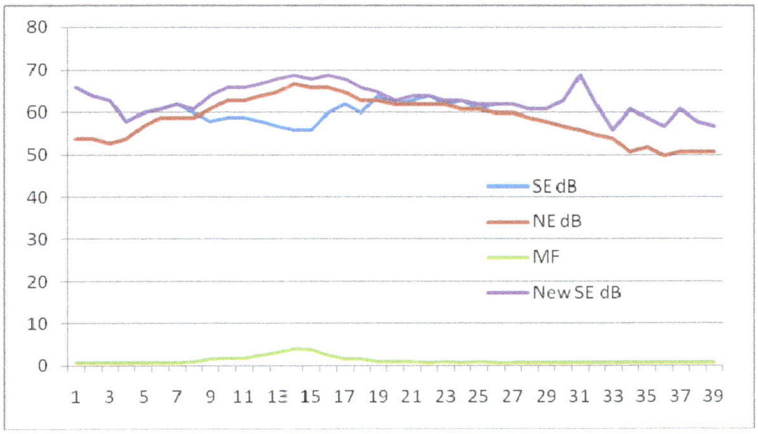

Fig. 9: Variation of speech, noise, enhanced speech signal and gain/MF using frequency domain approach.

The obtained results are plotted in figure 9. Number of frames is indicated in x-axis and y-axis indicates energy of the speech, noise & enhanced signal and variation of gain w.r.t. speech and noise signal. For a maximum gain of 4 the speech signal energy is enhanced by 13 dB.

4. Evaluation of Both the Methods

The gain derived from both the time and frequency domain (using psychoacoustic) approach is plotted in figure 10. Results indicate that later approach promises minimal gain required for speech enhancement, which is more realistic with reference to human audibility since psychoacoustic was involved. Complexity involved in the processing the samples will more in the latter approach because of frequency transformations (FFT) involved as compared to earlier approach. Algorithm enhances only the selected frequency components hence amplification required for enhancing speech signal is minimal which reduces the battery power as well as ears will not get fatigue.

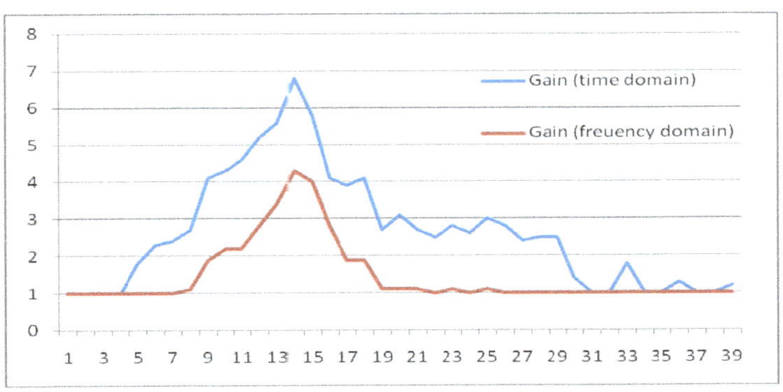

Fig. 10: Variation of gain/MF in time and frequency domain approach.

5. Conclusions

In this contribution we presented an efficient algorithm to enhance the speech signals in presence of noisy background noise. The paper discusses the implication of background noise on signal perception by the user and necessity of this problem needs to be solved for extended user experience and growth of mobile technology. The paper also discusses avenues to mitigate the effect of such noise by gain adjustment of speech signal according to noise variations. Experimental results are presented for a few combinations of parameters against the varying background noise, highlighting acceptable gain of the signal. Results indicate that gain obtained is adaptive and varies w.r.t. change in speech and noise signal. The proposed system is capable of tracking change in the user perception levels with time. The gain derived from frequency domain approach is more realistic because derived gain uses the masking properties of the human auditory system. Simulation results were verified using an audio editor tool Gold Wave v5.58, compiled with Eclipse C/C++ IDE and implemented on Beagle-Board-xM (OMAP-3530) platform.

ACKNOWLEDGEMENTS

The authors would like to thank everyone, just everyone!

REFERENCES

[1] S. F. Boll, "Suppression of acoustic noise in speech using spectral subtraction", IEEE Transaction on Acoustic, Speech, Signal Processing, vol. ASSP -27, pp. 113–120, April. 1979.

[2] Y. Ephraim and D. Malah, "Speech enhancement using a minimum mean square error short-time spectral amplitude estimator", IEEE Transaction on Acoustic, Speech, Signal Processing, vol. ASSP-32, pp. 1109–1121, Dec. 1984.

[3] N Virag, "Single channel speech enhancement based on masking properties of human auditory system", IEEE Transaction on Speech and Audio Processing, vol. 7, pp. 126–137, 1999.

[4] Jeon Yu-young and Lee Sang-min, "A speech enhancement algorithm to reduce noise and compensate for partial masking effect", Journal of Central South University of Technology, vol. 18, issue 4, pp. 1121-1127, August 2011.

[5] Malihe hassani and M. R. Karami mollaei, "Speech enhancement based on spectral subtraction in wavelet domain", IEEE 7th International Colloquium on Signal Processing and its Applications, pp. 366-370, March 2011.

[6] Bastian Sauert and Peter Vary, "Near End Listening Enhancement: Speech Intelligibility Improvement in Noisy Environments", in: Proceeding of International Conference on Acoustics, Speech, and Signal Processing (ICASSP), pp. 493–496, May 2006.

[7] Bastian Sauert, Gerald Enzner and Peter Vary, "Near End Listening Enhancement with Strict Loudspeaker Output Power Constraining", in: Proceeding of International Workshop on Acoustic Echo and Noise Control (IWAENC), pp. 01–04, Sept. 2006.

[8] Bastian Sauert and Peter Vary, "Near End Listening Enhancement Optimized with respect to Speech Intelligibility Index", in: Proceeding of European Signal Processing Conference (EUSIPCO) vol. 17, EURASIP. New York, Hindawi Publication, pp. 1844–1848, Aug. 2009.

[9] Bastian Sauert and Peter Vary, "Near End Listening Enhancement considering Thermal Limit of Mobile Phone Loudspeakers", in: Konferenez Elektronische Sprachsignal Verarbeitung (ESSV), Germany, vol. 61, pp. 333-340, 28-30th Sept. 2011.

[10] Jong Won Shin, Student Member, IEEE and Nam Soo Kim, Member, IEEE, "Perceptual Reinforcement of the Speech Signal Based on Partial Specific Loudness", IEEE signal processing letters, vol. 14, No. 11, pp. 887 – 890, Nov. 2007.

[11] Jong Won Shin, Woohyung Lim, Junesig Sung and Nam Soo Kim, "Speech Reinforcement based on Partial Specific Loudness", in 8th Annual Conference of International Speech Communication Association (ISCA), INTERSPEECH 2007, Antwerp, Belgium, 27-31st Aug. 2007. PP. 978-981.

[12] Jong Won Shin, Yu Gwang Jin, Seung Seop Park and Nam Soo Kim, "Speech Reinforcement based on Partial Masking Effect", in: Proceeding of IEEE Transaction on Acoustic, Speech, Signal Processing (ICASSP), pp. 4401 - 4404, April 2009.

[13] Yi Hu and Philipos C. Loizou, "Incorporating a Psychoacoustic Model in Frequency Domain Speech Enhancement", in IEEE signal processing letters, vol. 11 no. 2, February 2004.

[14] Eberhard Zwicker and Hugo Fastl, *Psychoacoustics, Facts and Models. New York:* Springer, 2007.

3

ENHANCING THE PERFORMANCE OF THE DNDP ALGORITHM

Hussein Al-Bahadili[1], Ghassan Issa[1], Alia Sabri[2]

[1]Faculty of Information Technology, Petra University, Jordan
[2]Faculty of Information Technology, Applied Science University, Jordan
hbahadili@uop.edu.jo, gissa@uop.edu.jo, a_sabri@asu.edu.jo

ABSTRACT

A novel dynamic noise-dependent probabilistic (DNDP) route discovery algorithm was recently developed to enhance the performance of the dynamic probabilistic algorithm in noisy mobile ad hoc networks (MANETs). In this algorithm, the mathematical model for calculating node retransmission probability (p_t) is calculated as a function of two independent variables: number of first-hop neighbors (k) and probability of reception (p_c). The model also shows another independent variable, namely, the maximum retransmission probability that can be assigned to the transmitting node ($p_{t,pcmin}$), which is assumed to be a fixed value. In this paper, we propose a new mathematical model for calculating p_t. In this new model $p_{t,pcmin}$ is calculated as a function of k. The performance of the DNDP algorithm using fixed and k-dependent $p_{t,pcmin}$ is evaluated through simulations. The simulation results showed that the new model enhances the performance of the DNDP algorithm as it significantly reduces the number of retransmissions at an insignificant reduction in the network reachability.

KEYWORDS

MANETs, dynamic routing protocols, route discovery, flooding optimization algorithms, probabilistic algorithm, the DNDP algorithm.

1. INTRODUCTION

A mobile ad hoc network (MANET) is defined as a collection of low-power wireless mobile nodes forming a temporary network without the aid of any established infrastructure or centralized administration [1, 2]. A data packet in MANET is forwarded to other mobile nodes on the network through reliable and efficient routing protocols, which are usually implemented in software as part of the network layer [3]. Dynamic routing protocols (DRPs), such as: ad hoc on-demand distance vector routing (AODV) [4], dynamic source routing (DSR) [5], zone routing protocol (ZRP) [6], and location-aided routing (LAR) [7], are widely used routing protocols in MANETs.

The DRPs consist of two phases: (i) route discovery, which is initiated when a node (source) desires to send a data packet to some node (destination) and does not have a valid route to that destination, and (ii) route maintenance, which is periodically initiated to maintain existing route; and if the route is broken for any reason, then the source either finds and uses other recognized route on its routing table or if it cannot find a route, it initiates a new route discovery procedure [3]. It is well approved that the cost of information exchange during route discovery is higher than the cost of point-to-point data forwarding after the route is discovered. Thus, significant efforts have been put forward by researchers to minimize the cost of route discovery [8].

In route discovery, the source initiates and broadcasts to its neighbours a small packet called route request packet (RREQ), which carries information on the source and destination addresses, packet sequence number (ID), and specific lifetime, i.e., time-to-live (TTL). Each recipient node checks to see if it is the destination; if not, then it appends its address to the payload of the RREQ and forwards it to its neighbours, and so on until the expiration of the

RREQ. If the RREQ reaches its destination, the destination sends a route reply packet (RREP) back to the source through the route from which it first received that particular RREQ [3, 5]. If the RREQ expires before reaching its destination, the node at which it expires, sends a route error packet (RERR) back to the source, once again through the route from which it first received that particular RREQ. As result of this route discovery failure, the source initiates a new RREQ and continues to do so until a route is found.

Pure flooding is one of the simplest and reliable mechanisms proposed in the literature for route discovery in MANETs [1, 9]. The main drawback of pure flooding is that it is very costly where it costs n transmissions in a network of n reachable nodes. In addition, it results in serious redundancy, contention, and collisions in the network; such a scenario has often been referred to as the broadcast storm problem (BSP) [10].

To eliminate the effects of BSP, a variety of flooding optimization algorithms have been developed, such as: probabilistic [1, 2, 11], multipoint relaying (MPR) [12, 13], counter/distance/location/cluster -based [3, 7, 10, 14] algorithms. They all try to minimize contention and collisions by limiting the number of retransmissions. As the number of retransmissions is decreased, the bandwidth and power are saved and contentions and collisions are reduced, and consequently the overall network performance is improved.

Probabilistic algorithm has been widely-used for route discovery in MANETs [1, 2]. However, it has been demonstrated that the performance of the probabilistic algorithm is severely suffered in presence of noise due to increases packet-loss (data or RREQ packets), and consequently reduces the overall network performance [15, 16].

A dynamic noise-dependent probabilistic (DNDP) algorithm was proposed by Al-Bahadili and Sabri in [17] for route discovery in noisy MANETs. In this algorithm, the retransmission probability of the transmitting node (p_t) is modeled as a function of two independent variable; these are: the number of first-hop neighbors (k) and the probability of reception (p_c). The model also shows another independent variable, namely, the maximum retransmission probability that can be assigned to the transmitting node ($p_{t,pcmin}$), which is assumed to be a fixed value.

In this paper, we propose a new mathematical model for calculating p_t. In this new model, $p_{t,pcmin}$ is calculated as a function of k. The new model is implemented on the MANET simulator (MANSim) [18], and the performance of the DNDP algorithm using fixed and k-dependent $p_{t,pcmin}$ is evaluated through simulations. The simulation results showed that the new model enhances the performance of the DNDP algorithm as it reduces the number of retransmissions at an insignificant reduction in the network reachability.

This section provides an introduction to the general domain of this paper. The rest of the paper is organized as follows. Related work is discussed in Section 2. The dynamic probabilistic and the DNDP algorithms are discussed in Sections 3 and 4, respectively. The proposed model is described in Section 5. Simulation results are presented and discussed in Section 6. Finally, in Section 7, conclusions are drawn and recommendations for future work are pointed-out.

2. PREVIOUS WORK

This section reviews some of the most recent and related work on probabilistic flooding in both noiseless and noisy MANETs. Probabilistic algorithm was first used by Haas et. al. [19] for route discovery in ad hoc networks, and they called it a gossip-based route discovery (GOSSIP1) approach. They used a predefined p_t to decide whether or not a node forwards the RREQ packets. GOSSIP1 has a slight problem with initial conditions. If the source has relatively few neighbors, there is a chance that none of them will gossip, and the gossip will die. To make sure this does not happen, Haas et. al. later developed a modified protocol, in which they gossip with p_t=1 for the first h hops before continuing to gossip with p_t<1. Their results showed that they can save up to 35% message overhead compared to simple flooding.

S. Tseng et. al. [10] investigated the performance of the probabilistic flooding for various network densities in noise-free environment. They presented results for three network parameters, namely, reachability, saved rebroadcast, and average latency, as a function of p_t and network density. Sasson et. al. [20] suggested exploring algorithms in which nodes would dynamically adjust their p_t based on local topology information. Because in their work they made the assumption that all nodes possess the same transmission range, they suggested another potential area for study which is to modify p_t according to the node radio transmission range.

Kim et. al. [21] introduced a broadcasting scheme in which a node dynamically adjusts its p_t according to its additional coverage area. The additional coverage is estimated by the distance from the sender. The simulation results showed this scheme generates fewer rebroadcasts than pure flooding approach. It also incurs lower broadcast collision without sacrificing high reachability.

Scott and Yasinsac [2] presented a dynamic probabilistic solution that is appropriate to solving BSPs in dense mobile networks. The approach can prevent broadcast storms during flooding in dense networks and can enhance comprehensive delivery in sparse networks.

Barret et. al. [22] introduced a probabilistic routing protocol for sensor networks. In this protocol, a sensor decides to forward a message with p_t that depends on various parameters, such as the distance of the sensor to the destination, the distance of the source sensor to the destination, or the number of hops a packet has already traveled. They proposed two protocol variants and compared the new methods to other probabilistic and deterministic protocols. The results showed that the multi-path protocols are less sensitive to misinformation, and suggest that in the presence of noisy data, a limited flooding strategy will actually perform better and use fewer resources than an attempted single-path routing strategy, also parametric probabilistic protocols outperforms other protocols. The results also suggested that protocols using network information perform better than protocols that do not, even in the presence of strong noise.

Viswanath and Obraczka [23] developed an analytical model to study the performance of plain and probabilistic flooding in terms of its reliability and reachability in delivering packets. Their results indicated that probabilistic flooding can provide similar reliability and reachability guarantees as plain flooding at a lower overhead.

Zhang and Agrawal [24] proposed a probabilistic scheme that dynamically adjusts p_t as per node distribution and node movement. The scheme combines between probabilistic and counter-based approaches. They evaluated the performance of their scheme by comparing it with simple flooding and fixed probabilistic algorithms. Simulation results showed that the new scheme performs better than the two algorithms.

Abdulai et. al. [25] studies the performance of the AODV protocol over a range of possible p_t. They focused on the route discovery part of the routing algorithm, they modified the AODV implementation to incorporate p_t; the RREQ packets are forwarded in accordance with a predetermined p_t. Simulation results showed that setting efficient p_t has a significant effect on the general performance of the protocol. The results also revealed that the optimal p_t for efficient performance is affected by the prevailing network conditions such as traffic load, node density, and node mobility. During their study they observed that the optimal p_t is around 0.5 in the presence of dense network conditions and around 0.6 for sparse network conditions.

Bani-Yassein et. al. [1] proposed a dynamic probabilistic flooding algorithm in MANETs to improve network reachability and saved rebroadcast. The algorithm determines p_t by considering the network density and node movement. This is done based on locally available information and without requiring any assistance of distance measurements or exact location determination devices. The algorithm controls the frequency of rebroadcasts and thus might save network resources without affecting delivery ratios.

Abdulai et. al. [26] proposed two probabilistic methods for on-demand route discovery, that is simple to implement and can significantly reduce the overhead involved in the dissemination of RREQs. The two probabilistic methods are: the adjusted probabilistic (AP) and the enhanced adjusted probabilistic (EAP) which address the broadcast storm problem in the existing OADV routing protocols.

Bani Yassein et. al. [9] combined probabilistic and knowledge based approaches on the AODV protocol to enhance the performance of existing protocol by reducing the communication overhead incurred during the route discovery process. The simulation results revealed that equipping AODV with fixed and adjusted probabilistic flooding helps to reduce the overhead of the route discovery process whilst maintaining comparable performance levels in terms of saved rebroadcasts and reachability as achieved by conventional AODV. Moreover, the results indicated that the adjusted technique results in better performance compared to the fixed one.

Khan et. al. [27] proposed a coverage-based dynamically adjusted probabilistic forwarding scheme and compared its performance with simple and fixed probabilistic schemes. The proposed scheme keeps up the reachability of simple flooding while maintaining the simplicity of probability based schemes.

Hanash et. al. [28] proposed a dynamic probabilistic broadcast approach that can efficiently reduce broadcast redundancy in MANETs. The algorithm dynamically calculates p_t according to k. They compared their approach against simple flooding approach, fixed probabilistic approach, and adjusted probabilistic flooding by implementing them in a modified version of the AODV protocol using GloMoSim. The simulation results showed that broadcast redundancy can be significantly reduced through their approach while keeping the reachability high.

Al-Bahadili [29] developed a new retransmission probability adjusting model, in which the neighborhood densities are divided into three regions (low, medium, and high). The performance of the new model was evaluated and compared with pure and other probabilistic algorithms. The model enhances the performance of probabilistic broadcast by reducing the number of transmissions while keeping almost the same network reachability.

Al-Bahadili and Kaabned [15] investigated the effect of noise-level on the performance of the probabilistic algorithm in MANETs. They investigated the effect of node density, node average speed, radio transmission range, p_t, and p_c on number of retransmissions, duplicate reception, average hop count, and reachability. Their results showed that the performance of the network is severely suffered as p_c increases, i.e. the noise-level increases.

3. THE DYNAMIC PROBABILISTIC ALGORITHM

In probabilistic broadcast, a node retransmits the packet with a certain p_t and with probability $(1-p_t)$ it discards the packet [1, 2]. A node is allowed to retransmit a given RREQ only once, i.e., if a node receives a RREQ, it checks to see if it has retransmitted it before, if so then it just discards it, otherwise it performs its probabilistic retransmission check. Nodes usually can identify the RREQ through its sequence number. The source node p_t is always set to 1, to enable it initializing a RREQ. On other hand, the p_t of the intermediate nodes (all nodes except the source and destination) can be set by using one of the following approaches:

(1) Static approach in which a pre-determined p_t is set for each node on the network and it can be expressed as: $p_t=P_t$, where P_t is a constant value $(0<P_t\leq1)$. $P_t=1$ for pure flooding.

(2) Dynamic approach in which each node on the network locally calculates its p_t as a function of k ($p_t(k)$), where $p_t(k)$ could be any linear/nonlinear function of k.

In this paper, we mainly concern with dynamic probability, therefore, in the next section, we shall provide a description of an efficient and flexible function that will be used for dynamically calculate $p_t(k)$ for intermediate nodes on the network.

3.1 Calculation of $p_t(k)$

Many functions have been developed for calculating $p_t(k)$ [1-2, 19-21, 24, 29]. However, in this paper, the discrete function presented in [29] is used for calculating $p_t(k)$. This function is shown in Figure 1. It demonstrates a satisfactory performance when used in various network conditions and it is mathematically expressed as:

$$p_t(k) = \begin{cases} p_{max} & for\ k \leq N_1 \\ p_1 - \frac{k-N_1}{N_2-N_1}(p_1 - p_2) & for\ N_1 < k < N_2 \\ p_{min} & for\ k \geq N_2 \end{cases} \quad (1)$$

Where p_{min} and p_{max} are the minimum and maximum retransmission probabilities, N_1 is the number of nodes at or below which $p_t(k)=p_{max}$, N_2 is the number of nodes at or above which $p_t(k)=p_{min}$, and p_1 and p_2 are the retransmission probabilities assign to intermediate nodes when they have N_1+1 and N_2-1 first-hop neighboring nodes. p_1 and p_2 should lie between p_{max} and p_{min}, and also p_1 is always $\geq p_2$.

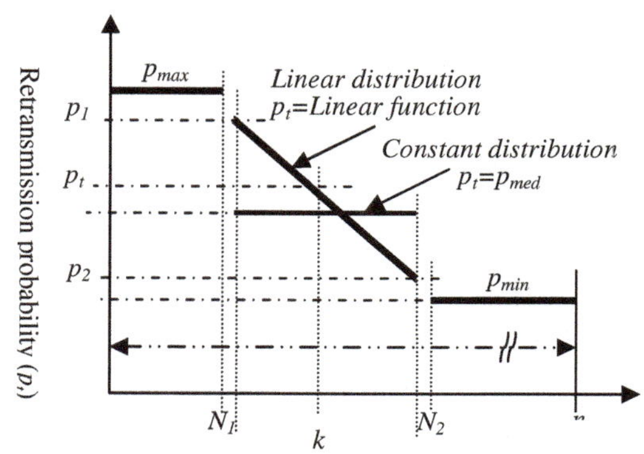

Figure 1. The k-dependent retransmission probability ($p_t(k)$)

In general, selection of a satisfactory distribution within the interval [N_1+1, N_2-1] and the values of p_{max}, p_{min}, p_1, p_2, N_1, and N_2 depend on a number of factors and need to be carefully selected. In this work, the values in [29] are used, which are 0.8, 0.5, 0.8, 0.5, 4, and 15 for p_{max}, p_{min}, p_1, p_2, N_1, and N_2.

Figure 2 outlines the main phases of the dynamic probabilistic algorithm in a noisy environment. To accommodate the effect of noise, it can be seen in Figure 2 that when the node is within the ratio transmission range of the transmitting node, a test is performed to find-out whether the RREQ packet is successfully delivered or not. This is tested by generating a random number ξ_1 and comparing it with p_c. If $\xi_1 \leq p_c$, then the packets is successfully delivered to the receiving node, otherwise, it is undelivered [15, 16].

4. The DNDP Algorithm

The DNDP algorithm was proposed to enhance the performance of dynamic probabilistic algorithm in noisy MANETs. In this algorithm, instead of calculating p_t as a function of k only, p_t is determined locally by the retransmitting nodes considering both k and p_c ($p_t(k, p_c)$) as:

$$p_t(k, p_c) = p_t(k) + p_t(p_c) \quad (2)$$

Where $p_t(p_c)$ is the noise-dependent p_t.

Dynamic Probabilistic Algorithm
If (*the receiving node is within the radio transmission range of the sender*) **Then**
ξ_1=rnd() {ξ_1 some random number between 0 and 1}
If ($\xi_1 <= p_c$) **Then** {Reception random test}
Update the node reception index
If (*The node has not retransmitted the received packet before*) **Then**
ξ_2=rnd() {ξ_2 some random number between 0 and 1}
p_t=function_p_t()
If ($\xi_2 \leq p_t$) **Then**
Retransmit the received packet
Update the node retransmission index
End if
End if
End if
End if
Function_p_t() {Determining p_t}
If (*Static probability*) **Then**
p_t=constant value
Else (*Dynamic probability*)
$p_t=f(k, p_c)$
End If

Figure 2. The dynamic probabilistic algorithm

4.1 Calculation of $p_t(k, p_c)$

It can be seen from Eqn. 2 that in order to calculate $p_t(k, p_c)$, we need to calculate $p_t(k)$ and $p_t(p_c)$. $p_t(k)$ can be calculated using Eqn. 1, and in this section we discuss the calculation of $p_t(p_c)$. The following main constraints should be considered when calculating $p_t(p_c)$,:

(1) The value of $p_t(p_c)$ should be ≥ 0 and $\leq 1-p_t(k)$, so that $p_t(k, p_c)$ will always be ≤ 1.

(2) The value of $p_t(k, p_c)$ lies between $p_t(k)$ and $p_{t,pcmin}$, which is a pre-adjusted maximum allowable p_t at a certain minimum value of p_c ($p_{c,min}$).

(3) The value of $p_{t,pcmin}$ should be ≤ 1 and $\geq p_t(k)$.

Considering the above constraints, $p_t(p_c)$ can be calculated as:

$$p_t(p_c) = \alpha(p_{t,pcmin} - p_t(k)) \tag{3}$$

Where α is called the noise-correction factor, which is a function of p_c, and has a value that lies between 0 and 1. Substituting Eqn. 3 into Eqn. 2 yields:

$$p_t(k, p_c) = p_t(k) + \alpha(p_{t,pcmin} - p_t(k)) \tag{4}$$

In [17], the value of α is calculated using the following equation:

$$\alpha = \frac{1 - p_c}{1 - p_{c,\min}} \tag{5}$$

Now, substituting Eqn. 5 into Eqn. 4 yields the following general equation for $p_t(k, p_c)$:

$$p_t(k, p_c) = p_t(k) + \frac{1 - p_c}{1 - p_{c,\min}}(p_{t,pc\min} - p_t(k)) \tag{6}$$

It can be seen from Eqn. 6 that $p_t(k, p_c)$ depends on $p_t(k)$, k, p_c, $p_{c,min}$, and $p_{t,pcmin}$.

In a noiseless environment p_c=1, α=0, and consequently $p_t(k, p_c)$=$p_t(k)$, i.e., p_t is a function of k only. In a noisy environment, when p_c=$p_{c,min}$, then α=1 and $p_t(k, p_c)$=$p_{t,pcmin}$. If p_c is any value between $p_{c,min}$ and 1, then $p_t(k, p_c)$ varies between $p_t(k)$ and $p_{t,pcmin}$ depending on p_c. According to the above discussion, $p_t(k, p_c)$ always lies between $p_t(k)$ and $p_{t,pcmin}$ as shown in Figure 3.

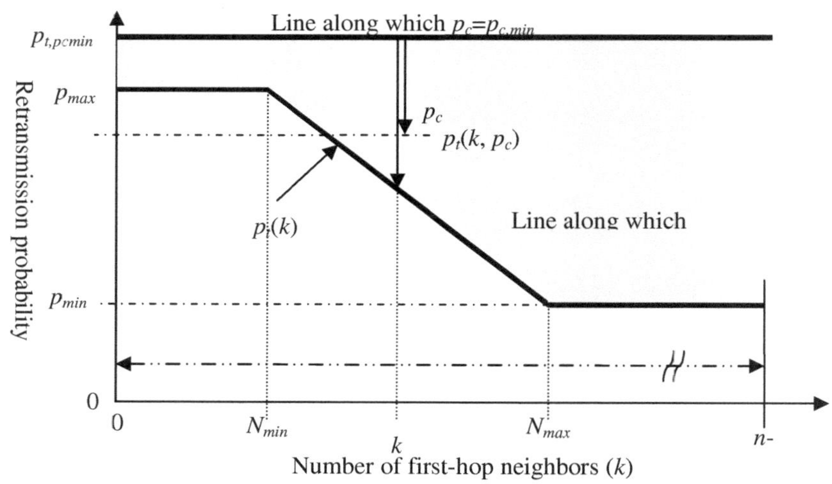

Figure 3. Variation of $p_t(k, p_c)$ (solid area) with k (constant $p_{t,pcmin}$)

5. ENHANCING THE DNDP ALGORITHM

In Eqn. 5, $p_{t,pcmin}$ is assumed to be a constant value, and as it can be seen in Figure 3 that the computed p_t is increased considerably with decreasing p_c. This may result in a significant increase in the number of retransmission, especially at high k. In order to alleviate the impact of this increase in p_t and to provide adjustable measure to control that increase, we introduce the following solution. The solution is simply based on using a k-dependent distribution for calculating $p_{t,pcmin}$, i.e., $p_{t,pcmin}$ can be set as a function of k ($p_{t,pcmin}(k)$) as shown in Figure 4.

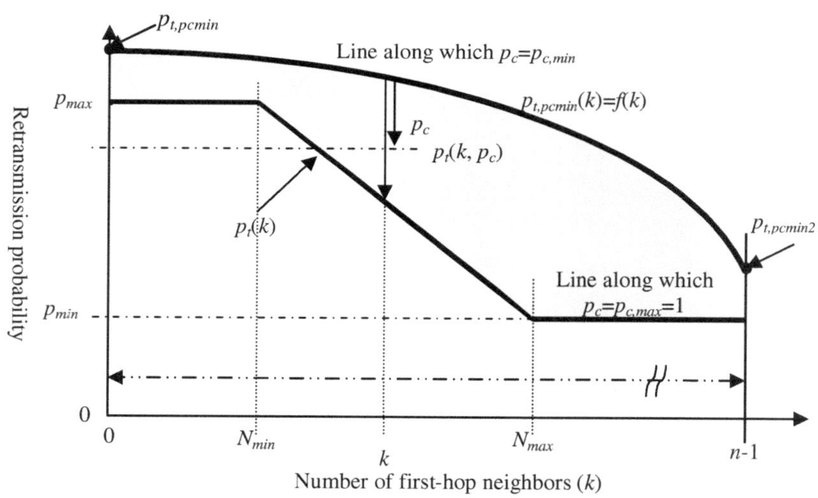

Figure 4. Variation of $p_t(k, p_c)$ (solid area) with k ($p_{t,pcmin}(k)$)

This distribution can be discrete similar to $p_t(k)$ or continuous distribution, such as linear, parabolic, exponential, etc. Thus, Eqn. 6 can be re-written as:

$$p_t(k,p_c) = p_t(k) + \frac{1-p_c}{1-p_{c,min}}\left(p_{t,pcmin}(k) - p_t(k)\right) \tag{7}$$

Mathematical representation of the linear, parabolic, and exponential distributions of $p_{t,pcmin}(k)$ can be given as follows:

Linear distribution:

$$p_{t,pc\min}(k) = p_{t,pc\min 1} + \frac{p_{t,pc\min 2} - p_{t,pc\min 1}}{n-1}k \tag{8}$$

Parabolic distribution:

$$p_{t,pc\min}(k) = p_{t,pc\min 1} + \frac{p_{t,pc\min 2} - p_{t,pc\min 1}}{(n-1)^2}k^2 \tag{9}$$

Exponential distribution:

$$p_{t,pc\min}(k) = p_{t,pc\min 1} + \frac{p_{t,pc\min 2} - p_{t,pc\min 1}}{1-e^{-\beta(n-1)}}(1-e^{-\beta k}) \tag{10}$$

Where $p_{t,pcmin1}$ and $p_{t,pcmin2}$ are the maximum allowable p_t in presence of noise at k=0 and k=n-1, respectively, and β is a constant ($\beta \leq 1$). It can be deduced from Figure 4 that due to the reduced size of the shaded area when compared with Figure 3, $p_t(k, p_c)$ can be tuned in a controlled way by choosing a proper distribution for $p_{t,pcmin}(k)$ for the sake of achieving optimum performance in terms of network reachability and number of redundant retransmissions.

6. Results and Discussion

In this section, we investigate the effect of $p_{t,pcmin}(k)$ on the performance of the DNDP algorithm through a number of simulations using the MANET simulator (MANSim) [18]. Also, in this section, the performance of the DNDP algorithm with different distribution for $p_{t,pcmin}(k)$ is compared against the performance of pure flooding, fixed and dynamic probabilistic algorithms, and the DNDP algorithm with fixed $p_{t,pcmin}$.

A number of network performance measures are calculated using MANSim, such as: network reachability (RCH), number of retransmission (RET), average duplicate reception (ADR), average hop count (AHP), saved rebroadcast (SRB), and disconnectivity (DIS). These parameters are recommended by the Internet Engineering Task Force (IETF) group to judge the performance of the flooding optimization algorithms. Definition of these parameters can be found in [13].

This paper presents results for two parameters only, these are: RCH and RET. RCH is defined as the average number of reachable nodes by any node on the network normalized to n; or the probability by which a RREQ packet successfully delivered from source to destination node. RET is defined as the average number of retransmissions normalized to n. The input parameters for these simulations are given in Table 1.

Table 1 - Input parameters.

Parameters	Values
Geometrical model	Random node distribution
Network area	600x600 m
Number of nodes (n)	100 nodes.
Transmission radius (R)	100 m
Average node speed (u)	5 m/sec
Probability of reception (p_c)	From 0.5 to 1.0 in step of 0.1
Simulation time (T_{sim})	1200 sec
Pause time (τ)	τ= 0.75*(R/u) = 15 sec
Size of the mobility loop ($nIntv$)	$nIntv = T_{sim}/\tau = 80$
Fixed: p_t	0.744
Dynamic: p_{min}, p_{max}, N_{min}, N_{max}, p_1, p_2	0.5, 0.8, 4, 15, 0.5, 0.8
Fixed: $p_{t,pcmin}$	1

Linear: $p_{t,pcmin}(k)$	$p_{t,pcmin1}$=1 and $p_{t,pcmin2}$=0.5
Parabolic: $p_{t,pcmin}$	$p_{t,pcmin1}$=1 and $p_{t,pcmin2}$=0.5
Exponential: $p_{t,pcmin}$	$p_{t,pcmin1}$=1 and $p_{t,pcmin2}$=0.5, β=1

The main outcomes of this scenario can be summarized as follows:

- For pure and fixed/dynamic probabilistic algorithms, RCH decreases as noise-level increases, due to the high packet-loss introduced by the high noise-level; and no measure is taken by the existing probabilistic algorithms to accommodate the negative effect of increasing noise-level.
- The DNDP algorithm with any of the suggested distribution functions for $p_{t,pcmin}(k)$ presents a better performance in terms of increasing RCH in presence of noise by effectively adjusting p_t based on both k and p_c. The results obtained demonstrated that the algorithm provides the highest RCH for various network noise-level, when compared with fixed/dynamic probabilistic algorithms.
- It can be seen from Figure 6 that the DNDP algorithm almost produces the same RCH using fixed, linear, and parabolic $p_{t,pcmin}(k)$ distribution, while the exponential distribution produces a lower RCH, but still it is higher than what can be obtained using fixed/dynamic probabilistic algorithms. However, enhancing RCH is paid by slight increase in RET as shown in Figure 7.
- In a noiseless environment, the fixed p_t is taken to be equal to $p_{t,avg}$ of the dynamic approach, which is 0.744. In a noisy environment, the average p_t ($p_{t,avg}$) depends on the area under the $p_{t,pcmin}(k)$ curve, and it can be seen in Figure 5 that the area under the curve can be arranged from the largest to the smallest as follows: fixed, parabolic, linear, and finally exponential. For the noiseless environment, $p_{t,avg}$ for the fixed and the dynamic probabilistic algorithms are equal and always less than that for the noisy environment. This comes in line with the results presented in Figure 8.
- The dynamic probabilistic algorithm and the DNDP algorithm with any $p_{t,pcmin}(k)$ distribution produced the same performance in noiseless environment (p_c=1). This is because for p_c=1, the noise-correction factor (α) (Eqn. 5) and the noise-dependent retransmission probability ($p_t(p_c)$) (Eqn. 3) are equal to zero. Consequently $p_t(k, p_c)=p_t(k)$ (Eqn. 7), which means all algorithms use the same distribution for p_t ($p_t=p_t(k)$).

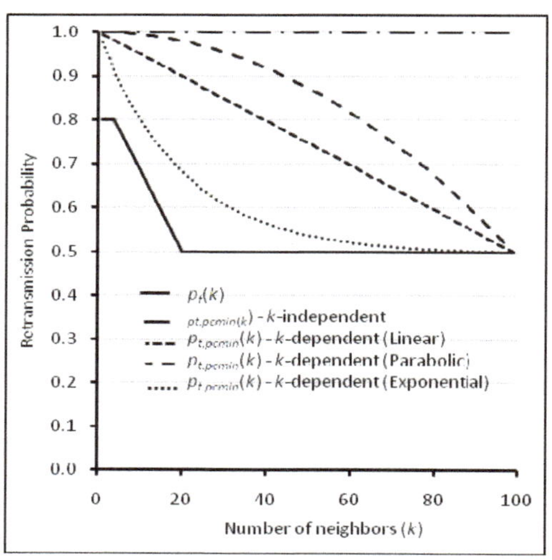

Figure 5. The variation of $p_{t,pcmin}(k)$ with k

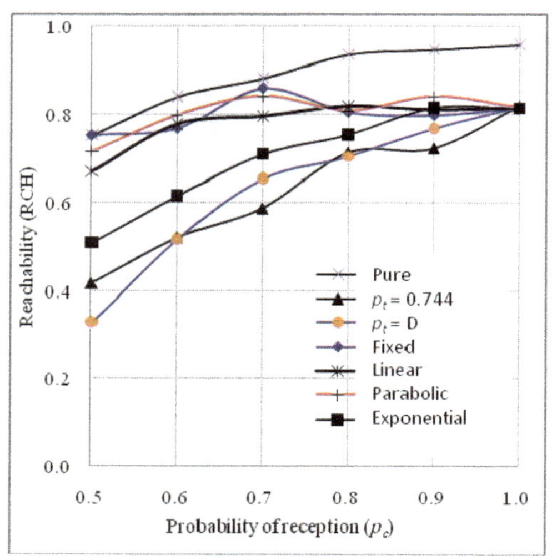

Figure 6. RCH against p_c for various algorithms

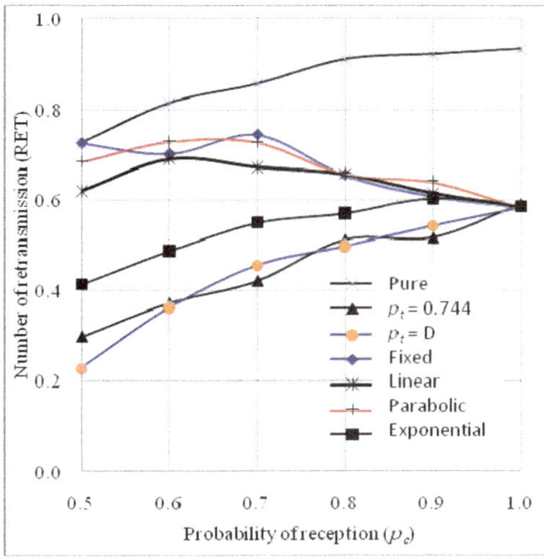

Figure 7. RET against p_c for various algorithms.

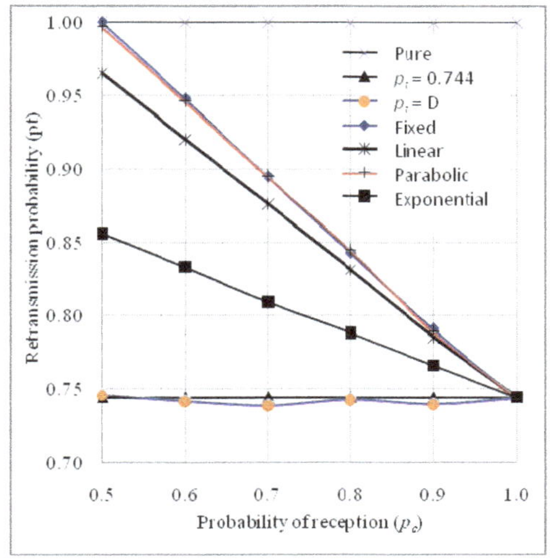

Figure 8. $p_{t,avg}$ against p_c for various algorithms.

7. CONCLUSIONS

The main conclusion of this work is that using a k-dependent $p_{t,pcmin}$ ($p_{t,pcmin}(k)$) enhances the performance of the DNDP algorithm, where the RCH can be improved against slight increases in RET or maintaining the same RCH against significant reduction in RET. The DNDP algorithm with linear $p_{t,pcmin}(k)$ provides the optimum performance when compared with other distributions. The results also demonstrated that the RCH of the DNDP algorithm is close to the RCH of pure flooding for various network noise levels at less cost.

It is highly recommended to investigate the effect of node densities and nodes speed and other network parameters on the performance of the DNDP algorithm with various distribution for $p_{t,pcmin}(k)$.

REFERENCES

[1] M. Bani-Yassein, M. Ould-Khaoua, L. Mackenzie, and S. Papanastasiou, "Performance Analysis of Adjusted Probabilistic Broadcasting in Mobile Ad Hoc Networks, International Journal of Wireless Information Networks, Vol. 13, No. 2, pp. 127-140, 2006.

[2] D. Scott and A. Yasinsac, "Dynamic Probabilistic Retransmission in Ad Hoc Networks", Proceedings of the International Conference on Wireless Networks (ICWN'04) (Editors: H. R. Arabnia, L. T. Yang, and C. H. Yeh), Las Vegas, Nevada, USA, CSREA Press, Vol. 1, pp. 158-164, 2004.

[3] E. Royer and C. Toh, "A Review of Current Routing Protocols for Ad Hoc Mobile Wireless Networks", IEEE Personal Communication Magazine, Vol. 6, No. 2, pp. 46-55, 1999.

[4] Z. Ismail and R. Hassan, "Evaluation of Ad Hoc on Demand Distance Vector Routing Protocol in HetMAN Architecture", Journal of Computer Science, Vol. 6, No. 7, pp. 830-836, 2010.

[5] D. Johnson and D. Maltz, "Dynamic Source Routing in Ad Hoc Wireless Networks", Mobile Computing, Kluwer Academic Publishers, pp. 153-181, 1996.

[6] Z. Haas and M. Pearlman, "The Zone Routing Protocol (ZRP) for Ad Hoc Networks", Mobile Ad Hoc Network Working Group, IETF, 1998.

[7] Y. Ko and N. Vaidya, "Location-Aided Routing (LAR) in Mobile Ad Hoc Networks", Journal of Wireless Networks, Vol. 6, No. 4, pp. 307-321, 2000.

[8] A. Rahman, W. Olesinski, and P. Gburzynski, "Controlled Flooding in Wireless Ad Hoc Networks. Proceedings of IEEE International Workshop on Wireless Ad Hoc Networks (IWWAN'04), University of Oulu, Finland, 2004.

[9] M. Bani-Yassein and M. Ould-Khaoua, "Applications of Probabilistic Flooding in MANETs", International Journal of Ubiquitous Computing and Communication, Vol. 1, No. 1, pp. 1-5, 2007.

[10] T. Tseng, S. Ni, Y. Chen, and J. Sheu, "The Broadcast Storm Problem in a Mobile Ad Hoc Network", Journal of Wireless Networks, Vol. 8, No. 2, pp. 153-167, 2002.

[11] H. Al-Bahadili, "Enhancing the Performance of Adjusted Probabilistic Broadcast in MANETs. The Mediterranean Journal of Computers and Networks (MEDJCN), Vol. 6, No. 4, 2010.

[12] A. Qayyum, L. Viennot, and A. Laouiti, "Multipoint Relaying for Flooding Broadcast Messages in Mobile Wireless Networks", Proceedings of the 35th Hawaii International Conference on System Sciences (HICSS'02), Island of Hawaii, pp. 3866- 3875, 2002.

[13] H. Al-Bahadili, R. Jaradat, "Performance Evaluation of an OMPR Algorithm for Route Discovery in Noisy MANETs", International Journal of Computer Networks and Communications (IJCNC), Vol. 2, No. 1, pp. 85-96, 2010.

[14] C. Bettstetter, "The Cluster Density of a Distributed Clustering Algorithm in Ad Hoc Networks", Proceedings of the IEEE International Conference on Communications (ICC'04), Paris, France, IEEE Computer Society Press, Vol. 7, pp. 4336-4340, 2004.

[15] H. Al-Bahadili and K. Kaabneh. Analyzing the Performance of Probabilistic Algorithm in Noisy MANETs. *The International Journal of Wireless & Mobile Networks* (IJWMN), Vo. 2, No. 3, pp. 83-95, 2010.

[16] H. Al-Bahadili and Y Jaradat, "Development and Performance Analysis of a Probabilistic Flooding in Noisy Mobile Ad Hoc Networks", *Proceedings of the 1st International Conference on Digital Communications and Computer Applications* (DCCA'07), Irbid, Jordan, pp. 1306-1316, 2007.

[17] H. Al-Bahadili and A. Sabri, "A Novel Dynamic Noise-Dependent Probabilistic Algorithm for Route Discovery in MANETs", International Journal of Business Data Communication and Networking (IJBDCN), Vol. 7, No. 1, pp. 52-67, 2011.

[18] H. Al-Bahadili, "On the Use of Discrete-Event Simulation in Computer Networks Analysis and Design", In Handbook of Research on Discrete-Event Simulation Environments: Technologies and Applications (Eds.: E. Abu-Taieh and A. El-Sheikh,), Information Science Reference, USA, pp. 414-442, 2009.

[19] Z. J. Haas, J. Y. Halpern, and L. Li, "Gossip-Based Ad Hoc Routing", IEEE/ACM Transactions on Networking, Vol. 14, No. 3, pp. 479-491, 2006.

[20] Y. Sasson, D. Cavin, and A. Schiper, "Probabilistic Broadcast for Flooding in Wireless Mobile Ad Hoc Networks. Proceedings of IEEE Wireless Communications and Networking (WCNC'03), Vol. 2, New Orleans, LA, USA, pp. 1124-1130, 2003.

[21] J. S. Kim, Q. Zhang, and D. P. Agrawal, "Probabilistic Broadcasting Based on Coverage Area and Neighbor Confirmation in Mobile Ad Hoc Networks", Proceedings of IEEE Global Telecommunications Conference Workshops (GlobeCom'04), Dallas, Texas, USA, IEEE Computer Society Press, pp. 96–101, 2004.

[22] C. Barrett, S. Eidenbenz, L. Kroc, M. Marathe, andL. Smith, "Parametric Probabilistic Routing in Sensor Networks", Journal of Mobile Networks and Applications, Vol. 10, No. 4, pp. 529-544, 2005.

[23] K. Viswanath, and K. Obraczka, "Modeling the Performance of Flooding in Wireless Multi-Hop Ad Hoc Networks", *Journal of Computer Communications*, Vol. 29, No. 8, pp. 949-956, 2005.

[24] Q. Zhang and D. P. Agrawal, "Dynamic Probabilistic Broadcasting in MANETs", Journal of Parallel and Distributed Computing, Vo. 65, No. 2, pp. 220-233, 2005.

[25] J. Abdulai, M. Ould-Khaoua, L. Mackenzie, and M. Bani-Yassin, "On the Forwarding Probability for On-Demand Probabilistic Route Discovery in MANETs", Proceedings of the 22nd Annual UK Performance Engineering Workshop (UKPEW'06), Poole, Dorset, UK, pp. 9-15, 2006.

[26] J. Abdulai, M. Ould-Khaoua, and L. Mackenzie, "Improving Probabilistic Route Discovery in Mobile Ad Hoc Networks", Proceedings of the 32nd Conference on Local Computer Networks (LCN '07), Dublin, Ireland, IEEE Computer Society, pp. 739-746, 2007.

[27] I. Khan, A. Javaid, and H. Qian, "Coverage-Based Dynamically Adjusted Probabilistic Forwarding for Wireless Mobile Ad Hoc Networks", Proceedings of the 1st ACM International Workshop on Heterogeneous Sensor and Actor Networks (HeterSanet '08), Hong Kong, China, 2008, pp. 81-88, 2008.

[28] A. Hanash, A. Siddique, I. Awan, and M. Woodward, "Performance Evaluation of Dynamic Probabilistic Broadcasting for Flooding in Mobile Ad Hoc Networks", Journal of Simulation Modeling Practice and Theory, Vol. 17, No. 2, pp. 364-375, 2009.

[29] H. Al-Bahadili, "Enhancing the Performance of Adjusted Probabilistic Broadcast in MANETs", The Mediterranean Journal of Computers and Networks (MEDJCN), Vol. 6, No. 4, 2010.

4

INTEGRATION OF 4G WIRELESS TECHNOLOGIES IN A TEST-BED ENVIRONMENT

Amir Esmailpour[1], Veljko Knezevic[2], Maria Gracias[3], and Gelareh Kokabian[4]

[1,4]Department of Electrical and Computer Engineering,
University of New Haven, West Haven, USA
([1]aesmailpour, [4]gkoka1)@newhaven.edu

[2,3]Department of Computer Networks, Ryerson University, Toronto, Canada
([2]vknezevi and [3]maria.gracias)@ryerson.ca

ABSTRACT

Amongst other mandates of the 4th Generation of Wireless Networks (4G), integration and unification of various technologies is a critical subject. In this paper an end-to-end wireless network deployment in a test-bed environment is established to demonstrate a practical model for integration of various wireless technologies used by ISPs to provide services to large and small enterprise customers. The focus of this research is on the Quality of Service (QoS) considerations in a WiFi-WiMAX integrated network. The results show that integrated deployment does not deteriorate throughput as compared to a standalone WiMAX system. The throughput of approximately 967.5 kbps was observed throughout the path and shows a steady performance across various parts of the network regardless of the technology involved. Numerical results also confirm that outcomes for effectiveness of the QoS routines defined in the proposed architecture are not severely affected by the integration.

KEYWORDS

WiMAX, WiFi, Wireless technology integration, IPerf

1. INTRODUCTION

In recent years, wireless and mobile applications have rapidly become an integral part of our everyday life, demanded by a large portion of the global population. These applications and their user demands have led to the development of innovative wireless research technologies at a fast pace. Researchers in this area work on sophisticated solutions that involve network integration, Quality of Service (QoS), security as well as all-IP networks with anywhere and anytime network availability.

Workplace mobility is essential for companies to perform their day-to-day activities. Mobile devices are equipped with business-like applications to meet employees' needs such as use of emails, make voice and video calls and utilize many other web based client applications. Several competing broadband wireless technologies exist to provide enterprises with wireless coverage on local and/or global scale. These wireless technologies include 3G/4G cellular networks such as Wideband CDMA, High Speed Packet Access (HSPA), Worldwide Interoperability for Microwave Access (WiMAX), Long Term Evolution (LTE) and Wireless Fidelity (WiFi) 802.11n. Majority of enterprises have adopted Wireless Local Area Network (WLAN) deployments for their infrastructure which are based on WiFi. WLANs are fixed networks, have relatively small coverage area but have very high throughput. Extending WLAN coverage to the Internet usually requires enterprises to connect their wireless Access Points

(APs) to the wired networks. Enterprises located in remote areas far from Internet Service Provider's (ISP) core have a choice of several technologies to use to backhaul their traffic. WiMAX is one of the contending 4G technologies that can extend long range coverage to WLANs [1].

Due to widespread adoption of WLAN, and level of consumer comfort resulting by its low cost and ease of maintenance, WiMAX is not introduced as a replacement, but rather a complement to the technology. WiMAX has been proposed as an ideal backhaul to WiFi networks, which could provide an end-to-end wireless network if deployed in an integrated WiFi-WiMAX infrastructure.

Wireless technologies are consistently improving many different aspects of their functionalities such as transmission speed, coverage, and QoS support. The prevalence of WLANs in the consumer and enterprise market made IEEE 802.11 the de-facto standard that has been deployed around the globe; give that there are limitations inherit to WiFi standard. Conventionally WiFi setups still would require a wired connection as their backend to be able to connect to Internet. This is not the desirable prerequisite in many instances such as rural or less developed areas which lack the required infrastructure. This need justifies a new wireless schema that would provide the required backhaul between the WiFi spots and the internet backbone in a reasonable and cost effective manner. One such solution is the IEEE 802.16 family of Wireless Metropolitan Area Network (WMAN) technologies that showcase an encouraging resolution to provide WLAN hotspots with required backend support. This hybrid approach would provide an ideal answer to our requirements for having a cost effective and broadband wireless network solution.

The purpose of this research is to design and implement an integrated WiFi-WIMAX network. Our design resembles an enterprise with two separate remote offices each with a local WLAN of its own that are connected to each other through a WiMAX backhaul network. We have established connection from one of these locations to the other through our WIMAX network. Our next goal is to define and study implementing a QoS system on our network. We created multiple scenarios with dissimilar usage patterns and analyzed the result of applying different classes of service on their access time and throughput.

The rest of this paper is organized in the subsequent arrangement. Section 2 is 4G technologies with emphasis on the background and comparison of how each wireless technology defines and implements QoS support. In Section 3 we discuss the design and define the requirements for the simulated network topology. In Section 4, we establish a test-bed environment to investigate the end-to-end connectivity and data rate flow. Section 5 covers the results and also provides analysis for the gathered data. Section 6 is conclusion to the study.

2. BACKGROUND

The transition to wireless network usage is increasing at an incredible rate. 3G/4G high-speed wireless technologies such as WiMAX and LTE, and WLANs' WiFi, will coexist, working in tandem to meet service provider and customer needs for truly mobile computing. No single technology can be ubiquitous since they all meet unique user requirements in a wirelessly connected world. The most robust wireless solutions will use a combination of technologies to enable increased mobility and eventually seamless roaming [1, 2].

2.1. New developments in 4G

New developments in 4G account for the success of recent wireless technologies such as LTE and WiMAX [2]. Those include new techniques in medium access, Orthogonal Frequency Division Multiple Access (OFDMA), and Multiple Input Multiple Output (MIMO) to name a few [3]. 4G networks go beyond the traditional wireless technology capabilities and explore many features that have not yet been explored comprehensively.

Important areas to this end include complete IP layer support using wireless technologies, end-to-end wireless connectivity, wireless network integration, security and QoS support at packet level. In this research we examine one of these areas namely wireless integration between WiFi and WiMAX technologies as well as the QoS considerations for such integrated networks [4].

2.2. Technologies involved in the wireless integration

Wireless technology integration is an important area in the race towards 4G. Different wireless technologies support wide variety of applications, features, and capabilities. By integrating these technologies, designers could implement solutions that utilize the capability of one technology to compensate the shortcoming of the other. In-order to accomplish the above, different types of cellular-WLAN combinations such as WiFi-LTE and WiFi-WiMAX have been proposed. [5].

As shown in Table 1, complementary features of WiFi and WiMAX make it a suitable choice for integrating these two technologies. WiFi is plug and play network while WiMAX is like a carrier transport network. WiFi is built into laptops and handsets, whereas WiMAX requires larger standalone receivers mounted on customer rooftops for optimum utility. An important benefit of WiMAX is its spectral efficiency, at least 50% more than 3G cellular networks, so it has much higher data-carrying capabilities in limited spectrum [6]. WiFi provides service over a range of 250m; whereas WiMAX can carry 100Mbps over a range of 50km between base-stations. WiFi provides a proven case for enterprise networks, while WiMAX is capable of carrier-class networks.

Table 1: Complementary capabilities of WiFi and WiMAX

Parameter	WiFi	WiMAX
Difficulty in deploying	Easy and cheap	Difficult and expansive
Mobility	Walking speeds	120 km/h
Range	250 m	50 km
Active Users	13	150
Data Rate	288.90 Mbps	100 Mbps
Typical Usage	WLANs	ISP Carrier-class networks

WiFi-backhaul has been proposed as one of the most popular derivers for WiMAX applications WiFi could benefit from WiMAX features such as long range coverage, security, QoS, and VPN capabilities. An integrated WiFi-WiMAX network is an ideal technology for service providers, to provide end-to-end wireless networks to customers in rural areas (Figure 1).

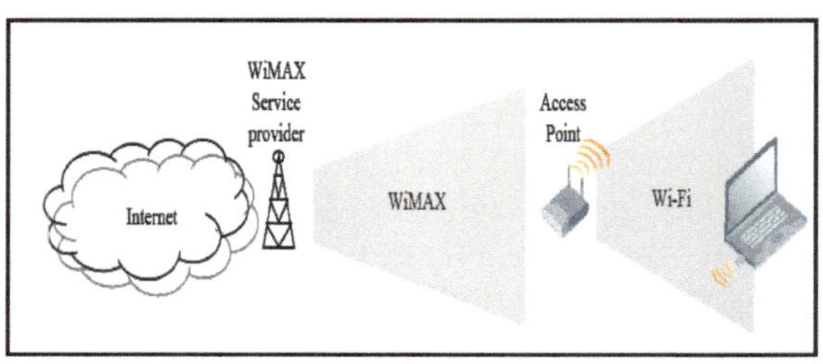

Figure 1: WiFi client with WiMAX Backhaul

2.3. QoS in Wireless Systems

Quality of Service is interpreted differently based on application and usage. Consequently one could define a set of parameters used to measure variety of radio resource levels related to the end user requirements [7]. WiMAX provides the ISPs with an excellent opportunity in allowing them to offer variety of different plans to customers depending on cost and usage model. They could potentially use different flow types offered in WiMAX to provide different levels of guaranteed service for voice, data, and video [8, 9]. This is highly desirable for both the enterprise and consumer as well. For example using a voice over IP (VoIP) program should promptly be recognized by the system and prioritized in a way that guarantees an unbroken and high quality conversation to be happening. Therefore we require a flexible QOS implementation that provides acceptable levels of latency and jitter which are the requirements of providing time sensitive services such as VoIP and Video Conferencing. The IEEE802.16d WiMAX standard provides five classes of prioritization of traffic which as follows: Unsolicited Grant Service (UGS), Real-Time Polling Service (rtPS), Extended Real-Time Polling Service (ertPS), Non-Real-Time Polling Service (nrtPS), and Best Effort (BE) [10].

WiMAX has two scheduling agents that work in tandem to make the system work; one in the Base Station (BS) and the other in the Subscriber Station (SS). Main scheduling unit would be the BS, which controls all system parameters and the burst profile and transmission periods for every connection. It is also responsible for selection of coding and modulation techniques based on the link quality, the demand, and the load on the network. On the SS side the main job of the scheduling agent is to categorize all the outgoing packets in to different classes.

3. Integrated Network Design and Implementation

The complete network architecture is comprised of an ISP core and several sites connecting to the core using a combination of WiMAX and WiFi and based deployments.In the current study we have designed a complete WiFi-WiMAX network, and demonstrated how service providers could capitalize on benefits of both technologies to provide an end-to-end wireless network, without use of infrastructure networks or cellular systems. We employed Cisco Unified Wireless Network [11] and AlvarionBreezeMAX Extreme [12] platform solutions respectively to implement the end-to-end wireless network, and measured network performance and traffic using software application Paessler Router Traffic Grapher (PRTG) network monitoring solution [13]. In this section, we illustrate the overall topology of our design and how we used integration techniques to create end-to-end WiFi-WiMAX-WiFi connectivity.

3.1. Design Overview

WiFi WLAN deployment including Access Points (AP) and Wireless LAN Controller (WLC); a WiFi-WiMAX network integration including base station, subscriber stations and ASN-gateway; and a Network Operation Center (NOC) side with the required equipment: routers, switches, severs, etc.

The design and implementation is divided into three distinct sections – a large enterprise, a small business, and an ISP core. The large enterprise network is scattered over multiple sites located in rural places, whereas the NOC, sales sites, and head office are located in urban areas. WiMAX is an apt choice to connect the enterprise's sites. This enterprise has its server site provisioned in the NOC directly connected to the ISP's core network.

The second enterprise is an example of a small business. It resides in an urban area and has connectivity to the service provider's network through wired networks. The small enterprise site does not require WiMAX. Instead it utilizes a wireless LAN controller and light weight

APs with in-house DHCP and authentication servers to provide a high quality wireless experience to its customers and employees.

The ISP is the core part of the network environment. It will provide wireless connectivity using WiMAX for the rural enterprise sites thus connecting them to the Internet and other enterprise sites. The ISP consists of core routers connected to the Base Transceiver Station (BTS) as illustrated in Figure 2.

The main objective of this study is to measure an end-to-end QoS support in a WiFi – WiMAX network model. To create this topology we have built a core wireless network, based on WiMAX technology which consists of a Base Station and two Subscribers. We have also used an internal DHCP and local authentication for assigning IP addresses and authenticating the subscribers.

The WiMAX segment is considered to be the ISP section of our design. We used AlvarionBreezeMAX Extreme (BTS 5Ghz RADIO and Embedded ASN Gateway, Version 1.5.1.72) as our base station and AlvarionBreezeMAX PRO 5000 CPE as our customer side WiMAX receiver.

On the WiFi side, we have simulated an enterprise which consists of two different locations. Both locations are based on wireless environments with Cisco's Wireless LAN Controllers (WLC) and Light weight Access Points (LWAP). More specifically the WLCs are Cisco 4402 IOS Version 7.01 and LWAPs are Cisco Aironet 3500 IOS Version 7.01. WLC on each site are using layer 2 authentications to validate the APs and Layer 3 authentication to verify clients. They are also using an internal DHCP server for assigning IP addresses to the clients and to the APs[14].

Our service provider network provides three different levels of service: Best Effort (BE), none Real Time (nRT), and extended Real Time (eRT) services. We will employ these levels of service for three types of traffic: management, data and VOIP in BS and will compare their differences.

We used PRTG Network Monitoring Solution software Version 8 [13] for monitoring the network based on predefined levels of service we have in the WiMAX BS. We monitored five different types of packets that go through the system, and compared their performance in different levels of service: Ping, HTTP, FTP, RDP and VOIP.

3.2. Enterprise WiFi-WiMAX Network Architecture

The large enterprise has two main sites, located miles away from each other. In order to provide connectivity to its sites, the company utilizes WiMAX service from the ISP. The ISP's access router is plugged into a LAN switch. The WiMAX BTS, RADIUS server and a management station are also connected to the same LAN network. The Customer Premise Equipment (CPE) is located directly at one of the sites and communicates with the BTS wirelessly.

The management station equipped with the Alvaricraft software is utilized to configure the BTS. The free RADIUS server is used by the BTS to authenticate a CPE connecting to it, thus preventing any rogue CPE's from unauthorized access. The BTS comprises of an inbuilt DHCP server which is used to assign IP addresses both to the CPE as well as the network behind the CPE.

The CPE is directly connected to an AP which provides wireless connectivity to all clients. The AP and all the clients receive IP addresses from the BTS. Client machines that possess a WiMAX card can directly connect to the BTS; however WiMAX NIC cards are not yet popular, hence we extend WiMAX connectivity by using WiFi wireless LANs. This provides more flexibility in terms of wireless SSID's and user security options, since the CPE can be managed locally from client side machines by authorized users as well as over the air from the management station connected to the BTS.

The hypothetical enterprise has two different locations (branch offices). Both locations are using Unified Wireless LAN Controller. To provide end-to-end connectivity and Internet connection for the two sites they both have a WiMAX network connection in their network.

In each location the service provider has installed a CPE to establish wireless connectivity from the customer site to the service provider's network. Since the company's systems are only equipped with WiFi adaptors, we used a switch to build a connection between WiFi and WiMAX network. In order to do this we plugged our CPE to the same switch which we have our Wireless LAN Controller and LWAP on it. The CPE is used as a Gateway to Service Provider.

The second site for the enterprise network is co-located with the NOC which houses the corporate servers. The NOC consists of an oracle database server which stores the entire inventory and the company's intellectual properties. This server needs to be accessible from the other sites, so that employees at the site can update information as and when required. It also includes a web server, which hosts the enterprise corporate site.

We have also implemented a simplified representation of a service provider's core network, which plays the role of connecting inter and intra company sites. The core is utilized to demonstrate end-to-end connectivity, and consists of three routers using RIP-v2 for routing. The main site, the off-shore site and the NOC are connected to different routers in the ISP's core network.

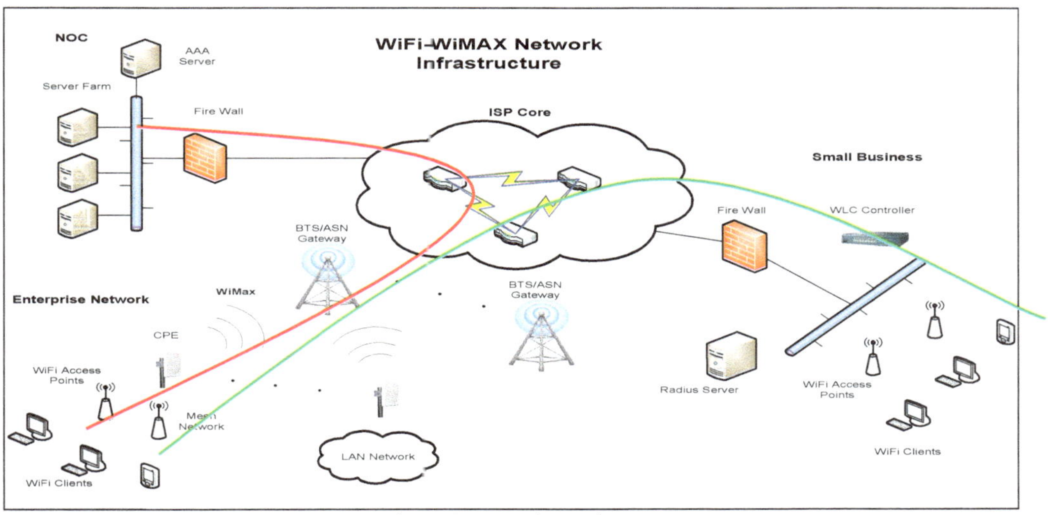

Figure 2 Network Topology
(a) Traffic flow from Enterprise Network to NOC
(b) Traffic flow from Enterprise Network to Small Business

3.3. WiFi Wireless LAN Solution

The small company resembles a coffee shop located in the city. The ISP's access router is connected to a switch in the shop. The small business is spread over two floors in a building connected through multiple APs for better wireless network coverage to all. For effective management of all APs we used Cisco's 2100 series wireless LAN controller (WLC) for AP management. The WLC and the APs are all connected to the same switch as the ISP.

The controller broadcasts two SSIDs called 'EE_Guest' and 'EE_Employees', which provide internet access to customers and the employees respectively. Since employees may need to access sensitive information, all users connecting to this SSID are authenticated by a radius server. The inbuilt DHCP server provides IP addresses to clients connecting to both SSID's. The access router from the ISP plays a router on stick role to provide inter-connectivity between the two SSID's.

3.4. WiMAX Solution Implementation

WiMAX technology is the solution for many applications transported across long distances with QoS support, longer reach, and high data rate. Typical applications for enterprise class networks are high speed Internet, VoIP telephony, Video conferencing and surveillance, and secured VPN.

WiMAX solution is provided to integrate various sites for the enterprise. In the proposed network scenario, we have two sites connected via the ISP's core network; the server farm site connected via an access router to the ISP's core network; and the client handling site, which utilizes WiMAX technology as a fixed broadband connection. The BTS and CPE are both provided and configured by the ISP to provide fixed WiMAX service. The CPE on the enterprise customer site can be plugged into a switch through which all employees can easily access the internet.

When a WiMAX service is used, clients need to be provisioned and a billing relationship established, in the same way as mobile phone or cellular data service. Rather than a SIM card, as in GSM networks, or a hardcoded equipment identifier as in CDMA, the WiMAX subscriber is identified by an X.509 certificate embedded in the client's device. The provisioning steps for the scenarios are required to identify the WiMAX client cards, and to activate them with the WiMAX network operator. All WiMAX trials and services thus far use operator-supplied client cards.

WiFi client cards are inexpensive as compared to WiMAX cards. The new WiFi -3G dual-mode phones are comparable in price to their cellular-only counterparts, while the simplest PCs today incorporate WiFi cards. Since WiFi entails no service subscription, it is a more economical connectivity option for all manner of consumer and business devices. Hence our enterprise extends wireless connectivity by converting the WiMAX connection to WiFi. This is achieved by plugging an AP to the Ethernet port of the CPE as shown in Figures 3.

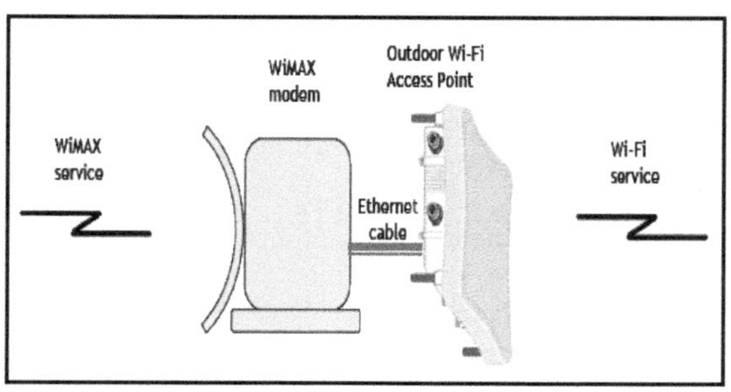

Figure 3 WiMAX - WiFi Integration

In the small business site we deploy WLANs to increase employee productivity, enhance collaboration, and improve responsiveness to customers. The company uses Cisco Unified Wireless Network Solution (CUWNS) which is a wired and wireless network solution that cost-effectively addresses the wireless network security, deployment, management, and control issues an enterprise faces with a low total cost of ownership. The modular architecture supports 802.11a/g/n, and enterprise wireless mesh for indoor and outdoor locations [15]. The small business network resembles a cafe, which offers WiFi connectivity to all its customers via Cisco's WLC. It also offers a secure wireless access for all employee wireless connections by utilizing a Cisco ACS RADIUS server.

4. Test-bed Analysis of End-to-end Traffic Flow

In this study, we designed and built a real-world WiFi-WiMAX integrated scenario depicting interoperability in a testbed environment. In this section we present the overall, end-to-end connectivity including both WiMAX and WiFi conversion signalling mechanism (inter- and intra-site conversions), as well as the bandwidth test results achieved using the IPerf tool. IPerf [16] is a tool to measure the bandwidth and the quality of a network link. The network link is limited to two hosts running IPerf. The quality of a link can be tested for latency, jitter and datagram loss. The bandwidth is measured through TCP tests. One host is set as a client and the other one as server using the commands 'iperf –c <server ip address>" and "iperf –s <client ip address>" respectively. The Iperf client connects to the Iperf server on the TCP port 5001 and the bandwidth displayed by Iperf is the bandwidth from the client to the server.

4.1. Connectivity between the enterprise, small business and the NOC site

We used laptops as end user wireless devices to connect to the WiFi network for both client sites. The traffic flows are shown in Figure 2. Traffic flow 2 (a), indicated in red, shows the packet originating from a client device in the Enterprise Network passing through the WiFi-WiMAX link then through the ISP core routers and finally reaching the NOC site. The traffic flow 2 (b), indicated in green, shows the same flow path of the packet to the ISP core after which it enters the Small Business' WiFi network and the destination client on the other side.

4.2. Bandwidth Test

We investigated the results for the performance of WiFi-WiMAX system as compared to a standalone WiMAX system. Our results indicate that system throughput of WiFi-WiMAX integration did not deteriorate. We performed the test using IPerf tool on all three sites and investigated the effect of network integration on the performance of the network. The traffic flowed seamlessly throughout the network between WiFi and WiMAX networks, and we did not observe any performance degradation or packet loss due to integration. We carried out several sets of tests between the clients in Enterprise and Small Business networks. We also tested intra-site connectivity within the Enterprise. The results are displayed in Figure 4.

Figure 4(A) depicts a bandwidth of 971 kbps over a WiFi-WiMAX-WiFi link for traffic flowing from the WiFi client at the small business network traversing over the ISP core and WiMAX link to reach the WiFi client at the large Enterprise. The traffic in the reverse direction over the same link is highlighted by Figure 4(B). The achieved data rate of the above mentioned link averages to approximately 967.5 kbps. For the WiMAX link we are using QPSK modulation with a code rate of ½ and channel bandwidth of 10 MHz. The WiMAX base station and the client machine are tested in Non Line of Sight (NLOS) conditions.

A study was carried out by A. Durantini et al. [17] which showed the performance of an end-to-end WiMAX system using CPE's that were placed at varying locations from the BS. They performed tests for different modulation schemes. Their results also show data rate values for

QPSK-1/2 modulation ranging from 950 kbps to 1000 kbps for the various CPE's, which is in coherence with our results. This indicates that a WiFi-WiMAX system still retains all the advantages of a basic WiMAX system while allowing enterprises to keep their WiFi architectures intact. It reinforces that integrating WiFi with WiMAX doesn't deteriorate WiMAX capabilities.

Lastly, Figure 4(C) represents a bandwidth over a WiFi–WiMAX link to the NOC. The observed bandwidth is 957 kbps which is slightly less as compared to a WiFi-WiMAX-WiFi link. The overall bandwidth can be significantly improved by using higher modulation schemes on the PHY WiMAX link [17].

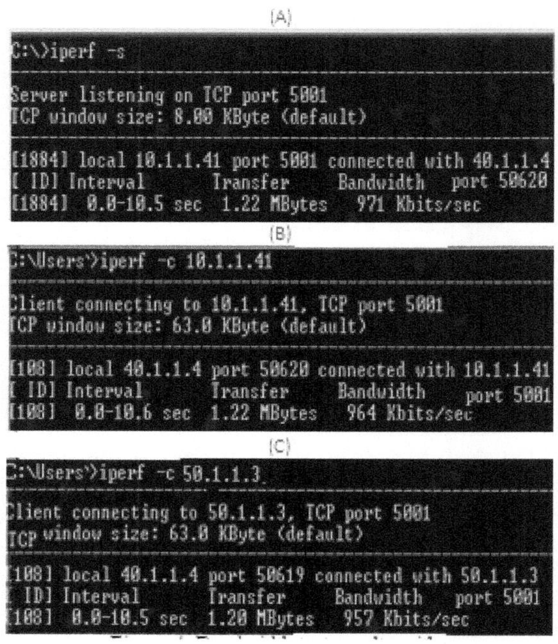

Figure 4: Bandwidth test results with:
(A) WiFi -WiMAX-WiFi link with small business site as server and large enterprise client
(B) Reverse of flow path from A
(C) Large Enterprise site to the NOC

In the next set of experiments, we setup the WiFi–WiMAX integration network, generated traffic for the network, and installed PRTG to further monitor the network performance, with the QoS parameters under surveillance. In the following section we explain different scenarios to monitor our network and further review all the results from these scenarios in the Result and Discussions section. The following scenarios are designed to monitor our wireless network using five installed sensors: Ping, FTP, HTTP, RDP and VOIP. The WiMAX BS supports three different predefined categories for shaping the traffic: nRT, eRT and BE. We will use these classes of service with our traffic and capture the results.

In the first Scenario we have a single type of traffic with each class of Service, including the traffic flow for Ping traffic with BE class of service. In the second scenario we generated traffic for all five types of packet which we have in our network with different class of service designated to all packets.

5. Results and Discussion

In this section we present the result of each scenario in the entire experiment and compare the result with the influence of each class of service in WiFi–WiMAX integrated network. Due to the limitation of the equipment we cannot take advantage of all WiMAXQoS capabilities and are limited to only three classes of service instead of the five classes identified by the standard.

5.1. Scenario one

In the first scenario we have a single type of traffic with each class of service. Due to similarity of topology diagram we are going to only illustrate PING traffic given different Class of Service. For other type of packet the diagram would stay the same and only the traffic type is going to change.

The graphs presented in Figures 5 through 9 are derived from our results in RTPG and showcase the responsiveness of the system given different types of flows (PING, FTP, HTTP, RDP, VOIP) matched with three types of QoS profiles (i.e. nRt, eRT, BE) in the BreezeMaxWiMAX setup [18]. Each packet flow is running individually in order to create a base line for the next scenarios.

Ping results in Figure 5 show that the average ping time stayed the same in three different types of QoS scenarios. The results show the evident exception that having only one type of traffic on the line would make the QoS prioritization mute, and the result will not change.

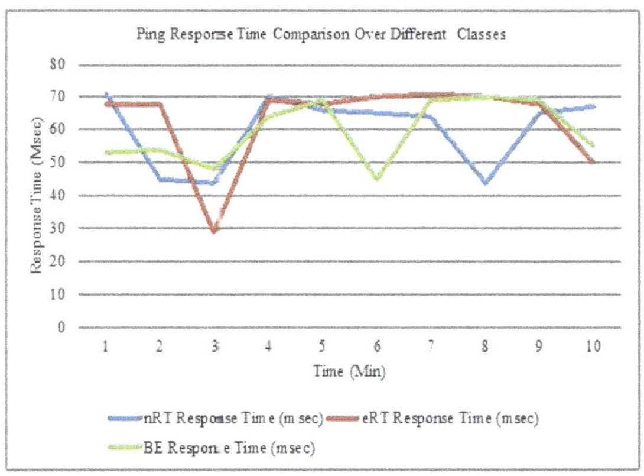

Figure 5: Response Time Comparison for Ping Traffic

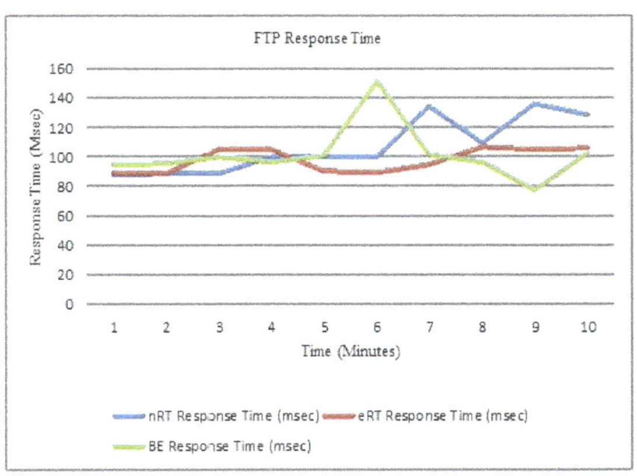

Figure 6: Response Time Comparison for FTP Traffic

Again we notice that in FTP, HTTP, and RDP traffic (Figures 6, 7 and 8) response time and page loading time stays the same and do not change by the type of QoS profile, whether we use nRT, eRT or BE.

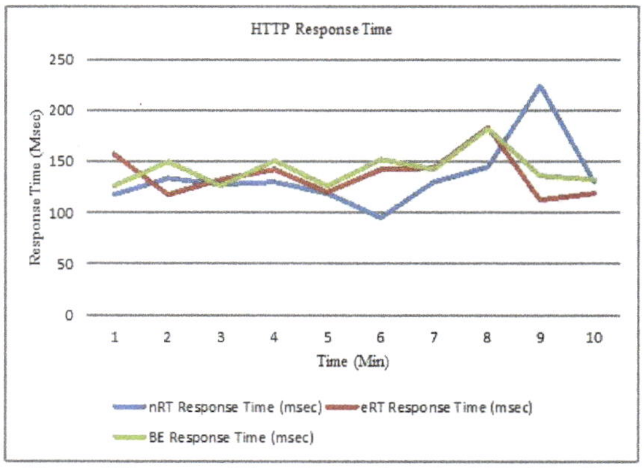

Figure 7 Response Time Comparison for HTTP Traffic

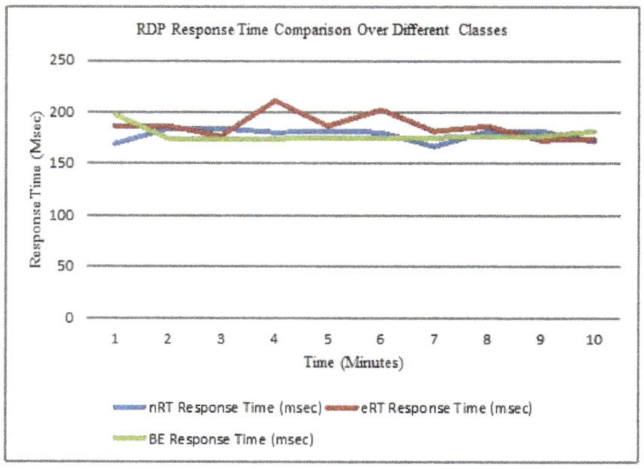

Figure 8 Response Time Comparison for RDP Traffic

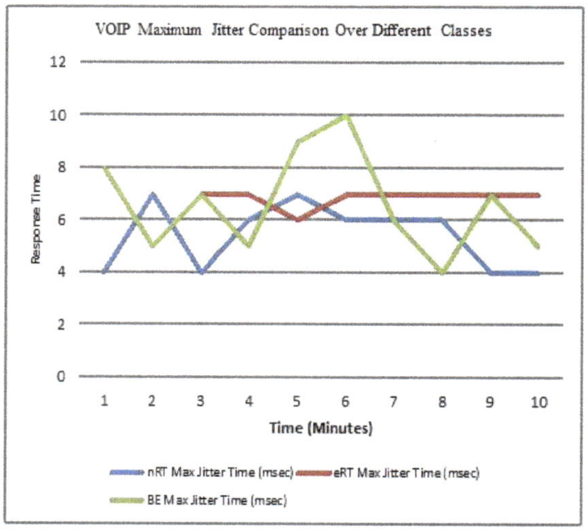

Figure 9 Response Time Comparisons for VOIP Traffic

In Figure 9, a comparison of VOIP traffic in different classes of service based on response time has been shown. We were expecting to see higher response time for the Best Effort class and this was evident in the result. In terms of response time however, nRT flows unexpectedly showed lower values than their RT counterparts.

5.2. Scenario two

Having captured a baseline of how different types of traffic response time are affected based on each class of QoS assigned in the WiMAX environment we further investigated the results by executing all of the flows at the same time and measure how this will be reflected on the overall system response time.

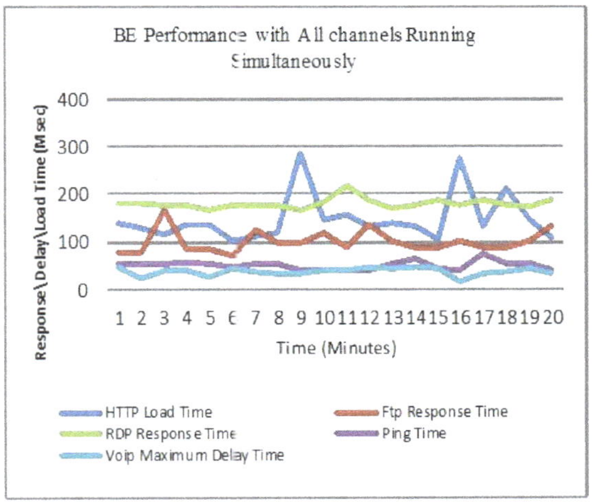

Figure 10: BE Performance With all Traffic Flows

Figures 10 and 11 are derived from the results in RTPG and showcase the responsiveness of the system given all type of traffic (PING, FTP, HTTP, RDP, VOIP) running simultaneously and matched with each types of QoS profile (nRt, eRT, BE) in the BreezeMaxWiMAX [19] setup on three separate instances.

Figure 10 QoS profile is set to Best Effort, which shows competition between packets trying to pass the media, consequently all the flows' response time are fluctuating. The winners are VOIP and Ping with response times around 50 ms. HTTP has the highest response time at almost 300 ms.

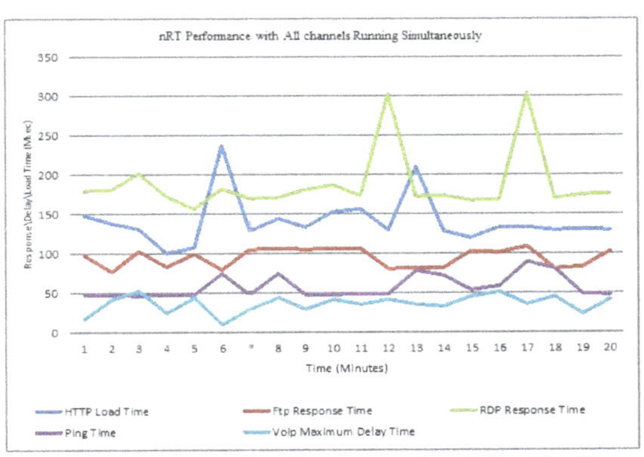

Figure 11: nRT Performance With all Traffic Flows

By changing the class of service from BE to nRT, we had a better response time as an overall trend. All five types of traffic flow had an improvement in their response time ranging from 50 to just under 200 Msec (Figure 11).

6. Conclusions

WiMAX has emerged as a prevailing technology for wireless metropolitan area services in licensed and unlicensed spectrums. It utilizes RF technology comparable to WiFi, while providing superior performance to current cellular standards in terms of data rates and wide coverage area. In this paper, we explored the WiFi-WiMAX integrated architecture. WiMAX equipment is now available for the 5.4-5.9 GHz unlicensed band. It is a profitable solution for enterprise networks to integrate their multiple sites. It also enables ISP's to integrate existing WiFi networks operating in similar bands with WiMAX as backhaul.

An ISP could use WiMAX to backhaul multiple WiFi hotspot sites. An enterprise could receive services from a WiMAX service provider and then combine the WiMAX terminal equipment with WiFi WLANs to achieve a complete mobile experience. We successfully demonstrated this solution by providing end-to-end WiFi-WiMAX connectivity with providing similar data rates to those of pure WiMAX network solution. Generally, WiMAX will be useful for providing intermediate-distance backhaul for WiFi access points, which could be a preferable arrangement to WiFi mesh technology.

In this study, we integrated the CPE with a basic access point; however advanced options using the CPE in conjunction with a Wireless LAN controller is an open area to be investigated further. True 802.16e mobility and handover testing using WiMAX cards and BS operation with an external ASN gateway are also areas that need to be investigated. Lastly, the bandwidth of integrated WiFi-WiMAX link can be explored by using different modulation schemes on PHY WiMAX link.

We further investigated the effect of integration of the technologies on one of the important aspects of WiMAX networks; namely the quality of service (QoS) support, and carefyully measured the QoS parameters like throughput, and delay values for the entire network, and compared it to those of the stand-alone networks.

We built the integrated network for a complete ISP/Enterprise deployment, and created multiple scenarios and configured different types of QoS settings in the laboratory environment. The results showed that in congested circumstances implementing any of QoS policies has a positive effect on the overall responsiveness of the integrated network just as in the stand-alone scenarios.

References

[1] WiMAX Forum, "Mobile WiMAX – Part II: Competitive Analysis", Available (Feb 12) online at:http://www.wimaxforum.org/technology/downloads/Mobile_WiMAX_Part2_Comparative_Analysis.pdf

[2] Krapichler, Christian, "LTE, HSPA and Mobile WiMAX a comparison of technical performance",LTE vs.WiMAX and Next Generation Internet, Institution of Engineering and Technology, Issue (27), page: 1 – 31. (Sept. 2007)

[3] Hassan Yaghoobi, "Scalable OFDMA Physical Layer in IEEE 802.16 Wireless MAN", Intel technology Journal, Volume 8, Issue 3, (2004)

[4] LoutfiNuaymi, "WiMAX Broadband Wireless Access Technology", available on e-tutorial WiMAX-parts 2 and 3 at: http://etutorials.org/Networking/wimax+technology+broadband+wireless+access (Oct 2010)

[5] Motorola," LTE: A Technical overview", available on-line at: www.motorola.com/.../a%20Driving_4G__WiMAX_and_LTE.pdf (July 2012)

[6] WiMAX Overview, available on-line at: http://www.airspan.com/products_wimax.html (Sept 2011)

[7] Adam Petcher, "QoS in Wireless Data Networks", available online at: http://www.cse.wustl.edu/~jain/cse57406/ftp/wireless_qos/index.html#1 (summer of 2011)

[8] Etemad, K, Lai, M, "WiMAX Technology and Network Evolution", Wiley-IEEE Press, pp. 245-290, 2010.

[9] IETF RFC 2212, Specification of Guaranteed Quality of Service, available in on line at: http://www.ietf.org/rfc/rfc2212.txt (summer of 2011)

[10] LoutfiNuaymi, "WiMAX Broadband Wireless Access Technology", available on line at: http://etutorials.org/Networking/wimax+technology+broadband+wireless+access (summer 2011)

[11] Cisco Unified Wireless Network, available on-line at: http://www.cisco.com/web/DE/pdfs/products/ccmigration_09186a0080184925.pdf (Aug 2010)

[12] Alvarion, "Comparing Mobile WiMAX, 3G and Beyond" available on-line at: www.alvarion.com, http://www.crystalcomltd.com/featured_whitepapers/010231032509Comparing_WiMAX_vs_3G_White_Paper.pdf. (Aug 2010)

[13] PRTG Network Monitor V8 User Manual, available on line at: http://download.cdn.paessler.com/download/prtgmanual.pdf (in summer of 2011)

[14] Wireless LAN Switches: Functions and Deployment, Cisco systems, available online at: http://www.cisco.com/web/about/ac123/ac147/archived_issues/ipj_93/wireless_lan_switches.html (summer 2012)

[15] Paul DeBeasi, "Burton Group, 802.11n: Enterprise Deployment Consideration", version 1, May 2008. Available online: http://podcast.burtongroup.com/ip/2008/05/80211n-enterpri.html (July 2012)

[16] Iperf network performance analyzer tool, available on-line at: http://www.softpedia.com/get/Network-Tools/Misc-Networking-Tools/Iperf.shtml (Dec 2010)

[17] A. Durantini, M. Petracca, and F. Ananasso, "Experimental evaluation of IEEE 802.16 WiMAX performances at 2.5 GHz band," in Proc. International Wireless Communications and Mobile Computing Conference 2008 (IWCMC 2008), pp. 338-343, Aug 2008.

[18] BreezeMAX Extreme Ver.1.5 System Manual, available on line at: http://www.scribd.com/doc/55408946/BreezeMAX-Extreme-Ver-1-5-System-Manual-100502 (summer of 2011)

[19] BreezeMax Pro 5000 CPE Manual Ver.4.6, available on line at: http://www.intelek.cz/db/repository.nsf/v/0721D9A675A5BE93C1257704002B4F52/$file/Manual_Alvarion_subscriber_unit_XTRM-SU-OD-1D_v4.6_090723.pdf (summer of 2011)

5

Performance Analysis of On-Demand Routing Protocols for Vehicular Ad-hoc Networks

A. Shastri[1], R. Dadhich[2], Ramesh C. Poonia[3]

[1]Prof. & Vice-Chancellor, Banasthali University, Banasthali, India.
adityashastri@yahoo.com
[2]Associate Prof. & Head Department of MCA, Govt. College of Engineering, Ajmer, India.
reena.dadhich@gmail.com
[3]Research Scholar, Apaji Institute of Mathematics & Applied Computer Technology, Banasthali University, Banasthali, India.
rcpoonia@yahoo.com

Abstract

Vehicular Ad Hoc Networks (VANETs) are a peculiar subclass of mobile ad hoc networks that raise a number of technical challenges, especially from the point of view of their mobility models. Currently, the field of VANETs has gained an important part of the interest of researchers and become very popular. More specifically, VANETs can operate without fixed infrastructure and can survive rapid changes in the network topology. The main method for evaluating the performance of routing protocols for VANETs by Network Simulator-2.34. This paper is subjected to the on-demand routing protocols with identical loads and evaluates their relative performance with respect to the two performance context: average End-to-End delay and packet delivery ratio. We investigated various simulation scenarios with varying pause times, connections and no. of nodes particularly for AODV and DSR. We will also discuss briefly about the feasibility of VANETs in respect of Indian automotive networks.

Keywords

VANETs, NS-2.34, V2V, V2R, AODV, DSR

1. Introduction

Ad-hoc network is an infrastructureless network with no fixed routers, hosts or wireless base stations. Vehicular Ad-hoc Networks (VANETs) represent a rapidly emerging, particularly challenging class of Mobile Ad Hoc Networks (MANETs) [1]. VANETs are distributed, self organizing communication networks built up by moving cars as nodes in a network to create a very high mobility of nodes and limited degrees of freedom in nodes movement patterns. It turns every participating car into a wireless router or node, allowing cars approximately 100 to 300 meters of each other to connect and in turn, create a network with a wide range. VANET integrates on multiple ad-hoc networking technologies such as Wi-Fi IEEE 802.11 b/g, WiMAX IEEE 802.16, Bluetooth, IRA, ZigBee for easy, accurate, effective and simple communication between vehicles on dynamic mobility [2]. Some Important features of VANETs are: Vehicles move in more predictable fashion, portability issues are not a problem (computational power, energy, antennas, etc.), location information can be assumed, the nodes in a VANET are vehicles and road side units, the movement of these nodes is very fast, the motion patterns are restricted by road topology, vehicle acts as transceiver i.e. sending and receiving at the same time while creating a highly dynamic network, Which is used as part of automotive electronics to identify an optimally minimal path for navigation with minimal traffic intensity. The system can also be used as a city guide to locate and identify landmarks in a new city. Vehicles are enabled to communicate among themselves (vehicle-to-vehicle, V2V) and via roadside access points (vehicle-to-roadside, V2R) [3]. Vehicular communication is expected to contribute to safer and more efficient traffic by providing timely information to drivers, and also to make travel more convenient, figure 1.

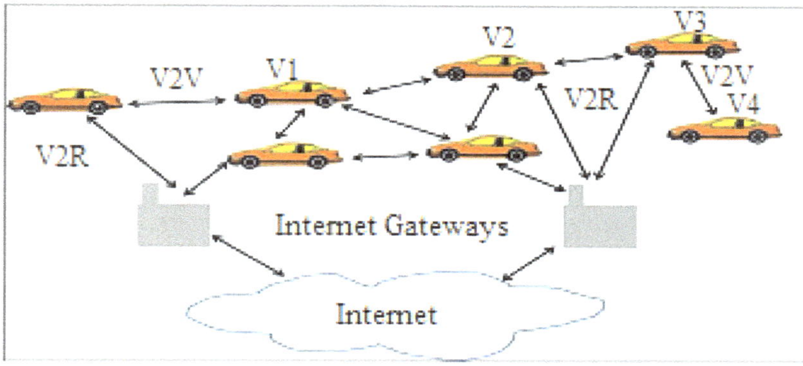

Figure 1. The scenario of VANETs

For communication in vehicular ad hoc networks, On Demand Routing Protocols is required. Currently there is lot of research work in the field of VANETs for several scenarios [4]. Recent research [5],[6] has focused on topology related problems such as range optimization, routing mechanisms, or address systems, as well as security issues like traceability or encryption. Concerning to these facts there is an urgent need of protocols and technologies to develop VANETs for the safety of human beings. The main goal of this paper is to focus on analyzing the performance of routing protocols in VANETs so that it will help to researchers to adopt the application for their contribution in VANETs.

The complete paper is organized as follows: In Section 2, we shortly provide some related work, while in Section 3, we will discuss in brief about the feasibility of VANTEs in Indian Automotive Networks, Section 4 presents the routing protocols in VANTEs. Evaluation parameters for routing protocols are discussed in Section 5. The simulation results and performance analysis is done in section 6and the paper concludes in Section 7.

2. RELATED WORK

In this section we will discuss in brief the initial development that were carried in creating a simulation that was aimed at testing VANETs , e.g. CanuMobiSim, NS-2 and NS-3, GlomoSim, QualNet, MOVE, TraNs, VanetMobiSim, NCTUns [7]-[9] and many more. The very first work related to the title of the paper was proposed by Ott and Kutscher's seminal in [10]. They discussed about the drive-thru Internet problem and performed a detailed experimental analysis of opportunistic Internet access in vehicles. Jaap et al. [11] evaluated AODV, DSR, FSR, and TORA in city traffic scenarios. The simulation shows that AODV has the best performance and lowest control overhead. It is followed by FSR, DSR, and then TORA. DSR suffers from a very high delay because source routes change continuously due to high mobility. Its route overhead is comparable to FSR yet higher than AODV since DSR keeps route information with in the packet header. The common characteristic among all four routing protocols is that performance degrades as network densities increase, indicating their scalability problem.

There are many papers [12],[13] focusing on new routing protocols to VANETs, in order to achieve increased communication reliability in some mobility scenarios (highway and Urban area) or particular application.

3. THE FEASIBILITY OF VANETs

The Automotive in India is one of the largest fields in the world and one of the fastest growing globally. India manufactures over 17.5 million vehicles (including 2 wheeled and 4 wheeled) and exports about 2.33 million every year. India's passenger car and commercial vehicle manufacturing industry is the seventh largest in the world, with an annual production of more than 3.7 million units in 2010. According to recent reports, India is set to overtake Brazil to become the sixth largest passenger vehicle producer in the world [14]. The feasibility of VANET in a certain city as well as on highways has to be justified. First, there needs to be an adequate number of vehicles in the city road and highway at any given time in order for ad hoc networking among vehicles to be possible. The density of vehicles is λ vehicles/km^2. Furthermore, any vehicle should be within range of at least one other vehicle in order for a link to be established and multihop routing to prove practical [15]. According to the practical situation, we make the line-of-sight (LOS) distance r and the geometrical area of the communication zone S is calculated using

$$S = \pi r^2 \qquad (1)$$

Using the Possion distribution, the probability of the number of vehicle n in the region S can be worked out using

$$p(n) = \frac{(\lambda S)^n}{n!} e^{-\lambda S} \qquad (2)$$

Thus, the probability of finding at least k vehicles in area S is

$$q(k) = 1 - \sum_{n=0}^{k-1} \frac{(\lambda S)^n}{n!} e^{-\lambda S} \qquad (3)$$

4. ROUTING PROTOCOLS IN VANETs

Standard As a special type of network, Vehicular Ad hoc Networks (VANETs) have received increasing research attention in recent years. There are many active research projects [19] concerned with VANETs. Vehicular ad hoc networks are wireless networks that use multi-hop routing instead of static networks infrastructure to provide network connectivity. VANETs have applications in rapidly deployed and dynamic military and civilian systems. The network topology in VANETs usually changes with time. Therefore, there are new challenges for routing protocols in VANETs since traditional routing protocols may not be suitable for VANETs [16],[17]. Researchers are designing new VANETs routing protocols, comparing and improving existing ones by using simulations. This work is an attempt towards a comprehensive performance evaluation of two commonly used mobile ad hoc routing protocols (DSR and AODV).

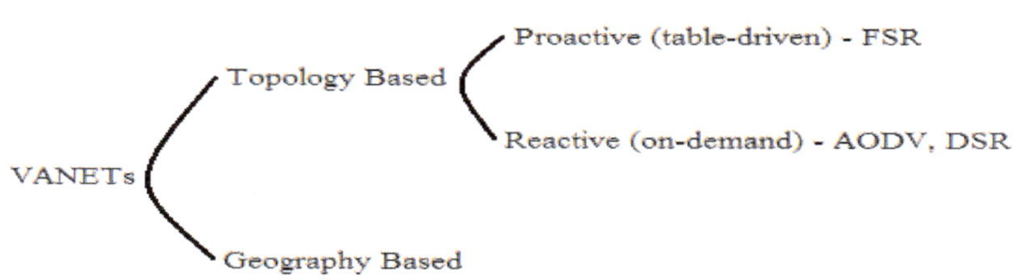

Figure 2. Taxonomy of Routing Protocols in VANET

Figure 2 illustrates the taxonomy of these VANET routing protocols [18],[19] which can be classified as topology-based and geographic (position-based) in VANET. Topology-based routing uses the information about links that exist in the network to perform packet forwarding. Geographic routing uses neighbouring location information to perform packet forwarding [20]. Since link information changes in a regular basis, topology-based routing suffers from routing route breaks.

4.1. Topology-based Routing Protocols

These routing protocols use link information that exists in the network to perform packet forwarding. They can further be divided into proactive (table-driven) and reactive (on-demand) routing [21].

4.1.1. Proactive (table-driven) Protocols

Table-driven routing protocols [22] try to maintain consistent, up-to-date routing information from each node to every other node. Network nodes maintain one or many tables for routing information. Nodes respond to network topology changes by propagating route updates throughout the network to maintain a consistent network view. The advantage of the proactive routing protocols is that there is no route discovery since route to the destination is maintained in the background and is always available upon lookup. Despite its good property of providing low latency for real-time applications, the maintenance of unused paths occupies a significant part of the available bandwidth, especially in highly mobile VANETs [23],[24].

4.1.2. Reactive (On-Demand) Protocols

Reactive routing opens [25] a route only when it is necessary for a node to communicate with another node. It maintains only the routes that are currently in use, thereby reducing the burden on the network. Reactive routings typically have a route discovery phase where query packets are flooded into the network in search of a path. The phase completes when a route is found.

4.2. Geographic Based

Geography based routing is basically a location based multicast routing. Its objective is to deliver the packet from source node to all other nodes within a specified geographical region (Zone of Relevance ZOR) [20]. In Geography based routing vehicles outside the ZOR are not alerted to avoid unnecessary hasty reaction. It is considered as a multicast service within a specific geographic region.

5. EVALUATION PARAMETERS FOR ROUTING PROTOCOLS

In this paper, we evaluated the performance of two widely used ad hoc network routing protocols using packet-level simulation using the latest simulation environment NS-2.34 [26],[27] under working Windows 7. The simulation characteristics used in our research, include parameters described as follows for detailed performance evaluation of any networking protocol and the simulation parameters are summarized in Table 1.

- Packet Delivery Fraction (PDF): It represents the ratio of the data packets received at destination and the delivered packets.
- End-to-end Average Delay: It indicates the propagation and transfer delay.
- Received packets: It measures the numbers of received packets.
- Reception time of the first packet: This parameter informs us about the time it takes to establish the connection for each of the protocols. Dropped Packets: It represents those packets which fail to reach their destination.

- Number of Hops: It refers to the number of hops that the packets need to reach their destination.

Table 1. Simulation Parameters

Parameters	Specifications
Network Simulator	NS-2.34
Routing Protocols	AODV, DSR
Simulation Area	500m x 500m
Propagation Model	Two-ray ground reflection model
Number of Vehicles	50
Antenna model	Omnidirectional
MAC Protocol	IEEE 802.11
Channel Type	Wireless channel
Traffic Type	Constant Bit Rate (CBR)
Source/Destination	Random
Max. Speed	20 m/s
Simulation Time	200 s
Data Rate	4 Mbps
Seed	1
Pause Time	50,100,150,200,250,300,350,400,500
Connections	10,20,30,40,50,60,70,80,90,100

In this table, the variables along with their appropriate values are specifying the simulation's parameters for all the density phases of movement model.

6. SIMULATION RESULTS AND PERFORMANCE ANALYSIS

Figure 3 shows the behaviour by throughput of AODV and DSR.

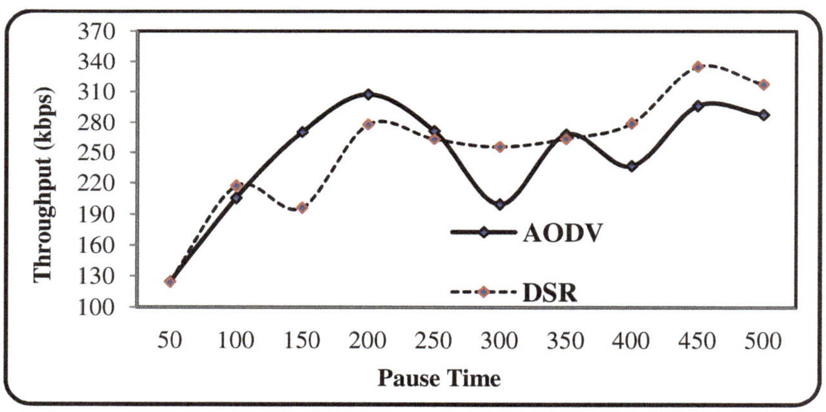

Figure 3. Throughput of AODV and DSR

First, we will compare two protocols under the above simulation environment. For the simulations, the same movement models are used, the number of traffic sources is fixed at 50, the maximum speed of the nodes is set to 20m/s and the pause time is vary as 50, 100, 150, 200, 250, 300, 350, 400, 450 and 500. Figure 4 highlights the relative performance of the two routing protocols. It shows that there is an increasing packets loss in AODV as compared to DSR when as the pause time increases.

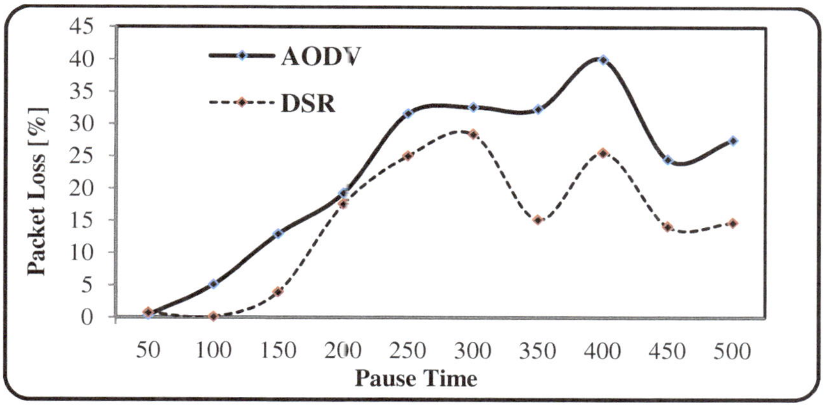

Figure 4. Packet Loss of AODV and DSR

6.1. Packet delivery Comparison

Figure 5 shows that Packet Delivery Fraction (PDF) for the same movement models are used, the number of traffic sources is fixed at 50, the maximum speed of the nodes is set to 20m/s and the pause time is vary as 50, 100, 150, 200, 250, 300, 350, 400, 450 and 500.The packet delivery fractions for AODV and DSR are almost similar with 50, 200, 250, 300 and 350 pause times shown in figure 5. However, with 100, 150, 200, 400, 450 and 500 pause times, AODV outperforms DSR by about 10 percent.

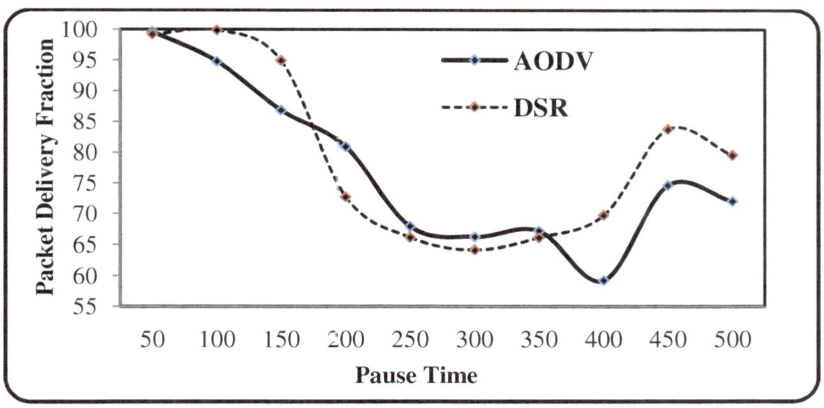

Figure 5. Packet Delivery Fraction (PDF) of AODV and DSR

6.2. Average End-to-End delay in Packet Delivery

The average end-to-end delay in packet delivery is higher in DSR as compared to AODV, figure 6. They are little and similar in the case of 100, 150, 400, 450 and 500 pause times. When the pause time increases to 200, the transmission delay of data packet increases greatly and that of AODV is obviously lower than DSR. The DSR becomes better than AODV when pause time becomes longer.

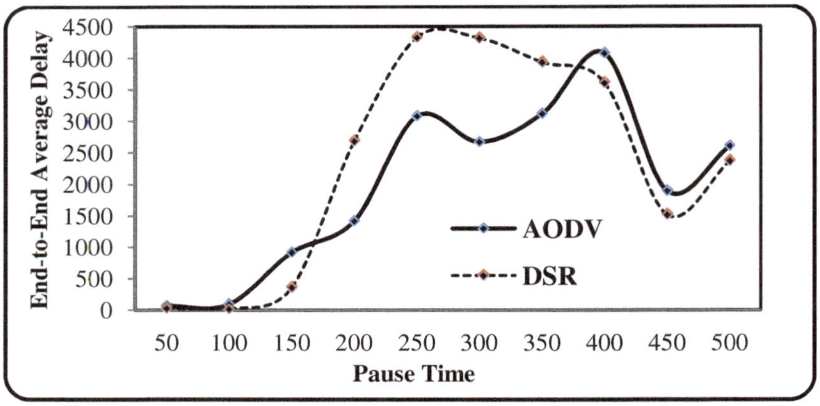

Figure 6. End-to-End average Delay of AODV and DSR

6.3. Received Packets

Similar to above observations for packet delivery ratio (PDR), elevate percentage means higher performance. In view of this fact, at all simulated pause times, the DSR delivers successfully a slightly greater performance of packets than the AODV so that DSR remained the highly acceptable approaches for the movement model. Figure 7 shows in all cases of received packet delivery with respect to pause times. It can be observed that the behaviour of DSR is better than AODV.

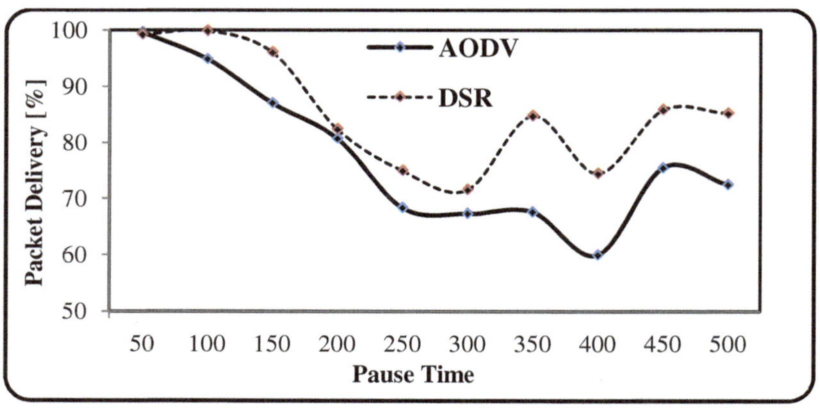

Figure 7. Received Packet Delivery of AODV and DSR

7. CONCLUSION AND FUTURE WORK

This paper presents a performance comparison between two reactive routing protocols for vehicular ad hoc networks: ad hoc on demand distance vector (AODV), dynamic source routing (DSR). Both protocols were simulated using the tool NS-2.34 and were compared in terms of packet loss ratio, end to end delay, throughput with 50 nodes, 20 speed and varying pause time 50 to 500 and making connections 10 to 100. From the simulation results in section V we can conclude that increase in the pause times increase both Packet Loss and Average End-to-End Delay but decrease the Packet Delivery Fraction. In the future, extensive complex simulations could be carried out using other existing performance metrics, in order to gain a more in-depth performance analysis of the ad hoc routing protocols. Other new protocols performance could be studied too. We can also analyse and evaluate in depth of VANETs in respect of Indian automotive networks.

REFERENCES

[1] Tarik Taleb, Ehssan Sakhaee, Abbas Jamalipour, etl.,(2007) "A Stable Routing Protocol to Support ITS Services in VANET Networks", *IEEE Transactions on Vehicular Technology*,56(6): 3337-3347.

[2] Pranav Kumar Singh, Kapang Lego, Themrichon Tuithung, (March 2011) "Network Centric Approach using MOVE &Application Centric Approach using TraNS for protocols and safety in VANET", *International Journal of Research and Reviews in Computer Science (IJRRCS)*, Vol. 2, No. 1.

[3] C2C-CC: CAR 2 CAR Communication Consortium, http://www.car-to-car.org/ Visited in May 2011.

[4] Xin Jin Weijie Su Yan Wei, (May 2011) "A study of the VANET connectivity by percolation theory", *Consumer Communications and Networking Conference (CCNC), IEEE.*

[5] Chin-I Lee, (May 2011) "Securing Vehicular Ad Hoc Networks: A Research Survey" *Ling Tung University.*

[6] Bhakthavathsalam, R. Nayak, S., (May 2011) "Operational inferences on VANETs in 802.16e and 802.11p with improved performance by Congestion Alert", *Consumer Communications and Networking Conference (CCNC), IEEE*

[7] CanuMobiSim available at http://canu.informatik.uni-stuttgart.de/mobisim/ Visited in May 2011.

[8] GloMoSim available at http://pcl.cs.ucla.edu/projects/glomosim/ Visited in Jan. 2011.

[9] VanetMobiSim available at http://vanet.eurecom.fr/ Visited in Jan. 2011.

[10] J. Ott and D. Kutscher, (2004) "Drive-thru Internet: IEEE 802.11b for Automobile Users", *In IEEE Infocom.*

[11] S. Jaap, M. Bechler, L. Wolf, (June 2005) "Evaluation of Routing Protocols for Vehicular Ad Hoc Networks in City Traffic Scenarios," *Proceedings of the 5th International Conference on Intelligent Transportation Systems (ITS) Telecommunications.*

[12] B. Ramakrishnan, Dr. R. S. Rajesh, R. S. Shaji, (November 2010) " An Intelligent Routing Protocol for Vehicle safety communication in Highway Environments", *Journal of Computing* , Volume 2, Issue 11.

[13] Carolina Tripp Barba, Karen Ornelas, Mónica Aguilar Igartua, (October 2010) "Performance Evaluation of a Hybrid Sensor and Vehicular Network to Improve Road Safety", *7th ACM PE-WASUN 2010*, Bodrum, Turkey.

[14] Automotive industry in India available at http://en.wikipedia.org/wiki/Automotive_industry_in_India Visited in May 2011.

[15] SUN Xi, LI Xia-miao, (Oct. 2008) "Study of the Feasibility of VANET and its Routing Protocols", *Wireless Communications, Networking and Mobile Computing*, 4th International Conference.

[16] J. J. Blum, A. Eskandarian, and L. J. Hoffman, (Dec. 2004) "Challenges of Inter vehicle Ad Hoc Networks", *IEEE Transactions on Intelligent Transportation Systems*, Vol. 5. No. 4".

[17] J.Blum, A. Eskandarian, and L. Hoffman, (2003) "Performance Characteristics of Inter- Vehicle Ad Hoc Networks", *The IEEE 6th International Conference on Intelligent Transportation Systems*, Shanghai, China, page 114-119.

[18] Yun-Wei Lin, Yuh-Shyan Chen and Sing-Ling Lee, (2010) "Routing in Vehicular Ad Hoc Networks: A Survey and Future Perspectives," *Journal of Information Science and Engineering.*

[19] Sajjad Ali & Asad Ali, (2009) "thesis - Performance Analysis of AODV, DSR and OLSR in MANET", *Department of Electrical Engineering with emphasis on Telecommunication Blekinge Institute of Technology*, Sweden.

[20] R.S. Raw, S. Kumar, S. Das, (2010) "An Efficient Position-Based Routing Protocol for Vehicular Ad Hoc Network in City Traffic Scenarios", *RISTEP-2010*, NIT Srinagar, India.

[21] Fan Li and Yu Wang, "Routing in Vehicular Ad Hoc Networks: A Survey," *IEEE Vehicular Technology Magazine*.

[22] Bilal Mustafa, Umar Waqas Raja, (June-2010) "Issues of Routing in VANET", *Thesis no: MCS-2010-20*, Sweden.

[23] WU Zhuang, YANG Yang, GUO Xun and AN Jianwei, (Jan. 2010) "Analysis of Collision Probability in IEEE 802.11 Based VANETs", *Chinese Journal of Electronics*, Vol.19.

[24] C. Hernández-Goya, P. Caballero-Gil, J. Molina-Gil, C. Caballero-Gil, (2009) "Cooperation requirements for packet forwarding in vehicular ad-hoc networks (VANETs)", *ACM International Conference Proceeding Series,* Vol. 433.

[25] Sandhaya Kohli, Bandanjot Kaur, Sabina Bindra, (Apr 2010) "A comparative study of Routing Protocols in VANET", *RIMT-IET,* 10.

[26] NS-2.34, available at http://www.isi.edu/nsnam/ns Visited in May-2011.

[27] Kevin Fall, Kannan Varadhan, (May 9, 2010) "The ns Manual", The VINT Project, available at http://www.isi.edu/nsnam/ns/ns-documentation.

6

Delay Analysis of IEEE 802.11 DCF with modified Binary Exponential Backoff

Rishipal Singh[1,2], D.K.Lobiyal[2]

[1]Department of CSE, Guru Jambheshwar University of Science and Technology, Hisar, India

[2]School of Computer and Systems Sciences, Jawaharlal Nehru University, Delhi, India

Abstract

In the design of wireless networks the medium access control (MAC) protocols have a very high impact on the performance of the network. The IEEE 802.11 is widely accepted technology for the Wireless LANs has been used by wireless networks. Delay is one of the important parameter to measure the performance of the wireless networks. We give a delay analysis of the Distributed Coordination Function of IEEE 802.11 under modified Binary Exponential Backoff (BEB) Algorithm using Frequency Hoping Spread Spectrum (FHSS). In the modified BEB, initially the contention window size increases with the factor of $\sqrt{2}$ *for first four collisions and after that for more collisions the size of contention window gets double. The size of contention window varies between* CW_{min} *and* CW_{max}*. Markov chain has been used for the analysis of IEEE 802.11 DCF for both modes basic and RTS/CTS. We have considered number of nodes and transmission probability to compute the delay for both the modes. The results show that the performance of RTS/CTS mode is better than the basic access mode. In this paper, we have numerically analyzed the IEEE 802.11 DCF by varying different parameters (i.e. number of nodes, transmission probability etc.).*

Keywords- Medium Access Control, Distributed Coordination Function, Markov chain, Delay analysis

I. Introduction

Wireless connectivity is becoming very popular now a days as compared to the other technology i.e. wired technology. Despite of the high speed data transfer rate of Wireless Local Area Networks (WLAN), the wireless connectivity covers limited geographical area. The wireless communication capability allows the devices stay connected even they are mobile. These devices include palmtop computers, personal digital assistants (PDAs), portable computers, digital cameras and printers. This paper focuses on the delay analysis of the MAC protocol of IEEE 802.11 protocol.

The wireless LANs at the physical layer level has more complexity in terms of design, because of the presence of hidden terminals or capture effect. The IEEE 802.11 standard used as WLAN have basic access protocol Distributed Coordination Function (DCF) and an optional Point Coordination Function(PCF). DCF is a asynchronous data transmission function, in this mode of transmission IEEE 802.11 can be used Ad hoc Network. Ad hoc network is infrastructure less network. The other mode PCF is used in infrastructure network and for the transmission of delay sensitive data.

The RTS/CTS scheme available in IEEE 802.11 helps to combat the hidden terminal problem [4]. The influence of hidden terminal problem has also been discussed in [6-7] along with capture effect. In references [7-9, 18,20] the performance evaluation of 802.11 has been carried out by means of analytical models with simplified Backoff, by employing a two dimensional Markov chain analysis. Bianchi [10] presents an easy systematic model to work out the saturation throughput performance assuming a finite number of stations and ideal channel conditions. Chatzimisios et al.[12] have analyzed throughput and delay of IEEE 802.11 DCF, further they have discussed the results for throughput but not for delay. The authors have analyzed the throughput with modified Binary Exponential Backoff (BEB) of IEEE 802.11 DCF [21]. In [19], a reservation based collision aware method has been used to improve the performance of the IEEE 802.11.
The paper is organized as follows: Section II discusses the working of Distributed coordination function of 802.11 MAC protocol and includes both basic access and RTS/CTS mechanism. Section III reviews the protocol presented in [12]. Section IV discusses the delay analysis results of DCF mode for both basic access and RTS/CTS mechanisms. Finally, section V concludes the paper and presents future work.

II. Distributed Coordination Function

The basic service set (BSS) is the primary building block of IEEE 802.11 WLANs. The area covered by BSS is called basic service area (BSA), just like a cell in cellular network. IEEE 802.11 supports both the ad hoc network and infrastructure network architecture.

The distributed coordination function (DCF) mode is a random access scheme and it is based on the carrier sense multiple access with collision avoidance (CSMA/CA) protocol. DCF consists two ways of transmission, among them one is basic access method and another method which is optional is called RTS/CTS method. The working of each method of medium access is explained below [15].

A. Basic access method

In this method the access to medium is controlled by the inter-frame space (IFS) time period between the transmission of frames. Inter-frame spacing refers to the time interval between the transmission of two successive frames by any station. There are four types of IFS: SIFS, PIFS , DIFS and EIFS, in order from shortest to longest. The IFS with shortest length has highest priority to access the medium, because the wait time to access the medium is lower. SIFS and DIFS are used for ad hoc network and PIFS for infrastructure network. After a SIFS only ACKs, CTS and data frames may be sent. The time interval after an idle DIFS is slotted and a station can transmit only at the beginning of a slot. The size of each slot is ,σ, which is equal to the time needed by any station to detect the transmission of a packet from any other station.

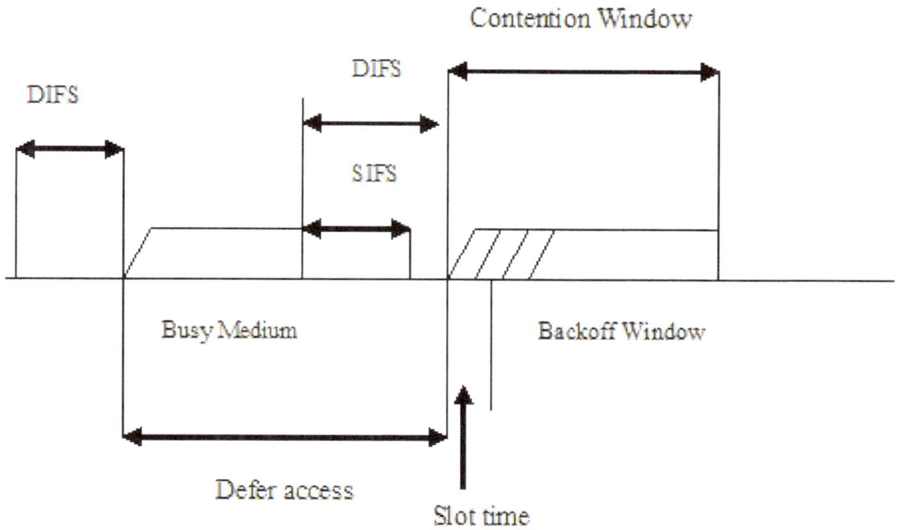

Fig. 1 Basic access mechanism in DCF

If the medium is sensed to be idle for a duration of DIFS, the node accesses the medium for transmission. If the medium is busy, the station defers until a DIFS is detected and then generate a random Backoff period before transmitting. The Backoff period is uniformly chosen within a contention window (CW). The contention window (CW) depends on the number of transmissions failed for the packet. At the first transmission attempt, CW is set equal to CW_{min} which is called minimum contention window. After each retransmission due to a collision, CW is increased as $(\sqrt{2}CW)$ for m'=4, after four retrials the size of contention window is doubled upto a maximum value $CW_{max} = 2^{m}.CW_{mid}$.

B. The RTS/CTS access method

The transmission of packets in WLAN is based upon CSMA/CA, because of CSMA/CA technique a source station can not hear its own transmission, thus collision takes place. Instead of transmitting a whole data packet of large size, small size control frames are transmitted in the form of RTS/CTS frames. RTS/CTS control frames can be used by a station to reserve channel bandwidth before the transmission of actual data packets. The sender node first transmits the RTS control frame to the receiver. The RTS frame contains the information about the expected duration of whole data transmission. After receiving RTS frame the receiver responds with the CTS control frame, after an SIFS idle period has elapsed. Station hearing the RTS or CTS frame read the duration field and update their network allocation vector (NAV) accordingly. Source station is allowed to transmit its packet only if it receives the CTS correctly. This updation of NAV(from the information) RTS/CTS frame of the neighboring stations helps to combat the hidden terminal problem [16-17].

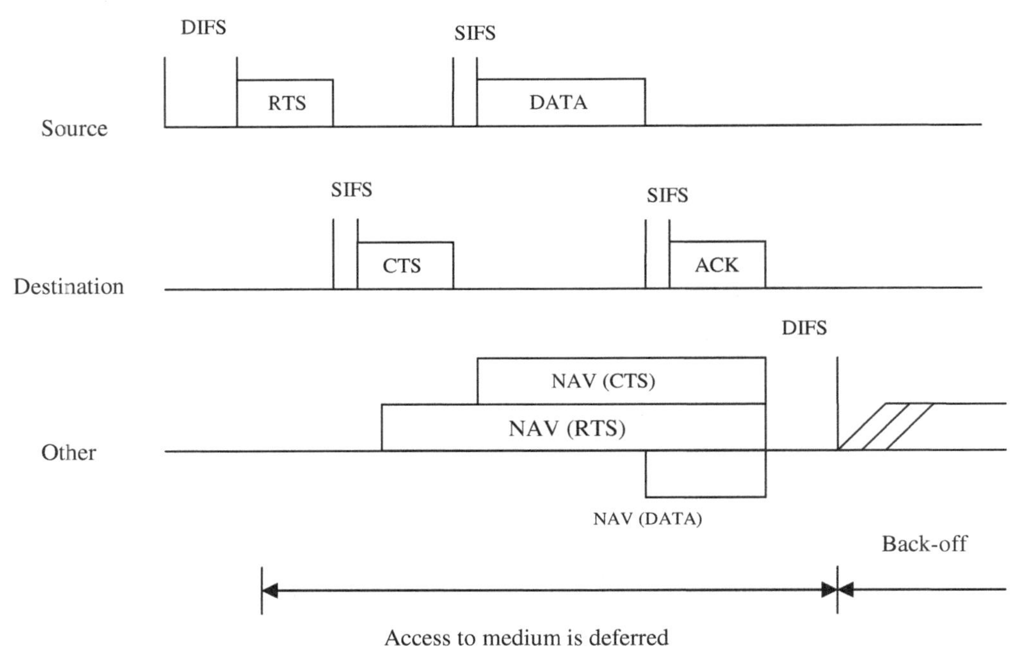

Fig.2 RTS/CTS access mechanism in DCF

Assumptions

Backoff algorithm in IEEE 802.11 is to minimize collisions during contention between multiple stations. Following assumptions are being made for the new modified backoff algorithm for IEEE 802.11.

1. There are collisions because of busy medium.
2. Whenever collision takes place for the resolution of the collision the backoff time to be selected from [0, CW-1] randomly.
3. The size of contention window is between ($CW_{\min}, CW_{\max}$).
4. Initial value of CW is set as $CW_{\min}$.
5. Initially the contention window is modified in the form of $(\sqrt{2}CW)$ then after four collisions at the 5th collision the size of contention window will be doubled in consecutive collisions.
6. The value of contention window (CW) will be set again after each successful transmission or CW reaches the maximum limit of transmission i.e. retry limit. This is the modified Binary Exponential Backoff algorithm.

III. PERFORMANCE ANALYIS FOR DCF

Initially if the collision is low then the size of window is increased in $\sqrt{2}$ factor. As the number of collisions will increase, it will indicate that the numbers of contending nodes are more. In this case the size of contention window will be increased by the factor 2 at every subsequent collision. The value of CW depends on the number of failed transmissions of a frame.

A. Markov Chain Model

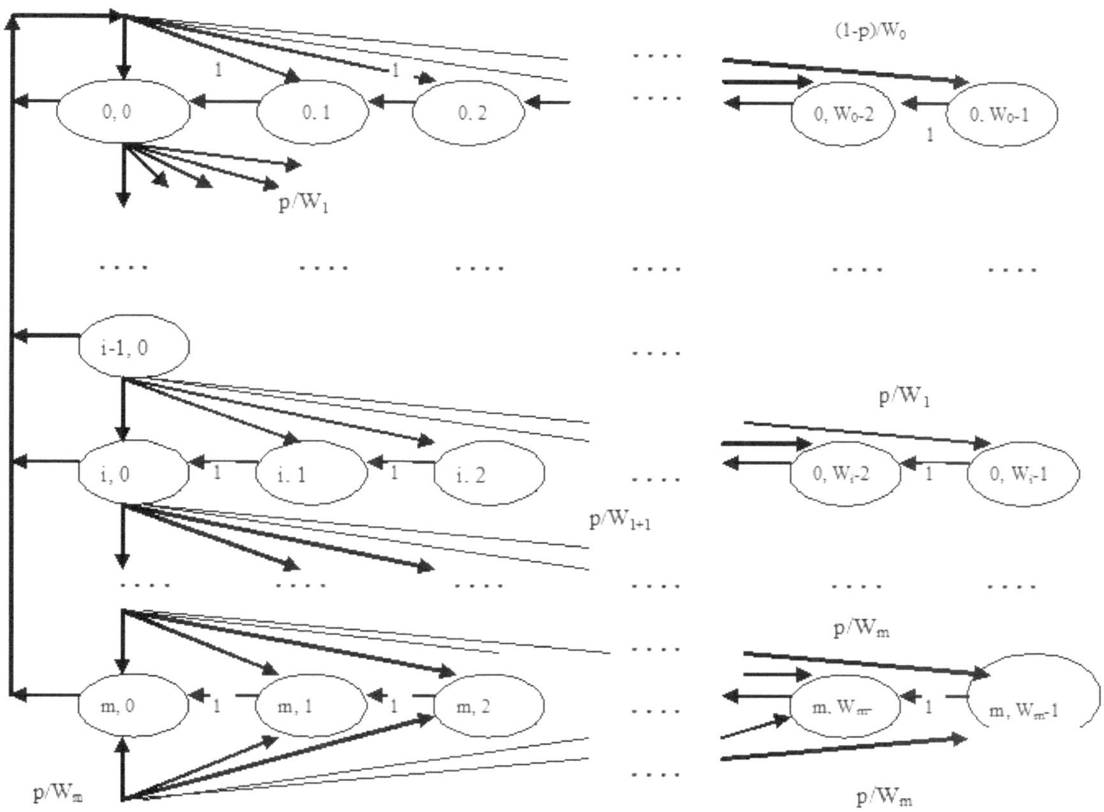

Fig 3 Markov Chain model for the IEEE 802.11 backoff window size

Transmission Probability

Let $b(t)$ be the stochastic process representing the backoff time counter and $s(t)$ be the stochastic process representing the backoff stage for a given station at slot time t. The bidimensional process $\{b(t), s(t)\}$ can be modeled with a discrete-time Markov chain given in the figure .

Let $b_{i,k} = \lim_{t\to\infty} P\{s(t)=i, b(t)=k\}, i \in (0,m), k \in (0, W_i - 1)$ be the stationary distribution of the chain then we can calculate the probability $b_{i,k}$. We have the following relations:

$$b_{i,0} = p.b_{i-1,0}, \quad 0 < i \le m \qquad (1)$$

$$b_{i,0} = p^i.b_{0,0}, \quad 0 \le i \le m \qquad (2)$$

$$W_i = \sqrt{2}.W \qquad i \le m'$$

$$W_i = 2^{m'}.W \qquad i > m' \qquad (3)$$

Where m represents the station short retry count and is equal to 7 according to IEEE 802.11 standard. Owing to the chain regularities, for each $k \in (1, W_i - 1)$, it is

$$b_{i,k} = \frac{W_i - k}{W_i} \cdot \begin{cases} (1-p).\sum_{j=0}^{m-1} b_{j,0} + b_{m,0} & ,i=0 \\ p.b_{i-1,0} & 0 < i \le m \end{cases} , \quad (4)$$

Using (2),(4) the simplified form will be as:

$$b_{i,k} = \frac{W_i - k}{W_i} b_{i,0} \quad i \in (0,m), \quad k \in (1, W_i - 1). \quad (5)$$

Thus, by relations (2) and (5) all the $b_{i,k}$ values can be expressed as function of $b_{0,0}$ and of collision probability p. By imposing the normalization condition and some simplification, we have:

$$1 = \sum_{k=0}^{W_i-1} \sum_{i=0}^{m} b_{i,k} = \sum_{i=0}^{m} b_{i,0} \sum_{k=0}^{W_i-1} \frac{W_i - k}{W_i} = \sum_{i=0}^{m} b_{i,0} \cdot \frac{W_i + 1}{2}$$

$$= \frac{b_{0,0}}{2} \left[W \left(\sum_{i=0}^{m'} \left(\sqrt{2}p\right)^i + \sum_{i=0}^{m'} p^i \right) + \sum_{i=m'}^{m} 2^{m'} p^i W \right]$$

From which

$$b_{0,0} = \frac{2.(1-\sqrt{2}p)(1-p)}{W.(1-(\sqrt{2}p)^{m+1}.(1-p)+(1-\sqrt{2}p)(1-p^{m+1})} \quad ,m \le m'$$

$$= \frac{2.(1-\sqrt{2}p)(1-p)}{W.(1-(\sqrt{2}p)^{m+1}.(1-p)+(1-\sqrt{2}p)(1-p^{m+1})+W.2^{m'}.p^{m'+1}(1-\sqrt{2}p)(1-p^{m-m'})} \quad ,m > m'$$

The probability of transmission τ, in a randomly chosen slot can be expressed as

$$\tau = \sum_{i=0}^{m} b_{i,0} = \frac{1-p^{m+1}}{1-p} b_{0,0} \quad (8)$$

Where $b_{0,0}$ can be expressed as in equation (7).

The transmission probability τ depends on the conditional collision probability p, which is derived next. The probability p is the probability that at least one of the n-1 remaining stations transmit in the same slot. If all stations transmit with probability τ, the collision probability p is:

$$p = 1-(1-\tau)^{n-1} \quad (9)$$

Equations (8) and (9) form a nonlinear system with two unknown variable τ and p, which can be solved by numerical results. Note that we must have $p \in (0,1)$ and $\tau \in (0,1)$.

B. Delay analysis

Let P_{tr} be the probability that there is at least one transmission in the considered slot time. For n contending stations for the network, P_{tr} is given by:

$$P_{tr} = 1-(1-\tau)^n \tag{10}$$

The probability P_s for a successful packet transmission is given by the probability that exactly one station transmits and the remaining n-1 stations defer transmission, conditioned on the fact that at least one station transmits:

$$P_s = \frac{n.\tau.(1-\tau)^{n-1}}{1-(1-\tau)^n} \tag{11}$$

Now we are able to express the delay E[X] as given in [12]:

$$E[D] = E[X]\cdot\sigma + E[N_{Fr}]\cdot(P_S \cdot T_S) + (1-P_S)\cdot T_C) + T_S \tag{12}$$

Where $E[N_{Fr}]$ is the mean number of times a station freezes its counter due to other transmissions before successfully transmitting a frame, σ is the size of a time slot and the time period E[X] that is required for empty slots before successfully transmitting a frame is given by:

$$E[X] = \sum_{i=0}^{m} (q_i \cdot E[BC_i]) \tag{13}$$

Where q_i is the probability that a station's transmission is in stage i and $E[BC_i]$ is the average Backoff counter value in stage i. Here, T_s and T_c are the average times that the medium is sensed busy due to a successful transmission or a collision respectively. The values of T_s and T_c depend on the channel access mechanism and in the basic access:

$$T_s^{bas} = H + E[P] + SIFS + \delta + ACK + DIFS + \delta$$
$$T_c^{bas} = H + E[P] + SIFS + \delta + ACK\_timeout + DIFS$$

For the RTS/CTS access mechanism, it is:

$$T_s^{rts} = RTS + SIFS + \delta + CTS + SIFS + \delta + H + E[P] + SIFS + \delta + ACK + DIFS + \delta$$
$$T_c^{rts} = RTS + SIFS + \delta + CTS\_timeout + DIFS$$

Where $H = PHY_{hdr} + MAC_{hdr}$.

After collision the colliding stations have to wait for the time equal to ACK_timeout and CTS_timeout.

Finally, the time period E[X] and $E[N_{Fr}]$, expressed in slot times, is equal to:

$$E[X] = \frac{b_{0,0}}{2\tau}\left[\frac{W((1-p)(1-2p)(1-(\sqrt{2}p)m') + (1-\sqrt{2}p)(1-p)(1-(2p)^{m-m'-1} - (1-p^m)(1-\sqrt{2}p)(1-p)}{(1-\sqrt{2}p)(1-p)(1-2p)}\right] \tag{14}$$

$$E[N_{Fr}] = \frac{E[X] - E[Idle]}{E[Idle]} \tag{15}$$

Where E[Idle] is the average number of idle slots before a transmission occurs and is equal to:

$$E[Idle] = \frac{idle}{busy} = \frac{1 - P_{tr}}{P_{tr}} \tag{16}$$

By putting (14), (15) and (16) into (12), the delay can be obtained and the outcome is given in the next section.

IV. Delay Analysis Results

The system parameters shown in the table I are being used for the analysis. The figure 4 below indicates the dependency of delay on the number of stations. The results indicate that the delay is lower for the RTS/CTS access mechanism then the basic access mechanism. The delay is lower for the small size networks for basic as well as RTS/CTS access mechanisms but the delay increases drastically for basic access mechanism for larger network size, as compared to the RTS/CTS access mechanism. For RTS/CTS and basic access mechanisms, the performance of the network is almost steady up to the size of network n=15.

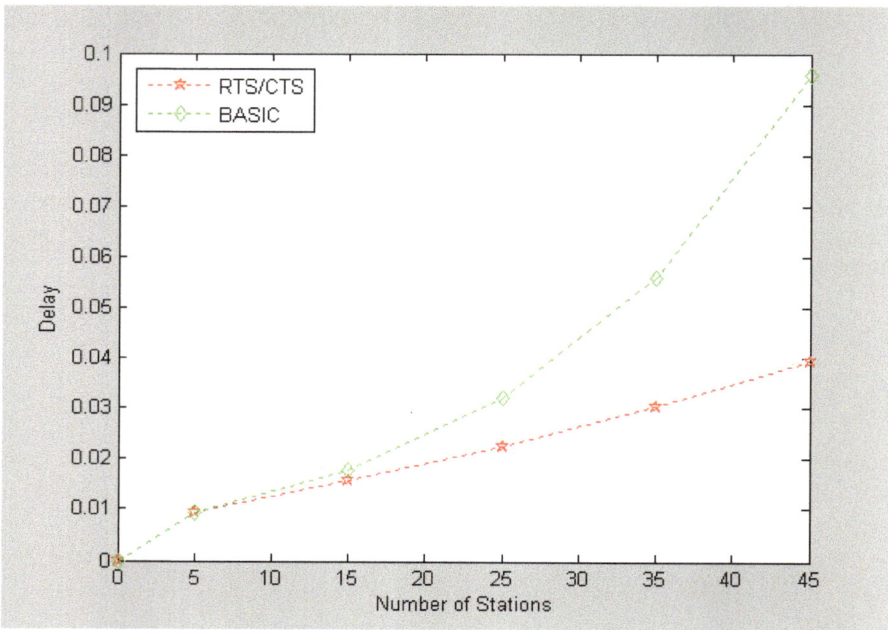

Figure 4 : Delay analysis versus Number of Stations Modified Exponential Backoff method for RTS-CTS and Basic access method

The delay performance of the IEEE 802.11 DCF (RTS/CTS and Basic access method) versus transmission probability has been shown in the figure 5, which indicates that the delay is almost same for both the access methods till the value of transmission probability (tau = 0.82), after that the performance of RTS/CTS is better than basic access method. There is steep rise in delay for basic access method than RTS/CTS method. Figure 6 indicates that the delay increases linearly (nearly) with the size of frame for both the access methods of IEEE 802.11DCF. There is not a considerable difference in terms of delay performance for both the access methods.

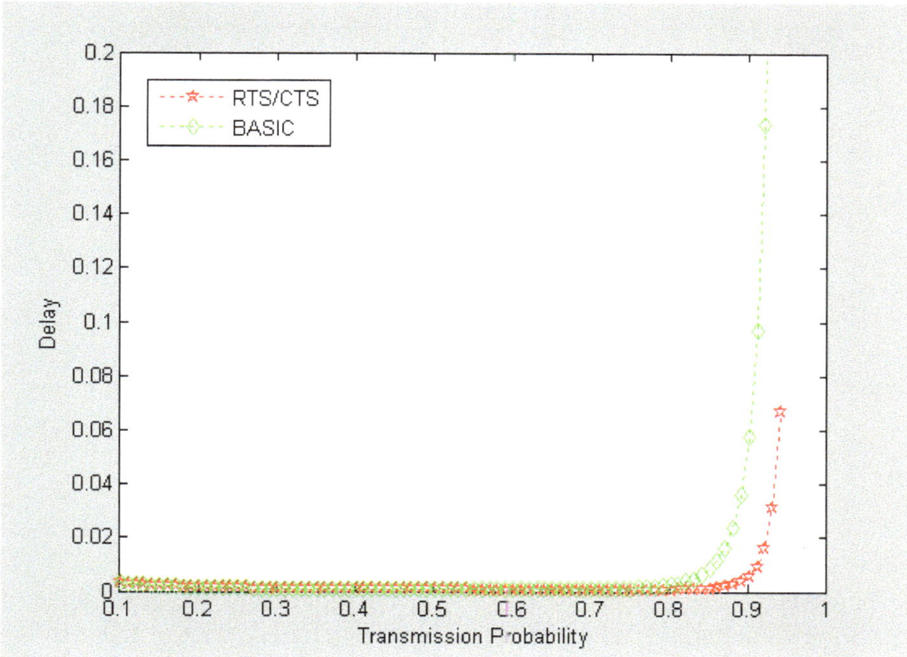

Figure 5 : Delay analysis versus Transmission Probability of Modified Exponential Backoff method for RTS-CTS and Basic access method

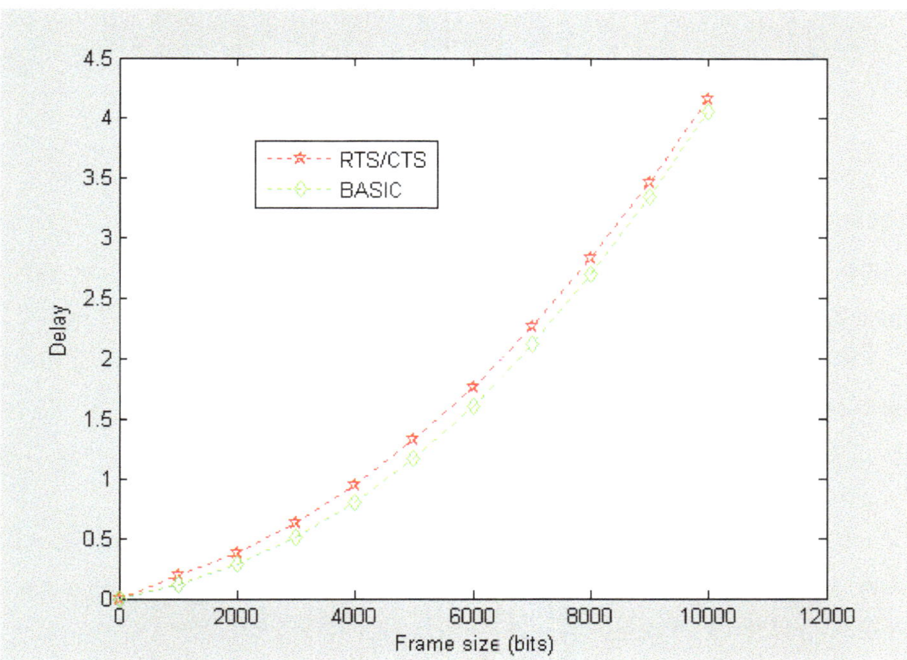

Figure 6 : Delay versus Frame size of Modified Exponential Backoff method for RTS-CTS and Basic access method

TABLE I System Parameters

Transmission Rate	1 Mbps
Packet Payload	8184 bits
PHY header	128 bits
RTS	160 bits + PHY header
CTS	112 bits + PHY header
ACK	112 bits + PHY header
DIFS	128 μs
SIFS	28 μs
Slot Time (σ)	50 μs
Propagation delay (δ)	1 μs
ACK_Timeout	300 μs
CTS_Timeout	300 μs

V. Conclusion

The paper presents a Markov chain based model with modified Exponential Backoff method. This model is used to evaluate the delay based performance of IEEE 802.11, the standard used for wireless LANs. The delay based performance of the IEEE 802.11 MAC protocol with modified Backoff algorithm has been evaluated, with respect to the frame size, network size and transmission probability. For number of stations the delay performance is better for RTS/CTS access method than the basic access method. The model is suitable for both access methods, separately or combined. The performance of both access methods is compared with respect to transmission probability. The performance of RTS/CTS access method is better than basic access method for all considered parameters. The delay of both the access methods is almost similar for frame size. However the performance of RTS/CTS method is far better for large network size.

VI. References

[1] IEEE standard for Wireless LAN Medium Access Control (MAC) and Physical Layer (PHY) specifications, ISO/IEC 8802-11:1999(E), Aug.1999

[2] B.P.Crow, I.Widjaja, J.G. Kim, "IEEE 802.11 Wireless Local Area Networks",IEEE Communication magazine, Sept.1997

[3] K. Huang, K. Chen, " Interference Analysis of Nonpersistent CSMA with Hidden Terminals in Multicell Wireless Data Networks", Proc. PIMRC, Toronto, pp. 907-911, 1995

[4] L. Kleinrock and F. Tobagi, "Packet switching in radio channels, Part II-The hidden terminal problem in carrier sense multiple access and busy tone solution," IEEE Trans. Commun., vol. COM-23, no 12,pp. 1417-1433, Dec. 1975

[5] G. Bianchi, L. Fratta, M. Oliveri, "Performance Evaluation and Enhancement of the CSMA/CA MAC Protocol for 802.11 Wireless LANs", proc. PIMRC'96. Taipei, Taiwan, pp.392-396, Oct.1996

[6] H. S. Chhaya, S. Gupta, "Performance modeling of asynchronous data transfer methods of IEEE 802.11 MAC protocol", Wireless Networks 3 (1997), pp.217-234,1997

[7] T. Ho, K. Chen, "Performance Analysis of IEEE 802.11 CSMA/CA Medium Access Protocol," proc. PIMRC'96 Taipei, Taiwan, pp.407-411, Oct.1996

[8] F.Cali, M.Conti, E.Gregori, "Dynamic Tuning of the IEEE 802.11 Protocol to Achieve a Theoretical Throughput Limit", IEEE/ACM Trans. On Networking, vol. 8, No. 6, Dec. 2000

[9] G.Bianchi, " IEEE 802.11 Saturation Throughput Analysis", IEEE Comm. Letters Vol 2, No. 12 , Dec. 1998

[10] G.Bianchi, " Performance Analysis of the IEEE 802.11 Distributed Coordination Function" ,IEEE Journal on selected Area in Comm. Vol 18 No. 3, March 2000

[11] H. Wu, Y. Peng, K. Long, S. Cheng, J. Ma, " Performance of Reliable Transport over IEEE 802.11 Wireless LAN: Analysis and Enhancement", IEEE INFOCOM'2002

[12] P. Chatzimisios, V. Vitsas, A.C. Boucouvalas, "Throughput and delay analysis of IEEE 802.11 protocol," Proc of IEEE International Workshop on Networked Appliances (IWNA), Liverpool, UK 2002

[13] R. Jain, " The art of Computer Systems Performance Analysis," John Wiley and Sons, 1991

[14] D. Bertsekas and R. Gallagar, "Data Networks", Prentice-Hall of India,2002.

[15] C. Siva Ram Murthy and B.S. Manoj, " Ad Hoc Wireless Networks" Pearson Education,2005

[16] H. Anousar and C. Bonnet, "Optimal Constant-Window Backoff Scheme for IEEE 802.11 DCF in Finite Load Single-Hop Wireless Networks" Proc of ACM MSWiM'06 Malaga, Spain Oct. 2006

[17] W.-T. Chen, " An Effective medium contention method to improve the performance of IEEE 802.11" Wireless Networks, Springer 2008.

[18] J. Choi and K.G.Shin, " QoS Provisioning for large scale Multi-ap WLANs" Elsevier Journal of Ad Hoc Networks 2010 (in press).

[19] L.Song and C.Yu, "Minimizing spatial and time reservation With Collision-Aware DCF in Mobile Ad Hoc Netwrks", Elsevier Journal of Ad Hoc Networks vol. 7 2009

[20] C. Hu, H. Kim and J.C.Hou, "Short term non-uniform access in IEEE 802.11compliant WLANs: A study on its impact on the saturation performance" Elsevier Journal of Computer Networks vol. 52, 2008.

[21] Rishi Pal Singh, D. K. Lobiyal, "Throughput Analysis of IEEE 802.11 DCF with Modified Binary Exponential Backoff in Mobile Ad Hoc Networks", IJMCMC 2(3): 61-70 (2010).

7

Optimized ZRP for MANETs and its Applications

M.N. SreeRangaRaju
Department of
Telecommunication Engg
Bangalore Institute of Technology
Bangalore, India
Email: mnsrr@rediffmail.com

Dr. Jitendranath Mungara
CMR Institute of Tech & Mgt
Bangalore, India
Email: jmungara@yahoo.com

ABSTRACT

In this paper, we unified two new algorithms, i.e., to selectively Bordercast the control packets in order to reduce the network load by limiting the number of control packets when ZRP searches for a new route. The second algorithm is proposed to optimize the performance of Query control mechanisms for the Zone Routing Protocol (ZRP) for mobile ad hoc networks for communication on an urban terrain. ZRP proactively maintains routing information for a local neighbourhood (routing zone), while reactively acquiring routes to destinations beyond the routing zone. This hybrid routing approach can be more efficient than traditional Zone Routing Protocol (ZRP) framework. However, without proper query control techniques, the ZRP cannot provide the expected reduction in the control traffic. Our proposed algorithm enhances the structure of the routing zone to provide enhanced detection and prevention of overlapping queries. These techniques can be applied to single or multiple channel mobile adhoc networks to improve both the delay and control traffic performance of ZRP. Our new algorithm allows ZRP to provide routes to all accessible network nodes, with less control traffic than purely proactive link state or purely reactive route discovery, and with less delay than conventional flood searching.

To further improve the efficiency of Enhanced ZRP, our future work shall propose a new algorithm to overcome the generation of periodic updates in the routing table to reduce the control traffic. Thus by the implementation of our proposed algorithm the time taken to receive the first packet will be minimized, which results in better throughput.

The performance evaluation via simulations using a well known simulator QualNet version 5.0 shows that the new approach has better performance than the traditional ZRP.

KEYWORDS

Adhoc network, Bordercast, hybrid routing, Proactive routing, query control, reactive routing, routing protocol, routing zone and zone routing protocol (ZRP).

1. INTRODUCTION

Manet is a collection of wireless nodes that can dynamically form a network to exchange information without using any pre-existing fixed network infrastructure with rapid configuration of wireless connections on-the-fly [1]. Due to rapidly changing connectivity, network partition, higher error rates, collision interference, bandwidth and power constrain together pose new problems in network control-particularly in the design of higher level protocol such as routing and in implementing applications with quality of service requirements. Routing in Manets is one of the primary functions that, each node must achieve network functions at any point of time. Routing enables connections between nodes those are not within each other's vicinity. To serve this purpose, various proactive, reactive and hybrid routing protocols are been proposed.

In proactive routing protocols viz., destination-sequenced distance vector (DSDV) [2] and optimized link state routing protocol (OLSR) [3], routes between source and destination are maintained regardless of the data traffic. Such a strategy avoids the need of finding routes for each message and is especially efficient when the nodes are relatively stationary and traffic is relatively heavy. Unlike proactive routing protocols, reactive routing protocols [4] viz., dynamic source routing (DSR) and ad hoc on demand distance vector (AODV), find route only when node attempts to send a message. It avoids the need of frequent link and route updates therefore substantially reduces energy consumption [5] when the traffic load is light or the network mobility is high. Hybrid routing protocol utilizes both reactive and proactive protocol mechanisms.

The rest of the paper is organized as follows: section- ii explains overview of DSR, AODV and ZRP for Manets. Section iii gives a detailed description of our proposed work such that we can have better efficiency. Section iv presents simulation based results, evaluation and performance comparison graphs of our previous work. Finally, conclusion and future work are presented in section v.

2. Overview of ZRP

2.1 Zone Routing Protocol

ZRP was proposed to reduce the control overhead of proactive routing protocols and decrease the latency caused by routing discover in reactive routing protocols. ZRP defines a zone around each node consisting of its k-neighbourhood (e. g. k=3). In ZRP, the distance and a node, all nodes within -hop distance from node belongs to the routing zone of node. ZRP is formed by two sub-protocols, a proactive routing protocol: Intra-zone Routing Protocol (IARP) [9] is used inside routing zones and a reactive routing protocol: Inter-zone Routing Protocol (IERP) [10] is used between routing zones, respectively. A route to a destination within the local zone can be established from the proactively cached routing table of the source by IARP, therefore, if the source and destination is in the same zone, the packet can be delivered immediately. Most of the existing proactive routing algorithms can be used as the IARP for ZRP. For routes beyond the local zone, route discovery happens reactively. The source node sends a route requests to its border nodes, containing its own address, the destination address and a unique sequence number. Border nodes are nodes which are exactly the maximum number of hops to the defined local zone away from the source. The border nodes check their local zone for the destination. If the requested node is not a member of this local zone, the node adds its own address to the route request packet and forwards the packet to its border nodes. If the destination is a member of the local zone of the node, it sends a route reply on the reverse path back to the source. The source node uses the path saved in the route reply packet to send data packets to the destination.

3. Related Work

This section reviews the some of literature pertaining to routing in wireless ad hoc networks. In [1], the work presents query controlled scheme to provide enhanced detection and prevention of overlapping queries. The work done in [2] presents a selective bordercasting zone routing protocol to reduce the network load by limiting the number of control packets when the protocol searches for new route. The work presented in [3], is a protocol called multipath distance vector zone routing protocol for mobile ad-hoc networks, which uses a topological map of the zone centered on a node to guarantee loop freedom and alternative paths in the case of route failure and disjoint paths. In [4] an algorithm is proposed to provide improved quality of service such as low end to end delay and high throughput via hybrid routing protocol ZRP. In paper [5] independent zone routing is proposed, which allows adaptive and distributed configuration for the optimal size of the each node's routing zone on the per node basis. In[6], it has given review about the current routing protocols in Ad-hoc networks.

4. Proposed Work

Basic idea of ZRP is based on the idea that querying can be done more efficiently than flooding, by directing route requests to target peripheral nodes. Since neighbouring routing zones heavily overlap, each node may forward same query messages multiple times. Thus leading to increase in the number of control packets. Also there are chances that these query messages may be forwarded again inward instead of moving towards the destination. This unnecessarily adds the delay in the system.

4.1 Detection of Query PacketId

To reduce this delay we can go for query control mechanisms. Query detection uses query source node's Id and query Id pair. Below Fig. 1 depicts query detection scheme. Where 'S' is the source node and destination node 'D' is not in S's local zone. Hence S Bordercasts the query message packet. Nodes J, K, I detects the query packet through QID Method-1 and node H detects it through evesdropping i.e QID Method-2.

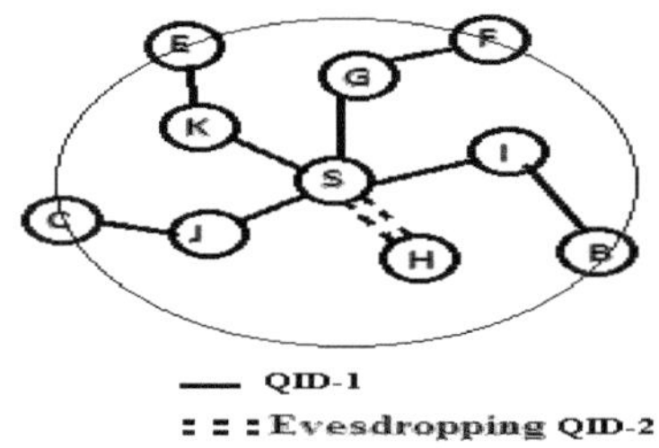

4.2 Premature Node Termination

With the knowledge of the QD obtained a procedure has to be followed in which covered nodes can be pruned by the Bordercast tree. This procedure is known as premature node termination. By referring to below Fig. 2, we can see that node 'S' Bordercasts the query message packet to its peripheral nodes. Node 'I' receives same copy of query message packet through Bordercast tree by the other node as it may be interior member of other routing zone. Now 'I' searches for the interior zone member of other node, and uses QD to detect is it necessary to relay that query message packet or no. 'I' identifies that nodes are been covered, it reconstructs the Bordercast tree. Prunes covered nodes from other node' Bordercast tree and deletes the query message received by other node using the criterion called premature node termination.

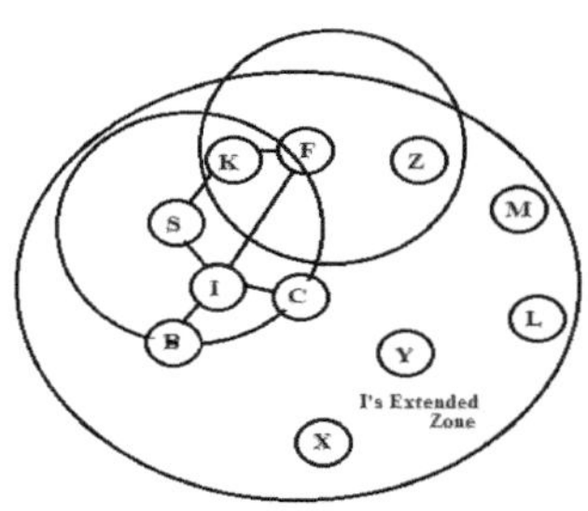

Fig. 2: Premature Node Termination

4.3 Random Query Analysing and Processing Delay

Bordercasting node initiates bordercasting to its peripheral nodes. We can observe from the below Fig. 3 that bordercasting node 'S' Bordercasts the query message packet to peripheral node 'F'. This node 'F' will border cast the query message packet to 'X' and 'Y'. 'X' and 'Y' receive query at the same time. If both of them Bordercasts the query then again there are chances that same node may relay the same query multiple times. Unless and until X and Y check for the QD information they both will not be knowing that the query Bordercasted by them is redundant.

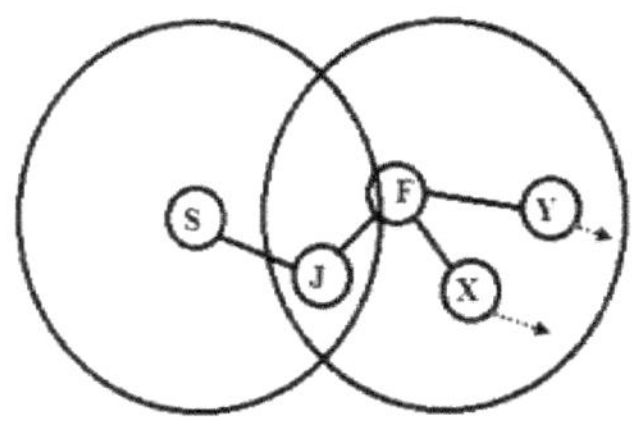

Fig. 3: Random Query Analyzing and Processing Delay

4.4 Selective Bordercasting Approach:

This approach is specially proposing for the minimization of control overhead packets for the ZRP protocol, which leads to congestion and delay in route acquisition The selective bordercasting mechanism use same IARP protocol as in ZRP, but use a new mechanism in IERP for interzone routing because the bordercasting procedure starts with IERP protocol.

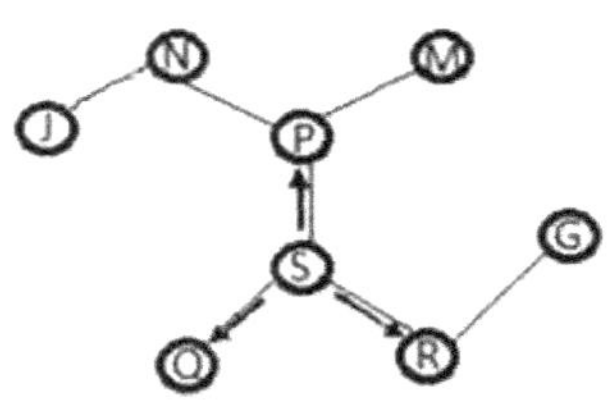

Fig. 4: IARP packet broadcasting.

The operation of the IARP is shown in Fig. 4. The source node S generates the IARP packet and broadcasts in its local routing zone, by using neighbor's information, the central node updates its routing table. When the destination is outside the routing zone, the interzone operation initializes. The IERP is slight different from IERP in original ZRP, to understand the mechanism of new IERP, consider the Fig. 5, which the route to the destination in unknown to the source. Then the IERP request packet is generated. The information in the new IERP packet is: SN=S, DN=D, and border-cast hop value (BH) =1 which is shown in Fig. 5(a) and Fig. 5(b). The generated IERP packet is sent to all border nodes which lie on the zone border, which have same zone radius.

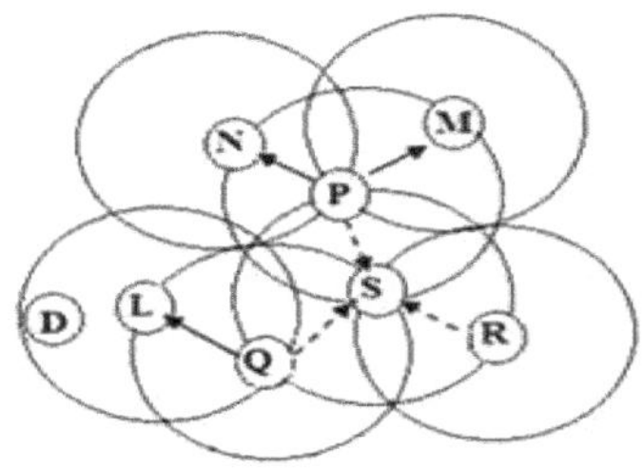

Fig. 5: IERP Packet transmission.

DN	SN	Intermediate Node	Packet Type	BH

Fig. 5(a): IERP Packet format.

DN	SN	Intermediate Node	Packet Type	BH
D	S	Null	REQ	1

Fig. 5(b): IERP Request packet format of Fig. 5.

The nodes in the zone boundary which receives the IERP request packet with increase in the BH value by 1. Then the node checks the destination in its routing table, if the destination address doesn't found in its routing table, then it should continue with the border casting procedure. When the nodes find the destination in its routing zone or itself a destination then the node will replays with IERP Replay packet as shown in the Fig. 6 and reply packet is shown in Fig. 6(a).

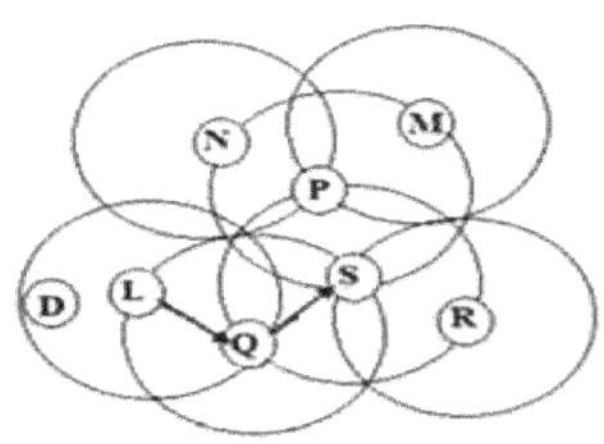

Fig. 6: IERP Reply Transmission

DN	SN	Intermediate BN	Packet Type	BH
D	S	L, Q	REP	2

Fig. 6(a): IERP reply packet format of Fig-6

In this Selective bordercasting approach the IERP Request packet's must save discovered routing information in a buffer for awhile and if there are requests for the same destination which the IERP REQ's are sent by nodes in the previous search. For example, when a destination moves inside the routing zone of Q, as shown in the Fig. 7, the previous recorded route information stored in the node Q and L, the node Q will generate the Reply packet to source S.

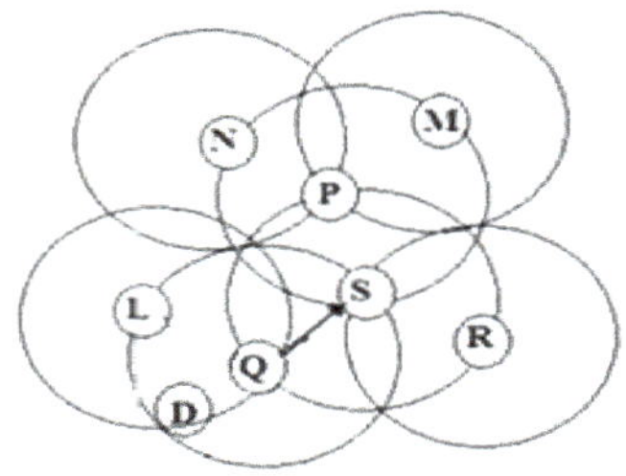

Fig. 7: a case when the destination node moves inside the zone.

In the case of destination which moves outside the zone of L as shown in the Fig. 8, it has no route to destination which is required for the query source, and then a new search is started from the node L.

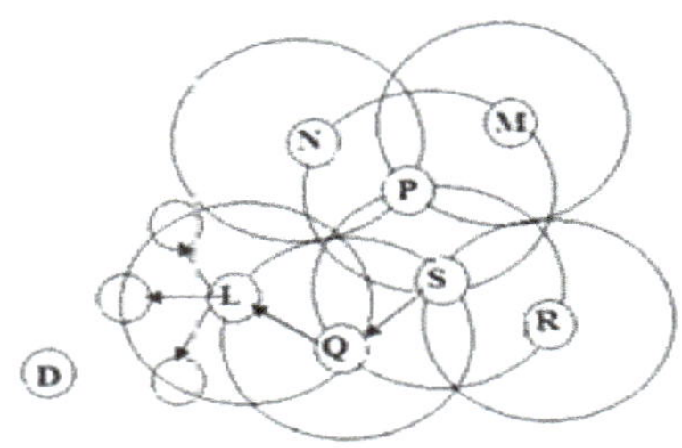

Fig. 8: A case when DN moves outside the zone

Thus the number of IERP Route Request packets is reduced, because the route reconfiguration is started from the destination failure reporting node instead of beginning from the source. Hence this reduces control overhead packets and end-to-end delay time, which results in higher throughput.

5. Simulation Results

In this section the methodology is presented .This methodology is used in order to isolate the impact on network performance. We used a well known network simulator QualNet version 5.0.

5.1 Simulation Platform, Models and Attributes

The image of the network as it appears in QualNet version 5.0 is presented in Fig. 9

Fig. 9: Topology of the Simulated Network on an Urban Terrain

The physical medium used is the well known 802.11DSSS PHY with a data rate of 2 mbps. The MAC protocol used is 802.11 MAC protocol, configured in a MANET mode on an urban terrain. More precisely we use only Distributed Coordination Function (DCF) of the protocol.

5.2 Results

Table-1 shows parameters we considered for simulations.

Table 1
Parameters we considered for simulations.

Parameters	Values
Routing Protocol	ZRP and Enhanced ZRP
Fading Model	Rayleigh
Shadowing Model	Constant
Pathloss Model	Cost-231 HATA
Energy Model	Mica Motes
Battery Model	Simple linear
Terrain File	DEM
Mobility Speed	15 to 20 meter/sec
Network Size-10-20 nodes	Area Considered-500 X 500
Network Size-50-100 nodes	Area Considered-1000 X 1000
Network Size-200 nodes	Area Considered-3000 X 3000
Node Placement	Random node placement under seed-1

After studying the operation of ZRP and Enhanced ZRP we can observe that, ZRP uses transmission of excess control messages there by resulting in poor QoS, by studying ZRP operation of route discovery we observe that ZRP uses additional time as it uses IARP, IERP and BRP. Thus leads to more number of control packets. When a destination node is not found in the local zone of the source node it initializes IERP. ZRP takes time for inter communication between IERP and IARP. Each node maintains routing table of their local zone. This adds unnecessary traffic in the network. This causes route acquisition delay. After discovering the route to the destination the data packets are encapsulated by two protocols. Hence it takes more time for data packet to reach to the destination.

The average end to end delay introduced is more for ZRP in our test case this is because of flooding more query control messages. Our proposed algorithm reduces this and there by promising reduced End to End delay. This is as shown in below Fig. 10.

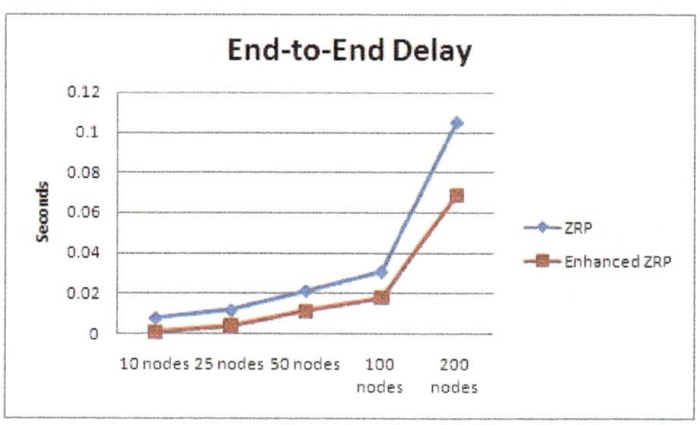

Fig. 10: Average End-to-End delay displayed in seconds for different runs on different network sizes

The time taken to receive the first packet in case of ZRP is higher when compared to Enhanced ZRP, this is because of the excess control messages flooded in case of IARP and IERP. This is as shown in below Fig. 11.

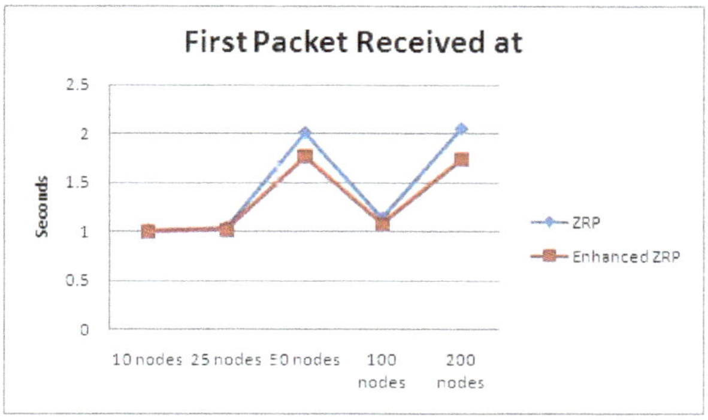

Fig. 11: The first packet received in seconds

In our simulation, time constraint is 100 sec, total number of packets and total numbers of bytes at the destination received within this simulation time are reduced for ZRP due to increased delay as compared with Enhanced ZRP. This is as shown in Fig. 12 and Fig. 13.

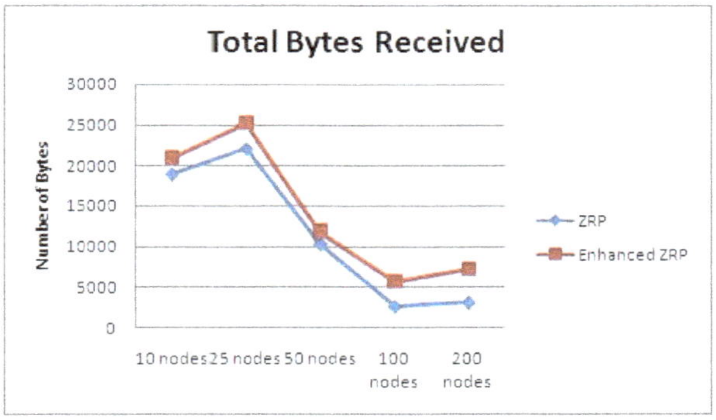

Fig. 12: Total Bytes Received on different network sizes.

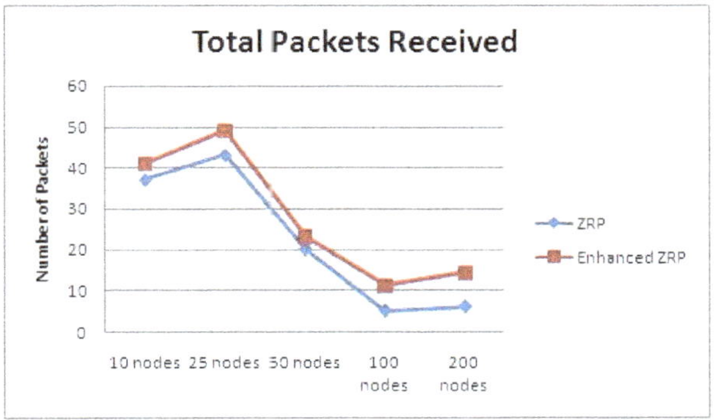

Fig.13: Total packets received for multiple runs with different node density.

In the above Fig. 14 we can observe that when network density is less i.e. for 10 nodes number of packets received are more as compared at high network density. This is because of number of nodes increase, number of overlapping zones increase, thus query messages.

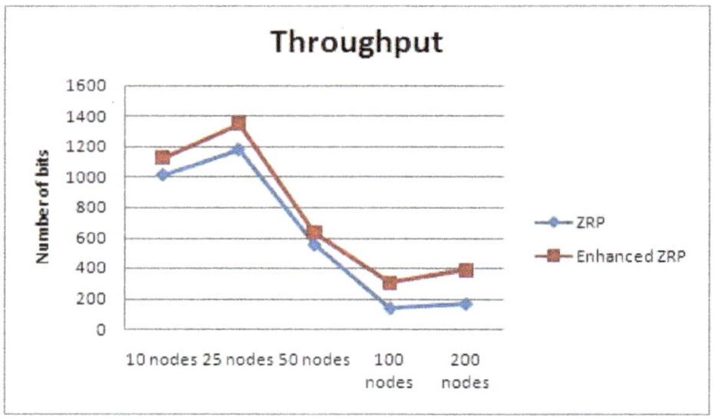

Fig.14: Throughput in bits per second obtained for different network sizes.

As the number of the nodes increases in the network, route discovery becomes more complicated, because centralized node routing zones will highly overlap with each other, hence the route request queries will be flooded in to the network, and the intermediate nodes will send same route request queries multiple times, hence the route acquisition delay will have higher percentage as the number of nodes increases. Therefore the zone routing protocol have lower throughput when compared to Enhanced ZRP. At 200 nodes Enhanced ZRP has low transmission rates but more than ZRP. This is as shown in Fig. 14.

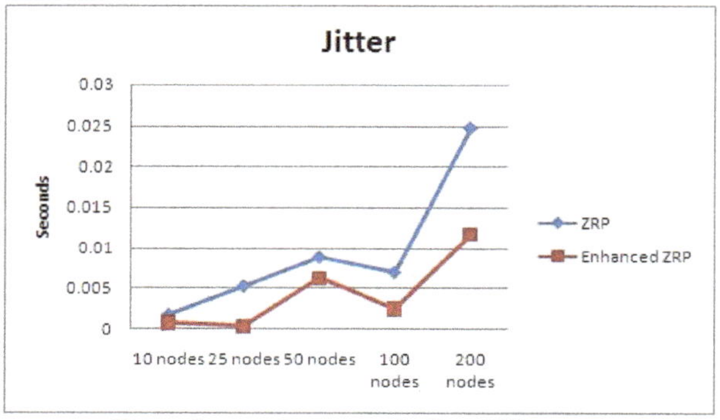

Fig. 15: The average jitter shown in seconds for different network sizes

Jitter is the variation in the time between packets arriving, caused by network congestion, timing drift, or route changes. As from the above Fig. 15 we can observe that at 10 nodes both the protocols have small jitter value. But at 25, 50, 100 and 200 nodes since query packets will be flooded throughout the network control overhead increases; it consumes more time to reconfigure the route if link failure occurs. Hence there will be more time variation between arrivals of packets results in more jitter value.

From above graphs of Fig. 10 thru Fig. 158 we can observe that Enhanced ZRP has performed better than ZRP to promise better throughput, end to end delay, jitter and number of bytes received. ZRP was not up to the task and it performed poorly throughout all the simulation sequences, hence putting itself out of competition.

6. Conclusions

With simulation results we compared ZRP with Enhanced ZRP by involving two enhanced algorithms to enhance the performance of ZRP for communication on MANETs on an Urban Terrain. By analyzing simulation results we can observe Enhanced ZRP had performed better in all the simulation sequences with an improvement of 15 to 20 percent when compared to ZRP. To further improve the efficiency of Enhanced ZRP, our future work shall involve a new algorithm to overcome the generation of periodic updates in the routing table to reduce the control traffic. Thus by the implementation of our proposed algorithm the time taken to receive the first packet will be minimized, which results in better throughput. So the number of bytes and total packets received at the destination will be increased.

The techniques can be applied to single or multiple channel MANETs to improve both the delay and control traffic performance of ZRP. In this scheme, we allow ZRP to provide routes to all accessible network nodes, with less control traffic than purely proactive link state or purely reactive route discovery, and with less delay. Extensive simulation results reveals that our scheme features better transmission delay, route convergence time, the last packet reception time, system efficiency and system through-put.

7. References

[1]. Jun-Zhao Sun, *"Mobile Ad-Hoc Networking: An Essential Technology for Pervasive Computing,"* Proceedings of International conference on info-tech and info-net, Vol-3, pp. 316-321, 2001.

[2]. Abdul Hadi Abd Rahman, Zuriati Ahmad Zukarnain, *"Performance Comparison of AODV, DSDV and I-DSDV Routing Protocols in Mobile Ad Hoc Networks,"* European Journal of Scientific Research ISSN 1450-216X, Vol.31 No.4, pp.566-576, 2009.

[3]. P. Jacquet, Ed. Project Hipercom, INRIA"Optimized Link State Routing Protocol (OLSR)," http://www.ietf.org/rfc/rfc3626.txt- DSR-09.txt, IETF Internet draft, Oct. 2003

[4]. Nasipuri and S.R. Das, *"On-demand multi-path routing for mobile ad hoc networks,"* IEEE ICCCN'99, pp. 64–70, 1999.

[5]. S. Singh, M. Woo and C. S. Raghavendra, *"Power aware routing in mobile ad hoc networks",* IEEE/ACM MobiCom-98, pp. 181-190. Oct.1998.

[6]. D. Johnson, D. Maltz and Yih-Chun Hu, *"The Dynamic Source Routing Protocol for Mobile Ad Hoc Networks,"* http://www.ietf.org/internet- drafts/draftietfmanet- DSR-09.txt, IETF Internet draft, Apr. 2003.

[7]. Vincent D. Park and M. Scott Corson, *"Adhoc On demand Distance Vector (AODV) version 4: Functional specification,"* Internet- Draft,draft-ietfmanet- AODV-spec-04.txt, July 2001.

[8]. Zygmunt J. Haas, Marc R. Pearlman,Prince Samar, *" The Zone Routing Protocol (ZRP) for Ad Hoc Networks,"* draft-ietf-manet-zone-zrp-04.txt, July 2002.

[9]. Zygmunt J. Haas, Marc R. Pearlman,Prince Samar, *"The Intrazone Routing Protocol (IARP) for Ad Hoc Networks,"* draft-ietf-manet-zone- iarp-01.txt, June 2001.

[10]. Zygmunt J. Haas, Marc R. Pearlman,Prince Samar, *"The Interzone Routing Protocol (IERP) for Ad Hoc Networks,"* draft-ietf-manet-zone- ierp 02.txt July 2002.

[11]. C. Perkins, E.M. Royer, S.R. Das, and M.K. Marina, *"Performance Comparison of Two On-demand Routing Protocols for Ad Hoc Networks,"* IEEE Personal Communications, pp. 16-28, Feb. 2001.

[12]. J. Broch et al., *"A Performance Comparison of Multihop Wireless Ad Hoc Network Routing Protocols,"* Proc. IEEE/ACM MOBICOM '98, pp. 85–97, Oct. 1998.

[13]. Pearlman MR, Haas ZJ, *"Determining the optimal configuration for the zone routing protocol,"* IEEE Journal on Selected Areas in Communications 1999, vol. 17, No.8, pp. 1395–1414, AUGUST 1999.

[14]. Zygmunt J. Haas, and Marc R. Pearlman, *"The Performance of Query Control Schemes for the Zone Routing Protocol"* IEEE/ACM TRANSACTIONS ON NETWORKING, VOL. 9, NO. 4, pp. 427-438, AUGUST 2001.

[15]. Idris Skloul Ibrahim A. Etorban Dr Peter J.B King, *"Multipath Distance Vector Zone Routing Protocol for Mobile Ad-Hoc Networks MDVZRP," PGnet 2008*, John Moores University, pages 171-176, July 2008.

[16]. Leonard Barolli, Yoshitaka Honma , Akio Koya, Arjan Durresi and Junpei Arai, *" A Selective Border-casting Zone Routing Protocol for Ad-hoc Networks,"* Proceedings of the 15th International Workshop on Database and Expert Systems Applications (DEXA'04), pp. 326-330, 2

[17]. Kavita Taneja and R.B.Patel," *Mobile Ad hoc Networks: Challenges and Future*", Proceedings National Conference on Challenges and Opportu-nities in Information Technology, RIMT-IET, Mandi Gobindgarh, PP.133-135, March 3, 2007

[18]. Prince Samar, Marc R. Pearlman and Zygmunt J. Haas,*Independent Zone Routing:An Adaptive Hybrid Routing FrameWork for Ad hoc Wireless Networks*",IEEE/ACM Transaction On Networking,Vol.12, No. 4, PP. 595-608, August 2004.

[19]. Moitreyee Dasgupta, S. Choudhury and N. Chaki," *Routing Misbehavior in Ad Hoc Network*", International Journal of Computer Applications, Vol.1, No. 18, PP.89-92, 2010

[20]. Ming Yu and Kin K. Leung, "*A trustworthiness-based QoS routing protocol for wireless ad hoc networks*", IEEE Transactions on Wireless Communications, Vol. 8, No. 4, April 2009

[21]. P. Papadimitratos and Zygmunt J. Haas, *Secure Message Transmission in Mobile Ad hoc Networks*",Journal Elsevier Ad hoc Networks, Vol. 1, No.1 PP. 193-209, 2003.

8

MODERNIZED INTRUSION DETECTION USING ENHANCED APRIORI ALGORITHM

Lalli[1] and Palanisamy[2]

[1]Department of Computer Science, Bharathidasan University, Trichy
Lalli_bdu@yahoo.co.in
[2]Department of Computer Science & Engineering, Alagappa University, Karaikudi
vpazhanisamy@yahoo.co.in

ABSTRACT

Communication networks are essential and it will create many crucial issues today. Nowadays, we consider that the firewalls are the first line of defense but that policies cannot meet the particular requirements of needed process to achieve security. Most of the research has been done in this area but we are lagging to achieve security needs. Already many models such as ADAM, DHP, LERAD and ENTROPHY are proposed to resolve security problems but we need an efficient model to detect new types of various intrusions within the entire network. In this paper, we proposed to design a modernized intrusion detection system which consist of two methods such as anomaly and misuse detection. Both are integrated and also used to detect novel attacks. Our system proposed to discover temporal pattern of attacker behaviors, which is profiled using an algorithm EAA (Enhanced Apriori Algorithm). This is experimented with a simple interface to display the behaviors of attacks effectively.

KEYWORDS

Intrusion Detection System, Data Mining, Security, Intrusions, Apriori Algorithm

1. INTRODUCTION

Intrusion detection is a device that monitors activity to identify malicious or suspicious events and also produce reports to a management station. Many new intrusions or attackers are coming out every day to abusing the computing system privileges. An intrusion detection system is a sensor, like smoke detector, that raises an alarm if any specific things occur. According to the alarms, the security analyst will be detected as positive and negative ones separately. If it is a smaller network, the intrusion detection systems perform simple pattern matching and report situations that match a pattern corresponding to a known attack type. But the detection of new attacks is not possible in a complex network. The quantity of traffic makes it obligatory for network administrators and network security experts to use specific tools, called intrusion detection systems (IDS), to prune down the monitoring activity they require to do. The IDS are evaluated on the basis of its accuracy, efficiency and usability. The characteristics used to evaluate the accuracy of the IDS are as follows. 1. Detection Rate: The number of attacks to be detected by the system. 2. False Positive Rate: It is the percentage of normal data that the system in-correctly determines to be intrusive. The accuracy of the different IDS can be compared with the detection rate against false positive rate. The comparison should be represented a plot named as Receiver Operating Characteristic (ROC) curve. The accuracy gives a measure of the ability of the IDS to handle noisy, discrete data and to adapt.

In general, detection mechanism used by IDS can be classified into two major categories. 1)Signature based detection: Models built from well known attack types, that is already known attack patterns. 2)Anomaly based detection: Modeled using normal traffic and deviation from this profile is considered anomalous. 3) Hybrid of both anomaly and misuse detection: Many systems are used anomaly detection and misuse detection to detect various types of intrusions effectively. To build powerful IDS, it is necessary to enumerate the desirable characteristics such as fault tolerance, imperviousness to subversion, scalability, adaptability, minimal overhead, configurability, graceful degradation of service.

The rest of the paper is organized as follows. In section 2, we briefly describe about Data Mining and their related work for IDS. We put forth a detailed description of our algorithm in section 3. We discuss about the experimental results in section 4. In section 5, we summarize the paper.

2. DATA MINING TECHNOLOGY

Data Mining is knowledge discovery in databases. Data Mining is knowledge discovery in databases. Data mining techniques are used to find out the hidden patterns automatically in an entire database. Data warehouse is an integration of multiple databases, data cubes and files. We have mined all the data, to predict it according to their relationship. Data mining is the search for the relationships and global patterns that exist in large databases but are hidden among vast amounts of data. It is also denoted as automated statistical analysis. The most commonly used techniques in data mining are: Neural networks, Genetic algorithms, Association rule mining, Clustering and Classification. The association rule mining technique is applied for our proposed work.

2.1. Related Work

Ye Changguo [1] has described about Network Intrusion Detection System (IDS), as the main security defending technique, are second guard for a network after firewall. Ye Changguo [1] has described about Network Intrusion Detection System (IDS), as the main security defending technique, are second guard for a network after firewall. The author described that the huge database log can be mined with the help of fuzzy rule mining algorithm. It gives the better results compared to Apriori algorithm.

Aly Ei-Senary et al[2] has integrated Apriori and Kuok's algorithms to capture the features of network traffic using fuzzy logic rules. Amir Azimi alasti et al[3]says that the reduction of false positive alerts and correlation between alerts are done by formalized SOM. The accuracy can be improved by Alert Filtering and Cluster Merging algorithms. Anderson J. P.[4] using data mining techniques and signature based tools for MINDS. The model can be developed by Denning D.E et al[5] to monitor the abnormal activities. Dewan M et al[6] has used ISABA for alert classification and also to reduce the number of false positives. Each and every user behavior can be identified by Neural network by Jake Ryan et al[7].

Jin-Ling Zhao[8] proposed to achieve the high detection rate and low false alarm rate through clustering and genetic optimizing stages. Norouzian M.R et al[9] described the neural network concepts to detect various attacks and also to classify six groups of two hidden layers in Multi-Layer Perceptron (MLP). All the system calls to be traced out by Host based Intrusion Detection System. OSwais. S et al[10] proposed the membership function can be tuned by genetic algorithm, it is performed by using approach based IDS and also implemented by Gas. Alan Bivens et al[11] has proposed the NIDS model to classify self organizing maps for data clustering. The MLP neural network is one of the best ways to creating uniform, grouping of inputs and they are detected effectively when a dynamic number of inputs are presented. S. Sathyabama et al[12] has described about clustering techniques and outliers for grouping the user's behaviors and also detecting different behaviors easily.

Teng.H.S [13] applied sequential rules to capture a user's behavior over time. According to the User activity patterns, the specific rules are used to find out the specific activities. High quality sequential patterns are automatically generated using inductive generalization and lower quality patterns are eliminated. An automated strategy for generation of fuzzy rules obtained from definite rules using frequent items. Taeshik Shon et al [14] proposed an enhanced SVM approach framework for detecting and classifying the novel attacks in network traffic such as packet profiling using SOFM, packet filtering using PTF, field selection using Genetic Algorithm and packet flow-based data pre-processing. SOFM clustering was used for normal profiling. The SVM approach provides false positive rate similar to that of real NIDSs.

Stephen F. Ownes, R. R. Levary[15] has stated that the expert system techniques are used for intrusion detection systems. The fuzzy sets are used to find out various types of attacks. Depends up on the types of threat it is easy to implement the expert system technology.

In that specific problem, we wish to tackle through our IDS is, to generate a modernized intrusion detection which will effectively detects both anomaly and misuse detections using EAA (Enhanced Apriori Algorithm) and also will reduce false positives effectively.

3. ARCHITECTURE FRAMEWORK AND PROPOSED ENHANCED APRIORI ALGORITHM

It is necessary to detect the potential attacks for anomalous behavior from the existing patterns because we predict the future actions of the attacker's. Sequential pattern mining techniques are used to analyze various IDS alerts. In this paper, we propose a model modernized intrusion detection system consist of two methods, anomaly detection and misuse detection to detect intrusions. We have integrated both anomaly and misuse detection due to its ability to detect novel attacks. Our system is used to find cut the behaviors of various attacks using EAA.

In this paper we proposed to design and develop an intelligent data mining intrusion detection system and its core part a composite detection engine with anomaly detection and misuse detection features and the two detection engines work serially to detect the user's activity in turn.

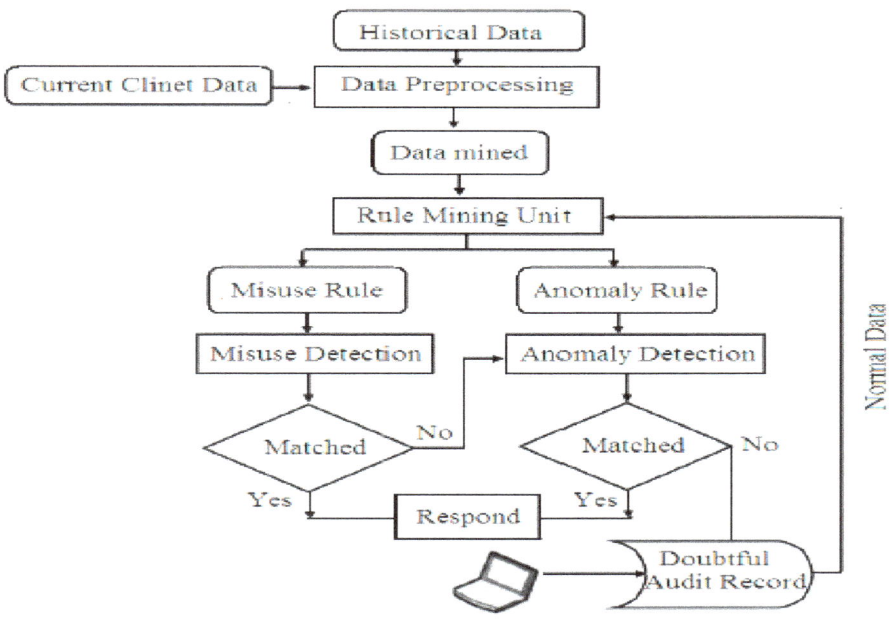

Figure 1. Design Architecture

The real time data's are collected from the database audit system. We have analyze the audit data and to find out its normal and abnormal behaviors. As a result, identify the abnormal activity through statistical analysis.

The design of architecture framework is mainly composed of the following parts: data collecting and pre-processing module, association rule mining module and intrusion detection analysis module, etc as shown in Figure 1.

3.1 The Proposed Algorithm - EAA

In this section, we proposed the Enhanced Apriori Algorithm(EAA). It is an extension of basic Apriori Algorithm. Several rule sets are framed as follows.

Rule 1 : If both anomaly and misuse module detected an attack then the specific attack can be classified by misuse module.

Rule 2 : If misuse module can detect an attack then the specific attack can be classified by misuse module.

Rule 3 : If anomaly module can detect an attack then the specific attack can be declared as unclassified attack.

```
L1= {frequent candidates};
for (k= 2; Lk-1 !=Ø; k++) do begin
Ck= candidates are generated from Lk-1 (cartesian product Lk-1 x Lk-1 and
eliminating any k-1 size itemset is not a frequent one);
While LK-1≠ϕ
Ck <- Generate(LK-1)
For transactions t € T
Ct, <- Subset(Ckt)
For candidates c € Ct
count[c] <- count[c] + 1
LK <- { c € Ck | count [c] >= € }
K <- K+1
Return µ Lk
        for each transaction t in database do
                increment the count of all candidates in
                 Ck that are contained in t
        Lk = candidates in Ck with min_sup
        end
Return µ Lk;
        Let's define:
                Ck  candidate itemset of size k
                Lk  frequent itemset of size k


Main steps of iteration are:
        Find frequent set Lk-1
        Join step: Ck is generated by joining Lk-1 with itself
                   ( cartesian product Lk-1 x Lk-1 )
         Prune step (apriori property):
            Any (k – 1) size itemset  is not a frequent subset
            of k size itemset, then it should be removed
        Frequent set Lk has been achieved
```

Figure 2. EAA Algorithm

Association Mining aims to extract interesting correlation, frequent patterns, and associations among sets of items or objects in transaction databases. Association rules satisfy both a minimum support threshold (min_sup) and minimum confidence threshold (min_conf). Audit data can be obtained from various databases, it consist of number of rows (audit record) and columns (system features). Association rules are used to capture the consistent behaviors and then it is implemented through aggregate rule sets. Our proposed algorithm is in pseudo code format.

4. Experimental Results

The proposed EAA is experimented to find out the efficiency of detecting false positives and false negatives that are occurred in IDS. The new type of attacks can be detected effectively by EAA and then automatically knowledge base can be updated. The experimental result shows, that it was better to find normal packet performance, faster in terms of execution time and also detecting the effectiveness and accuracy of false positives and false negatives.

The input parameters are data size (The size of the each audit record in database) and Data Length (Processing time of each audit record measured in seconds). Our proposed technique EAA approach gives better performance in evaluating large size audit data when compared to the existing techniques.

False Positive Rate = Number of false positives / Total number of abnormal labels.

False Negative Rate = Number of false negatives / Total number of normal labels.

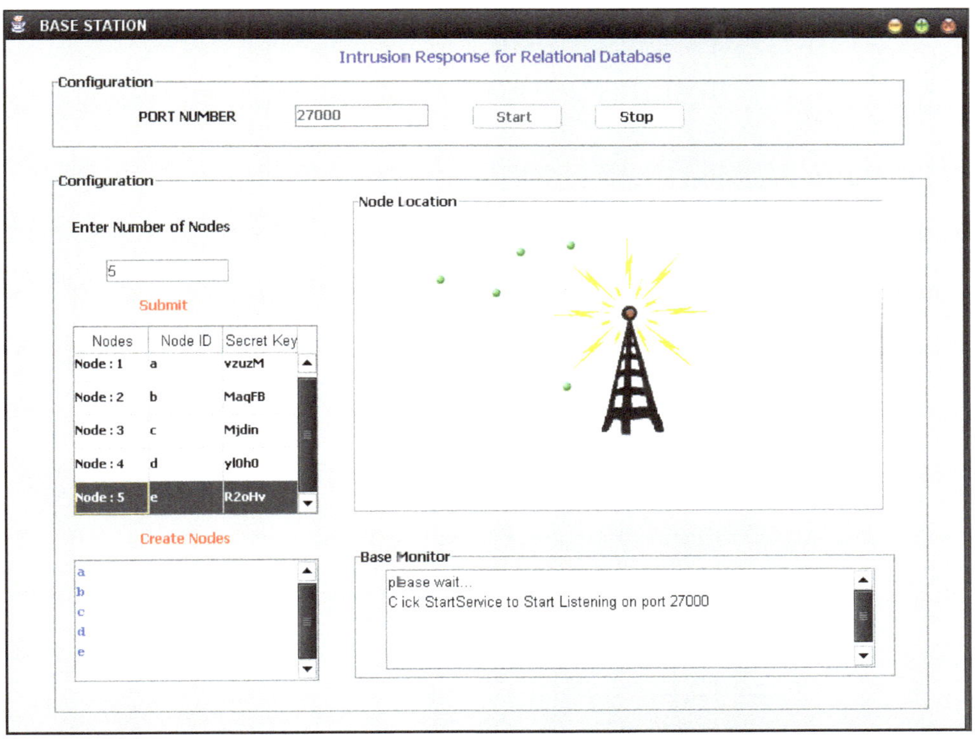

Figure 3. Nodes without intrusion

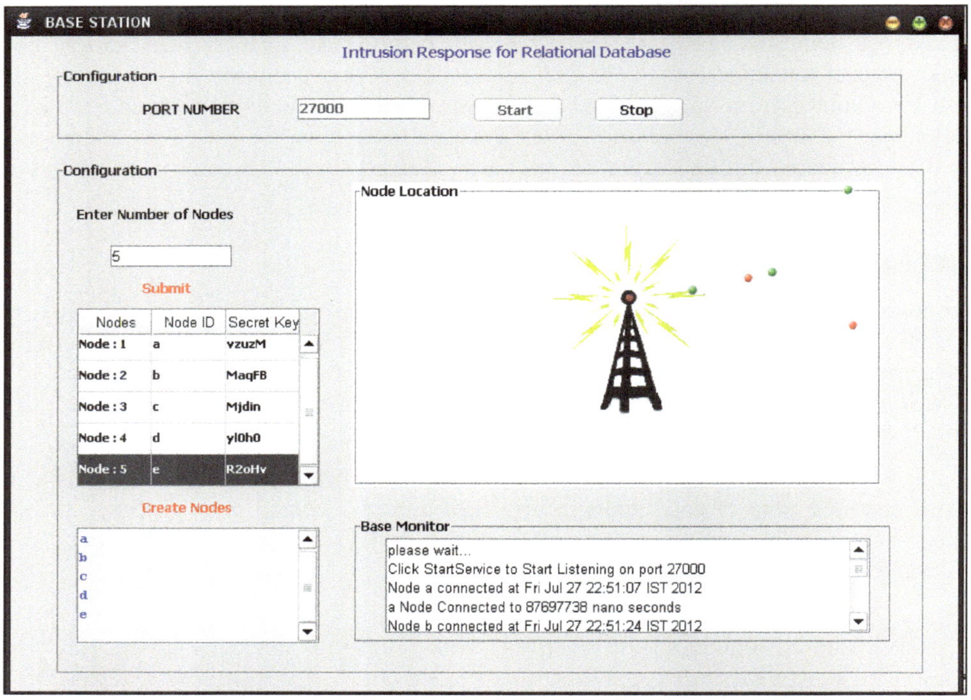

Figure 4. Nodes with intrusion

Table 1. Evaluation between existing and proposed Approach

Data Size	Data Length (Existing)	Data Length (Proposed)
1876	34	10
876	56	13
4526	87	20
5435	35	27
987	65	11
1234	27	16
675	74	08
2345	55	22
5432	33	25
5261	23	23
2176	22	19
165	11	02
765	65	09
763	23	09
343	25	10
7845	64	33

This paper improves the detecting speed and accuracy as a goal, and proposed a more efficient EAA (Enhanced Apriori Algorithm) to abnormal detecting experiment to be based on network.

In this approach we shows that new type of attack can be detected effectively in the system and the knowledge base can be updated automatically.

Finally the experimental results are compared and analyzed. Our proposed algorithm is effectively identify malicious attacks and also increase the efficiency. This section will shows the result based on the inputs as, Data Size: The size of the each audit record in database and Data Length: Processing time of each audit record measured in seconds. From the Figure 5. We can conclude that our proposed technique called EAA approach gives better performance in evaluating large size of audit data when compared to the existing techniques.

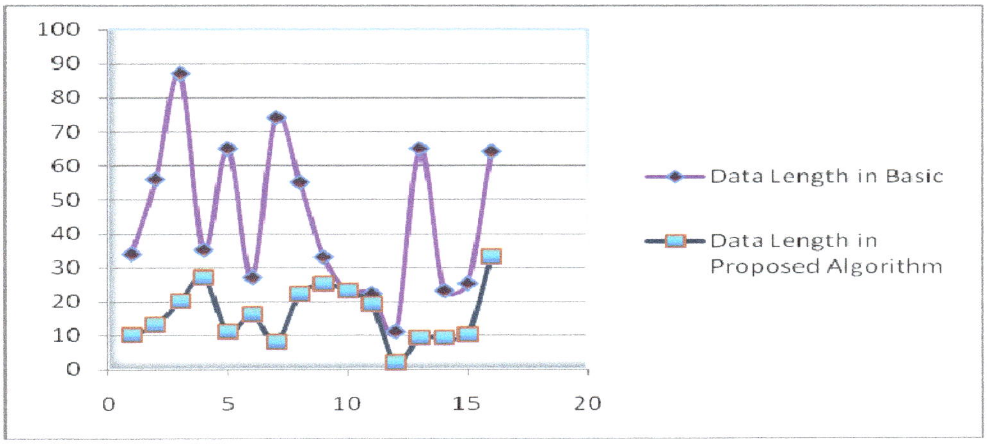

Figure 5. Performance Analysis

The IDS are evaluated on the basis of its accuracy, efficiency and usability. The characteristics used to evaluate the accuracy of the IDS are, *Detection Rate*: It is the percentage of attacks that a system detects and *False Positive Rate*: It is the percentage of normal data that the system incorrectly determines to be intrusive.

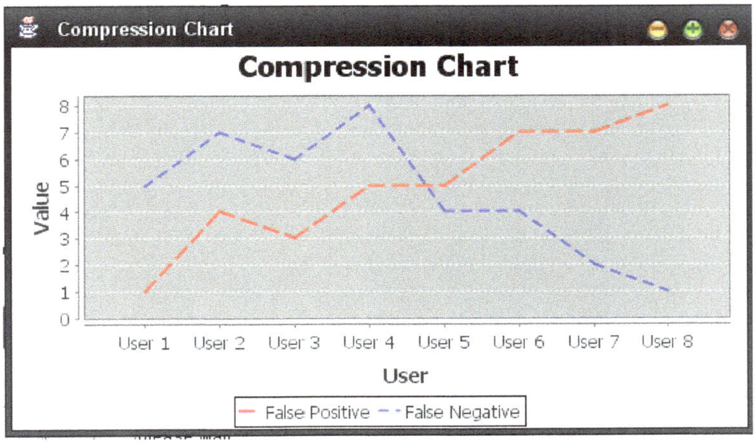

Figure 6. Compression chart for false positives and false negatives

Different types of IDS can be analyzed according to the detection rate and it is plotted as a Receiver Operating Characteristic (ROC) curve. The accuracy gives a measure of the ability of the IDS to handle noisy, discrete data and to adapt. The compression range of false positives and false negatives are shown in Figure 6.

The experimental results shows that the security range between existing and proposed technique based on the ratio (59:67) is in Figure 7.

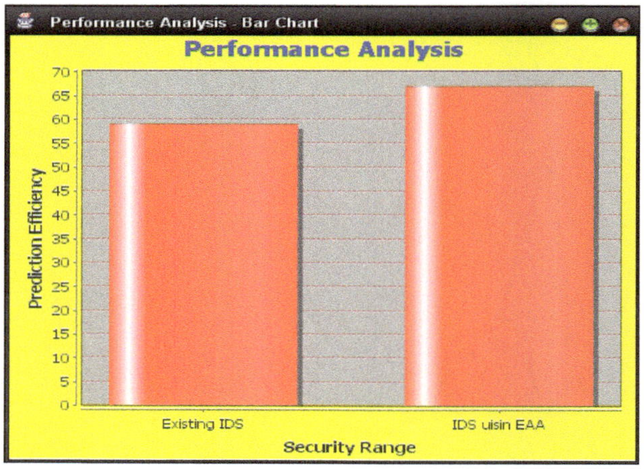

Figure 7. Performance Analysis for Existing and Proposed system

An existing research, it is based on simple and basic association algorithm to enhance the efficiency of association rule mining, which is not sufficient one to generate frequent item set. Another problem is in fuzzy set theory, it is often requires several analysis before the number of frequent item set produce. It can be very sensitive to the choice of initial analysis and it does not yield the same result with each run, since the resulting frequent item set depends on the initial random assignments. So it is not sufficient for runtime intrusion detection efficiency.

Reasons for Supremacy over other algorithms:-

- Proposed Design will be better than existing to find normal packet performance.
- Proposed Design will be faster than existing in terms of execution time.
- Proposed Design will be smaller than existing and easy to understand and implement.
- It will do not contain complex structure, control flow will be well defined and looping structure will be minimized. Due to the above facts it will take very less time for execution.

It is a modern Intrusion Detection with Intelligent Analysis that meets the challenges that traditional intrusion detection systems failed to meet:

- need for accuracy
- need for active responses
- need for speed and reliability
- need for usability

5. CONCLUSION & FUTURE WORK

Using advantages of characteristic extraction of data mining theory in dealing with large amounts of data, the new model MIDS which combines both misuse detecting and anomaly detection is proposed in this paper. In this model we applying Enhanced Apriori (EAA) algorithm to the training data set that containing a large number of intrusions to establish and update the 'normal' and 'abnormal' behavior rules. It is, therefore, concluded that the given methods can be easily adapted to the configuration of MIDS to reduce the manual efforts used to configure them, reduce alert generation and hence improve their effectiveness. And the results of simulation manifested that the new model has a trait of high detection rate and low false positive rate. Besides, unknown intrusion can be detected at the same time effectively.

In Future Enhancement, we investigate the methods and benefits of combining multiple simple detection models. In Future Enhancement, we investigate the methods and benefits of combining multiple simple detection models. We enhance this model by using multiple audit data sets from multiple data streams to improve the accuracy and efficiency.

REFERENCES

[1] Ye Changguo , "The Research on the Application of Association Rules Mining Algorithm in Network Intrusion Detection" Transactions on Software Engineering, IEEE Communication Magazine.

[2] Aly Ei-Semary, Janica Edmonds, Jesus Gonzalez-Pino, Mauricio Papa, "Applying Data Mining of Fuzzy Association Rules to Network Intrusion Detection", in the Proceedings of Workshop on Information Assurance United States Military Academy 2006, IEEE Communication Magazine, West Point, Y,DOI:10.1109/IAW.2006/652083.

[3] Amir Azimi, Alasti, Ahrabi, Ahmad Habibizad Navin, Hadi Bahrbegi, "A New System for Clustering & Classification of Intrusion Detection System Alerts Using SOM", International Journal of Computer Science & Security, Vol: 4, Issue: 6, pp-589-597, 2011.

[4] Anderson.J.P, "Computer Security Threat Monitoring & Surveilance", Technical Report, James P Anderson co., Fort Washington, Pennsylvania, 1980.

[5] Denning .D.E, "An Intrusion Detection Model", Transactions on Software Engineering, IEEE Communication Magazine, 1987,SE-13, PP-222-232,DOI:10.1109/TSE.1987.232894.

[6] Dewan Md, Farid, Mohammed Zahidur Rahman, "Anomaly Network Intrusion Detection Based on Improved Self Adaptive Bayesian Algorithm", Journal of Computers, Vol 5, pp-23-31, Jan 2010, DOI:10.4.304/jcp 5.1.

[7] Jake Ryan, Meng - Jang Lin, Risto Miikkulainen, "Intrusion Detection With Neural Networks", Advances in Neural Information Processing System 10, Cambridge, MA:MIT Press,1998,DOI:10.1.1.31.3570.

[8] Jin-Ling Zhao, Jiu-fen Zhao ,Jian-Jun Li, "Intrusion Detection Based on Clustering Genetic Algorithm", in Proceedings of International Conference on Machine Learning & Cybernetics (ICML),2005,IEEECommunicationMagazine,ISBN:0-7803-9091-DOI:10.1109/ICML.2005.1527621.

[9] Norouzian.M.R, Merati.S, "Classifying Attacks in a Network Intrusion Detection System Based on Artificial Neural Networks", in the Proceedings of 13th International Conference on Advanced Communication Technology(ICACT), 2011,ISBN:978-1-4244-8830-8,pp-868-873.

[10] Oswais.S, Snasel.V, Kromer.P, Abraham. A, "Survey: Using Genetic Algorithm Approach in Intrusion Detection Systems Techniques", in the Proceedings of 7th International Conference on Computer Information & Industrial Management Applications (CISIM), 2008, International Journal of Computer Applications (0975 – 8887) Volume 35– No.8, December 2011 56 IEEE Communication Magazine,pp-300-307,ISBN:978-0-7695-318-7,DOI:10.1109/CISM.2008-49.

[11] Alan Bivens, Chandrika Palagiri, Rasheda Smith, Boleslaw Szymanski, "Network-Based Intrusion Detection Using Neural Networks", in Proceedings of the Intelligent Engineering Systems Through Artificial Neural Networks, St.Louis, ANNIE-2002, and Vol: 12, pp- 579-584, ASME Press, New York.

[12] Sathyabama.S, Irfan Ahmed.M.S, Saravanan.A,"Network Intrusion Detection Using Clustering: A Data Mining Approach", International Journal of Computer Application (0975-8887), Sep-2011, Vol: 30, No: 4, ISBN: 978-93-80864-87-5, DOI: 10.5120/3670-5071.

[13] Teng.H.S, Chen.K and Lu.S.C, "Adaptive Real-Time Anomaly Detection using Inductively Generated Sequential Patterns, in the Proceedings of Symposium on research in Computer Security & Privacy, IEEE Communication Magazine,1990, pp-278-284.

[14] Taeshik Shon, Jong Sub Moon, "A Hybrid Machine Learning Approach to Network Anomaly Detection", Information Sciences 2007, Vol: 177, Issue: 18, Publisher: USENIX Association, pp- 3799-3821, ISSN:00200255,DOI:10.1016/j.ins-2007.03.025.

[15] Stephen F. Owens, Reuven R. Levary, "An adaptive expert system approach for intrusion detection", International Journal of Security and Networks, Vol: 1, No: 3/4, pp: 206-217, 2006.

[16] Tcptrace software tool, www.tcptrace.org.

[17] SNORT Intrusion Detection System. www.snort.org.

[18] Sans Institute. The SANS Security Policy Project. http://www.sans.org/newlook/resources/policies/policies.htm

PERFORMANCE EVALUATION OF MULTIMEDIA STREAMING OVER MOBILE ADHOC NETWORKS USING MAPPING ALGORITHM

N.Gomathi[1], Dr.P.Seethalakshmi[2] and Dr.A.Govardhan[3]
[1]JNTU, Hyderabad
gomathi1974@gmail.com
[2]Anna University, Triuchirappalli
seetha@tau.edu.in
[3]JNTUH College of Engg., Hyderabad
govardhan_cse@yahoo.co.in

Abstract

The main objective of this paper is to propose a novel method for enhancing the Quality of Service (QoS) of multimedia applications in wireless adhoc networks. The enhancement is achieved by implementing a cross layer mapping algorithm, between application layer and Medium Access Layer where Connectionless Light Weight Protocol (UDPLite) is used in transport layer that supports multimedia applications. The Proposed method achieves 17% improvement in reduction of delay and 5% improvement in PSNR as compared to the conventional methods under heavy traffic conditions.

Keywords

Enhanced Distributed Channel Access, MANETs, PSNR, UDPLite, Video Streaming.

1 INTRODUCTION

Recent advancements in computing techniques have become an integral part of wireless communication networks. Mobile Ad hoc networks (MANETs) have emerged amid the unprecedented growth of Internet and are increasingly attracting attention because of its ability to connect across nodes without relying on pre-existing network infrastructure. The widespread emergence of real-time voice, audio and video applications, stimulates the successful development of viable technologies to provide these multimedia applications over mobile adhoc networks. The performance of MANET is affected by various factors such as mobility of node, battery life and routing protocols, topology change etc... Hence providing Quality of service for multimedia applications in adhoc networks is difficult.

Quality of Service requirements of multimedia applications in adhoc networks have been supported by IEEE 802.11e standard. 802.11e defines four Access Categories(ACs) with different transmissions priorities. The transmission priority is the probability of successfully earning the chance to transmit when individual ACs are competing to access the wireless channel. Higher the transmission priority, better is the opportunity to transmit. But in a wireless channel the unavoidable burst loss, due to excessive delays and limited bandwidth there are challenges for good transmission over wireless network.

In this paper we argue that for 802.11e based adhoc networks, a partial checksum approach at the transport layer along with an adaptive cross layer mapping scheme between application layer and Medium Access Layer, can improve the performance of video transmission. In this paper an attempt has been made to get benefits of UdpLite along with cross layer approach. The rest of the paper is organized as follows. In this section 2 we discuss the aspects of anUdpLite and in section 3 enhanced Distributed

Channel Access(EDCA)and in section 4 adaptive cross layer mapping algorithm. In section 5 we discuss about proposed system. Section 6 establishes system simulation model and Section 7 gives results to illustrate the performance while conclusion are drawn in section 8.

2 UDPLITE

The quality of video can be increased by enabling the application layer to specify about the importance of packets and those packets can be preserved by the UDPLite protocol [1] in the transport layer along with Cross Layer Mapping approach. The notion of application-layer over transport-layer protection is not new and hence traditional real- time multimedia services have been realized as Realtime Transport Protocol (RTP) over User Datagram Protocol (UDP). User Datagram Protocol (UDP) is an unreliable protocol that is suitable for delay sensitive applications such as real-time media applications that are sensitive to network delays. UDPLite is an extension to UDP that even needs damaged data to be delivered rather than discarded by networks, so it allows partial checksums on multimedia data by enabling the applications to specify, the sensitive and insensitive parts of the multimedia stream on a per-packet basis. Errors in the sensitive part cause a packet to be discarded whereas an error in the insensitive part allows it to be delivered. The check sum is carried out on the sensitive part of the packet. UDP has a strict checksum where corrupted packets will be discarded if they contain any transmission errors. The UDPLite protocol allows the application to receive the corrupted packets instead of dropping them altogether. This is achieved by a partial checksum which only covers a fixed amount of sensitive data. Integrating UDPLite into existing UDP framework is simple. The length field in the UDP header is replaced by the coverage field, which signifies the number of bytes of the packet that are to be checksummed. With a checksum coverage value replacing the packet length, UDPLite packets are treated as classic UDP packets with the checksum enabled. To address security concerns and handle the multiplexing of other transport level flows, the packet header should always be checksummed. If corruption occurs in the Sensitive region or in the header, the packet is dropped at the receiver otherwise the packet is passed up to the application through the interface.

<table>
<tr><td colspan="3">Source Address</td><td rowspan="3">IP Pseudo-header</td></tr>
<tr><td colspan="3">Destination Address</td></tr>
<tr><td>Zero</td><td>Proto</td><td>UDP Length</td></tr>
<tr><td colspan="2">Source Port</td><td>Destination Port</td><td rowspan="2">UDP lite header</td></tr>
<tr><td colspan="2">CheckSum Coverage</td><td>Checksum</td></tr>
</table>

Figure 1: The UDPLite Header

The UDPLite protocol headers are shown in Figure 1. Shaded fields are the fields of the pseudo header provided by the IP layer, and white fields belong to the UDPLite header. The UDPLite checksum covers the conceptual IP pseudo-header in order to protect against misrouted packets.

The Checksum Coverage field in the UDP Lite header denotes the number of octets (counting from the first octet of the header) that are covered by the checksum. The value of Checksum Coverage is zero indicating that the entire UDP-Lite packet is covered by the checksum. This explains that the value of the Checksum Coverage field MUST be either 0 or at least 8. The UDP Lite header and the IP pseudo-header are always verified by the checksum, which means that the least acceptable value of the coverage field is eight (the number of bytes in the UDP Lite header). A UDP-Lite packet with a Checksum Coverage value of 1 to 7 MUST be discarded by the receiver.

3. EDCA

IEEE 802.11e supports quality of service by introducing priority mechanism. All types of data traffic are not treated equally as it is done in the original standard, instead, 802.11e supports service differentiation by assigning data traffic with different priorities based on their QoS requirements. Furthermore, four different Access Categories (ACs) have been defined each for data traffic of a different priority. Access to the medium is then granted based on the priorities of data traffic, such that each frame with a particular priority is mapped to an Access Category, and service differentiation is realized by using a different set of contention parameters to contend for the medium, for each AC.

In IEEE 802.11e, the AP and STA that provides QoS services are referred to as QAP (QoS Access Point) and QSTA2 (QoS Station) respectively, and the BSS they are operating in is called QBSS (QoS Basic Service Set). IEEE 802.11e introduces a new coordination function, called Hybrid Coordination Function (HCF), to provide QoS support. Subsequent sections describe HCF together with the detailed description of its service differentiation mechanism.

IEEE 802.11e defines a new coordination function called Hybrid Coordination Function (HCF). HCF is a centralized coordination function that combines the aspects of DCF and PCF with enhanced QoS mechanisms to provide service differentiation. HCF provides both distributed and centrally controlled channel access mechanisms similar to DCF and PCF in the original standard. The distributed, contention-based channel access mechanism of HCF is called Enhanced Distributed Channel Access (EDCA), and the centrally controlled, contention-free channel access mechanism is called HCF Controlled Channel Access (HCCA).

IEEE 802.11e introduces Transmission Opportunity (TXOP), defined as the time period during which a QSTA has the right to transmit. In other words, in 802.11e when a station gets access to the medium, it is said to be granted the TXOP. TXOP is characterized by a starting time and a maximum duration, called TXOP Limit. As a QSTA gets the TXOP, it can then start transmitting frames such that the transmission duration does not exceed the TXOP limit. TXOP Limit is specified by the QAP.

The EDCA provides differentiated, distributed access to the medium using different priorities for different types of data traffic. The detailed description of the components and operation of EDCA is presented next.

3.1 Access Categories (ACs)

EDCA defines four Access Categories (ACs) for different types of data traffic, and service differentiation is introduced such that for each AC, a different set of parameters is used to contend for the medium. These parameters are referred to as EDCA parameters and are described in the next subsection.

Frames from different types of data traffic are mapped into different ACs depending on the QoS requirements of the traffic/application the frames belong to. The four Access Categories are named AC_BK, AC_BE, AC_VI and AC_VO, for Background, Best Effort, Video and Voice data traffic, respectively. To simplify the notations, we assign AC_VO as AC3, AC_VI as AC2, AC_BE as AC1, and AC_BK as AC0. Each AC has its own buffered queue and behaves as an independent backoff entity, i.e., an AC, where each queue has its own AIFS and maintains its own Backoff Counter (BC). When there is more than one AC finishing the backoff at the same time, the collision is handled in a virtual manner. That is, the highest priority frame among the colliding frames is chosen and transmitted, and the others perform a backoff with increased CW values. The priority among ACs is then determined by AC-specific parameters, called the EDCA parameter set.

Each frame from the higher layer arrives at the MAC layer along with a priority value. This priority value is referred to as User Priority (UP) and assigned according to the type of application/traffic the frame belongs to. There are four different priority values ranging from 0 to 3.

Priority	**User Priority (UP)**	**Access Category (AC)**	**Designation**
Lowest	0	AC_BK	Background
-	1	AC_BE	Best Effort
-	2	AC_VI	Video
Highest	3	AC_VO	Voice

Table 1: User Priority (UP) to Access Category (AC) Mapping

At the MAC layer, a frame with a particular UP is further mapped to an AC. AC are derived from the UPs as illustrated in Table 1.

3.2 EDCAF (Enhanced Distributed Channel Access Function)

Every station maintains four transmit queues one for each AC, and four independent EDCAFs (Enhanced Distributed Channel Access Function), one for each queue, EDCAF is an enhanced version of DCF, and contends for the medium on the same principles of CSMA/CA and back off, but based on the parameters specific to the AC it is contending for. Next section discusses these parameters, referred to as EDCA parameters.

EDCA Parameters

An EDCAF contends for medium based on the following parameters associated to an AC. The set of EDCA parameters are:

- **AIFS -** The time period the medium is sensed idle before the transmission or backoff is started.
- **CWmin, CWmax -** Size of Contention Window used for backoff.
- **TXOP Limit -** The maximum duration of the transmission after the medium is acquired.

3.2.1 AIFS (Arbitration Inter-Frame Space)

The minimum time period for which the medium must be sensed idle before an EDCAF/station may start transmission or backoff process. It is not the fixed value DIFS, as it is in DCF, but is a variable value, AIFS, that depends on the AC for which the EDCAF is contending for. AIFS is derived from the following equation:

AIFS = AIFSN x aSlotTime + aSIFSTime,

where aSlotTime is the slot time, aSIFSTime is the SIFS time period and AIFSN (Arbitration Inter-Frame Space Number) is used to determine the length of the AIFS. AIFSN specifies the number of time slots in addition to the SIFS time period the AIFS consists of. Different AIFSN values are used for different ACs such that the high priority ACs use smaller values compared to the low priority ACs. The minimum possible value of AIFSN is 2.

The default AIFSN values for all four ACs can be seen in Table 1. Figure 2 further explains how priority is given to different ACs based on the AIFS time periods.

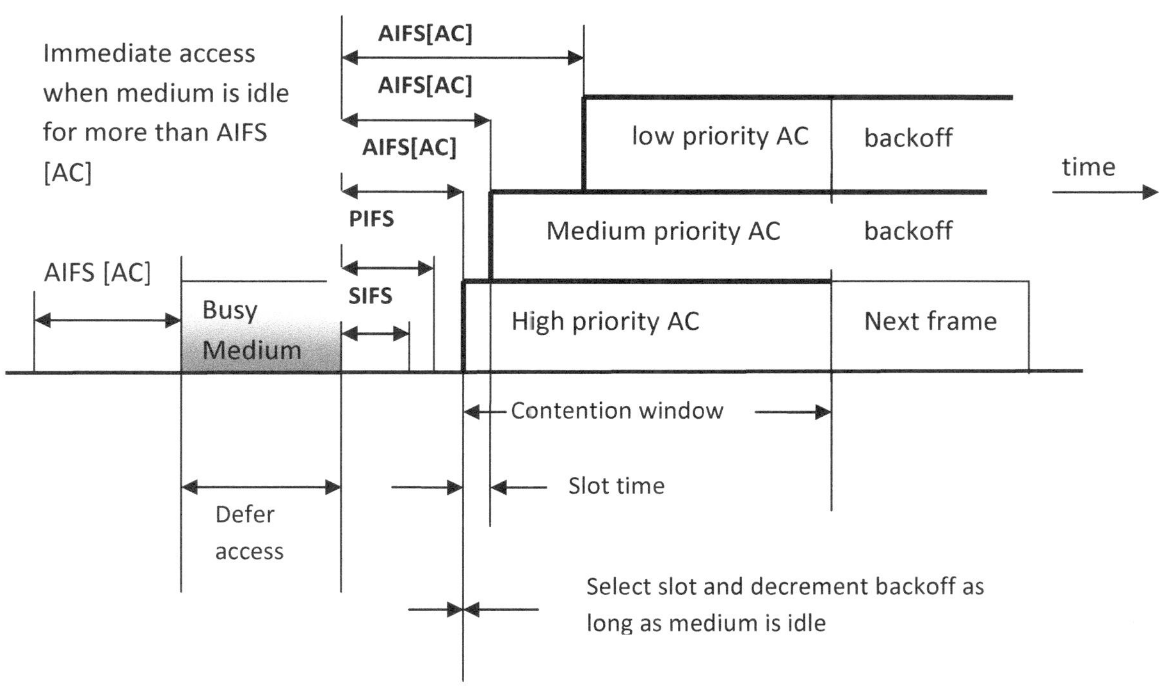

Figure 2: Prioritization based on AIFS

The smaller AIFSN value for a higher priority AC explains that the corresponding EDCAF has to wait shorter time period before it can start transmission or counting down its backoff timer compared to the EDCAF for a low priority AC. In this way, the higher priority ACs are guaranteed greater share of the bandwidth. Moreover, smaller AIFS lengths ensure that the higher priority ACs will not suffer from long delays, which are very critical for the delay-sensitive applications/traffics. The lower priority ACs may suffer from longer delays because of the larger AIFS durations they have to wait, but since these ACs are designed for delay-tolerant applications/traffics, certain amount of delays do not degrade their performance beyond an acceptable limit.

3.2.2. CWmin and CWmax

The minimum and maximum Contention Window size limits are not fixed as it is in DCF, but are variable depending on the AC. The higher priority ACs has smaller CWmin and CWmax values compared to lower priority ACs A smaller Contention Window for an AC will cause the corresponding EDCAF to choose smaller random backoff values, and thereby waiting shorter time period in addition to AIFS as the medium becomes idle. It gives such an AC priority over the AC with a larger Contention Window, which results in larger backoff values and thereby longer delays.

As seen in Table 1, for the commonly used Physical layer DSSS, the CWmin values for lower priority ACs, AC_BE and AC_BK, are same as it is for the legacy 802.11 DCF, but these values for higher priority ACs, AC_VO and AC_VI, are as small as one half or quarter of those of the lower priority ACs. This results in smaller backoff values for the high priority ACs and thereby shorter medium access delays. The negative aspect of small Contention Window sizes for higher priority ACs is that they suffer from higher number of collisions. The reason is, that the probability of choosing the same backoff values or counting the backoff timers to zero at the same time increases with the decreasing size of Contention Windows. CWmax values for high priority ACs are also set such that they are equal or less than the CWmin values for the lower priority AC This shows that after doubling the Contention Window size in case of an unsuccessful transmission, i.e., collision, its size still remains smaller than the CWmin size of lower priority ACs. Furthermore, it also indicates that while a low priority AC has to double its CW size after each unsuccessful transmission, until it reaches the CWmax, and with higher probability, has to choose a bigger backoff value for each retransmission, the Contention Window size of a high priority AC becomes constant after fewer retransmissions, allowing it to consistently choose smaller backoff values and thereby winning access to the medium. In this way, high priority AC is given consistent and greater share of the bandwidth in the situations when the network has become congested. On the other hand, this may severely degrade the performance of the low priority ACs since they might not be able to decrement their backoff timers because of the smaller post backoff durations of the higher priority ACs.

3.2.3 TXOP (Transmission Opportunity)

TXOP is the time duration an EDCAF may transmit after winning access to the medium. TXOP is characterized by a maximum duration, called TXOP Limit. As an EDCAF gets the TXOP, it can then start transmitting frames such that the transmission duration does not exceed the TXOP Limit. The transmission duration covers the whole frame exchange sequence, including the intermediate SIFS periods and ACKs, and the RTS and CTS frames if RTS/CTS mechanism is used.

Table shows the default TXOP limits for different ACs. A non-zero value of TXOP Limit indicates that the EDCAF may transmit multiple frames in a TXOP, provided that the transmission duration does not exceed the TXOP Limit and the frames belong to the same AC. This is then referred to as Contention Free Bursting (CFB).

The consecutive frame transmissions in a TXOP are then separated by SIFS time periods instead of AIFS plus the post backoff periods, as illustrated in Figure 3. It is important to note that the multiple frame transmission is granted to EDCAF (or AC) and not to the station, i.e., it is only allowed for the transmission of frames of the same AC as of the frame for which the TXOP was obtained.

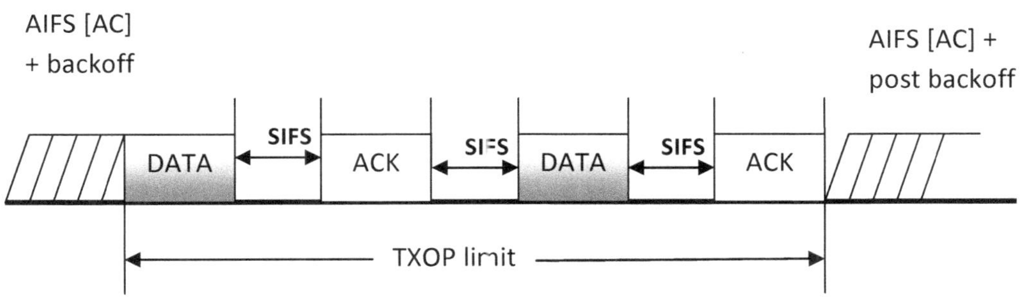

Figure 3: Contention Free Bursting

3.2.4 ARCHITECTURE OF IEEE 802.11e

Together with HCF and its two access mechanisms EDCA and HCCA, IEEE 802.11e also includes the two coordination functions from the original 802.11, DCF and PCF, in order to provide backward compatibility. Figure 4 illustrates the architecture of 802.11e MAC.

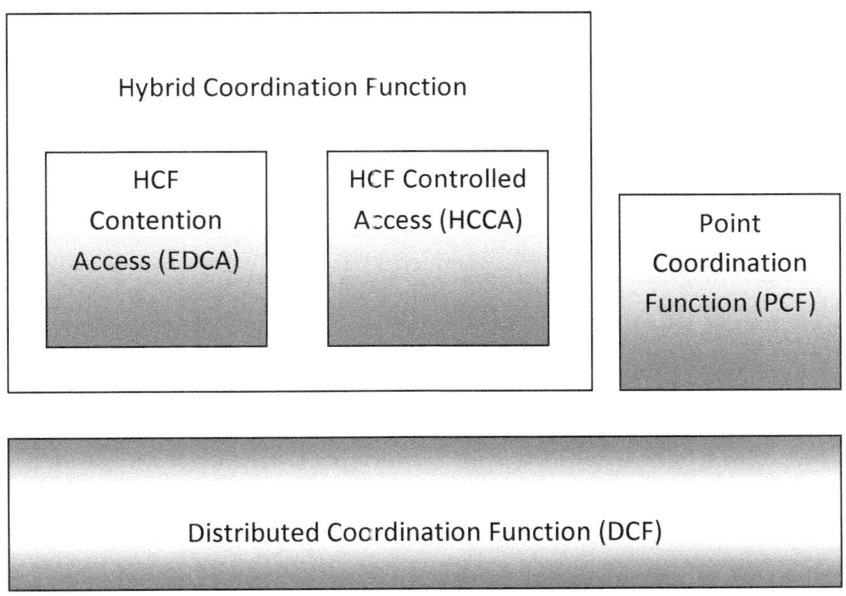

Figure 4: IEEE 802.1e MAC Architecture.

The centrally controlled, contention-free channel access mechanism of HCF, i.e., HCCA, uses a centralized coordinator called HC (Hybrid Controller), which is collocated in QAP. HC operates concurrently with the EDCA just like in the legacy 802.11, i.e., a Contention Free Period (CFP) is followed by a Contention Period (CP), such that the EDCA operates in CP while HC operates both in CP and CFP. This is in contrast with legacy 802.11 where PC can only operate in CFP. It indicates that HC is capable of polling QSTAs both in CP and CFP, and explains why it is referred to as Hybrid Controller.

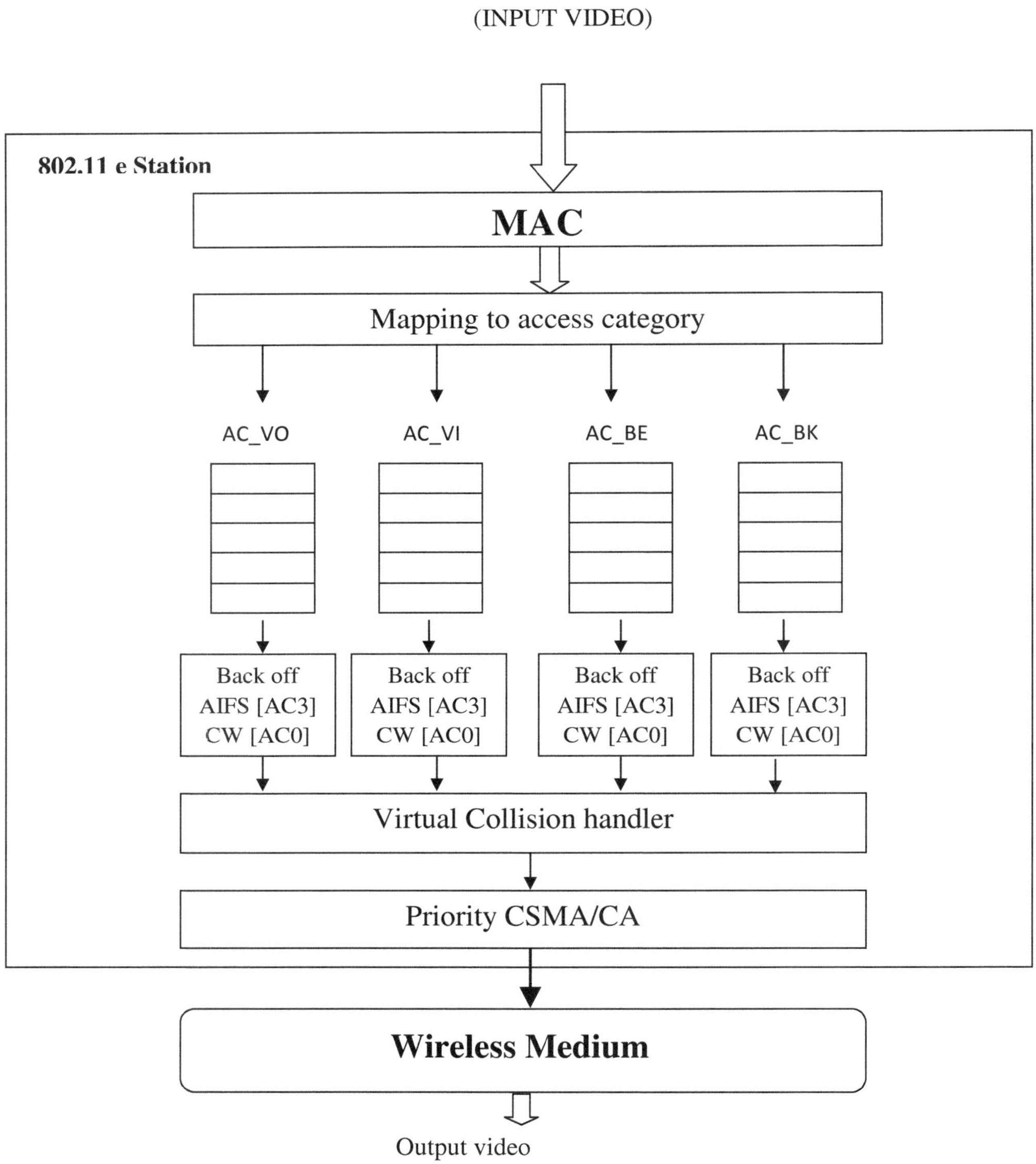

Figure 5: Four Access Categories in 802.11e

4 Cross Layer Approach

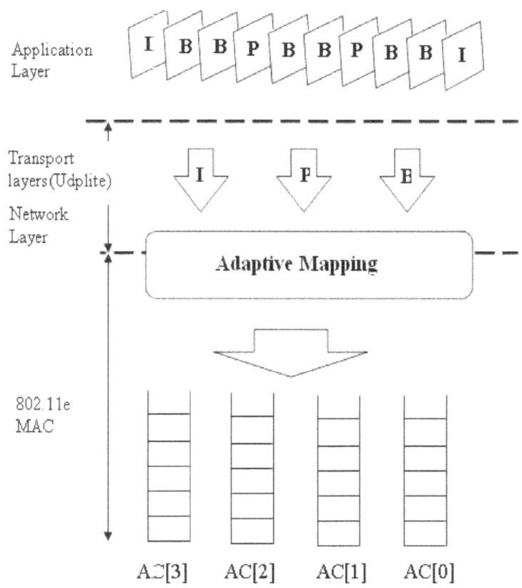

Figure 6: Architecture of adaptive mapping

From the Application Layer the video significance information is generated and transmitted to Enhanced Distributed channel access used in Medium Access Layer. In EDCA, packets arriving from Application layer and Transport layer(UDPLite layer) packets are tagged with four different user priorities and each priority is mapped to one of four Access Categories (ACs). The four different Access Categories (ACs) are Voice traffic, Video traffic, Best Effort traffic and Back Ground traffic that are represented as AC0, AC1, AC2 and AC3 respectively. Each AC maintains a local queue and an independent back off instance with a specific set of contention parameters. All ACs contend independently for access to the channel and internal collisions may occur, but are solved by allowing the AC with the highest priority to gain access to the channel.

To achieve differentiation between Access Categories(ACs), Contention Window(CW) parameters such as CW_{min} and CW_{max} and Arbitrary Interframe Space(AIFS) are used. to Instead of waiting for the normal Differentiated Inter Frame Space(DIFS) time, each AC waits a specific AIFS time. Higher priorities have lower values of the Contention Window(CW) parameters and AIFS. This leads to a higher fraction of the capacity and lower delays since the channel access frequency is increased. An additional parameter is the Transmission Opportunity (TXOP) that specifies the length of time the channel is occupied by a station. Depending on this limit, one or several packets may be transmitted when an AC has acquired the channel. Priority differentiation used by EDCA ensures better service to high priority class while offering a minimum service for low priority classes.

Priority	Access Category	Designation	AIFSN	CWmin	CWmax	TXOPlimit
3	AC_VO	Voice	2	(CWmin+1)/4-1	(CWmin+1)/2-1	0.003008
2	AC_VI	Video	2	(CWmin+1)/2-1	CWmin	0.006016
1	AC_BE	Best Effort	3	CWmin	CWmax	0
0	AC_BK	Background	7	CWmin	CWmax	0

Table 2: Default EDCA parameter set values

The AC with the smallest AIFS has the highest priority, and a station needs to defer for its corresponding AIFS interval. From Table 2 it is inferred that the smaller the parameter values (such as AIFS, CWmin and CWmax) the greater the probability to access the medium. Each Access Category within a station behaves like an individual virtual station, it contends for access to the medium and independently starts its backoff procedure after detecting the channel being idle for at least an AIFS period. When a collision occurs among different access categories within the same station, the higher priority access category is granted the opportunity to transmit, while the lower priority access category suffers from a virtual collision, similar to a real collision outside the station. This concept suits well when the traffic is very high, but the delay is still more and there occurs losses in packets though the traffic is very less since we have employed static mapping algorithm. If we employ dynamic mapping algorithm there will be unnecessary delays and high packet loss

In the cross-layer approach, the frames of the MPEG-4 video packets are dynamically mapped to the appropriate Access Category based on both the significance of the video frame and the network traffic load. Based on this Access Category the MPEG4 video packets are mapped dynamically to the appropriate Access Category(AC). Typically a MPEG video contains B-Frames ,I-Frames and P-Frames. Loss of Important frames in MPEG4 video stream would degrade the delivered video quality, whereas loss of B-Frame doesn't affect all the frames of Group of Pictures (GOP) but itself. Loss of I-Frame would cause all frames in Group of Pictures (GOP) to be undecodable. Based on the significance of the video frame, the channel access priorities are used to prioritize the transmission opportunity at the MAC layer are set with the I frame as the highest; the P frame below I but above B's priority, and the B frame set at the lowest priority. Mapping probability defined as Prob_TYPE is assigned according to its coding significance of video data. This way the important video data is alloted to high priority AC queue in 802.11e MAC layer which has been discussed in [2]. If allocating a frame into a lower priority queue is inevitable, the transmission allocating probability of lower significant frames is higher than that of important video frames. When larger Prob_TYPE is assigned to less important video frames the MPEG4 downward mapping probability relationship of the video frame types become Prob_B> Prob_ and Prob_I, values lying between 0 and 1. Moreover, to support dynamic adaptation to changes in network traffic loads, MAC queue length has been used as an indication of the current network traffic load. According to the IEEE 802.11e specification, when transmitted over an IEEE 802.11e wireless network, MPEG-4 video packets are placed in AC2 category which has better opportunity to access the channel than lower priority Access Categories (ACs). The tradeoff is, when the video stream increases, this queue rapidly jams and drops occur. So an adaptive, cross-layer mapping algorithm approach has been implemented and it is already discussed in [3]. This mapping algorithm re-arranges most recently received video packets into other available lower priority queues, while the AC2 queue is getting filled. Two parameters, ***threshold_low*** and ***threshold_high*** which was denoted in *[2]* has been used predicatively to avoid the upcoming congestion by performing queue management in advance. The integrated function to introduce these two parameters in the cross layer mapping approach is in the following expression:

$$\text{Prob\_New}=\text{Prob\_Type} * \frac{\text{qlen (AC[2])} - \text{threshold\_low}}{\text{threshold\_high} - \text{threshold\_low}}$$

In this function, the original predefined downward mapping probability of each type of video frame, ***Prob_TYPE***, will be adjusted according to the current queue length and threshold values, and about the result is a new downward mapping probability, ***Prob_New***. The higher ***Prob_New***, the greater the opportunity for the packet to be mapped into a lower priority queue.

In the cross layer mapping approach when a video packet arrives, first the queue length of AC2 is checked and it is compared with values of threshold_high and threshold_low. If queue length is lower than the threshold_low (light load), the video data is mapped to AC[2] irrespective of the type of video data being transferred. But if the queue length is greater than the threshold_high (heavy video traffic load) the video data is directly mapped to lower priority queues, AC[1] or AC[0]. However if queue length of AC[2] is between threshold_high and threshold_low, the mapping decision considers both the mapping probability (Prob_TYPE) and the current buffering size condition of the queue. Hence, the video data packet will be mapped to different AC's according to the calculated downward mapping probability. With such a priority scheme in mac layer along with UDPLite in Transport layer the transmissions are prioritized and the drop rate of video is minimized.

5 Related Work

EDCA has been improved by adjusting the parameter adaptively to channel state or congestion level in [4]. adaptive EDCA had been implemented, where the access point adopted the contention window based on the network congestions was discussed in [5]. a two level protection had been applied for voice and video traffic by distributed admission control. The Budget calculation had been done in EDCA to protect existing video streams and the issue of bandwidth allocation for video streams had been investigated in [6]. The cross layer architecture was used which is based on the data partitioning and they have been associated to each partition within the access layer categories of EDCA was discussed in [7]. The macro and micro rate control schemes had been used at the application layer and network layer which uses bandwidth estimation and adaptive mapping of packets using video classifications was discussed in [8]. A wireless video system had been built using the error resilient low bit rate video coder by implementing UDPlite and PPP lite in transport and link layer protocols for cellular video was discussed in [9]. The H.264 had been transmitted for video over an adhoc scenario using Udplite which has reduced retransmission using unequal error protection was discussed in [10]. A multimedia network asic design had been implemented which includes the characteristics of H.264 with Udplite to reduce packet loss was discussed in [11]. A distributed algorithm had been implemented for channel time allocation among multiple video streams, and they had investigated several heuristic packet pruning schemes for rate adaption of high definition video streams were discussed in [12]. A distributed rate allocation scheme had been implemented with a goal of minimizing the total video distortion of all peers without excessive network utilization was discussed in [13]. This scheme relied on cross-layer information exchange between MAC and application layers. A work had been done on 802.11e, where the parameters of MAC layer had been adjusted to provide very good quality by adding UDPlite in transport layer was discussed in [14].

6 Proposed System

The figure 7 depicts the main components of the system architecture for wireless media streaming. The media source which can be a real time encoder or pre compressed media file, generated media packets that are initially sent to the application layer buffer. Subsequently the UDPLite header is added by the transport layer, IP header is added by the network layer. The IP packet is sent to the 802.11e MAC layer and creates an MAC protocol data unit for wireless transmission. All packets are stored in link layer buffer. Similarly the above process is been reversed at the receiver end.

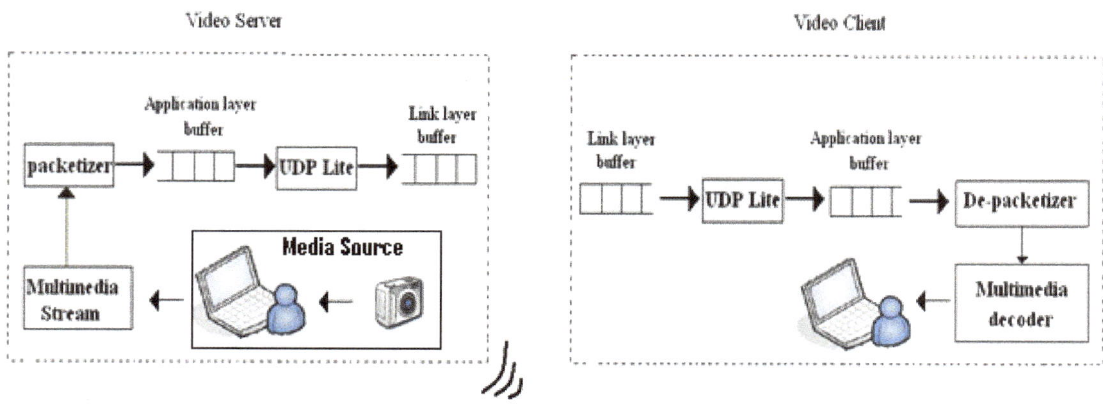

Figure 7: Multimedia Streaming System

7 Simulation Results and Discussions

A.Implementation and simulation setup of Cross Layer Mapping along with UDPLite .

The simulation of Cross Layer Mapping approach along with UDPLite using NS2 simulator has been done , and the performance of the received video at the receiver has been examined. We simulate a Cross Layer Mapping approach along with UDPLite over a ad-hoc network using the NS2 simulator. By exploiting the cross layer mapping approach, prioritization is done in the transmission of essential video data that improves the queue space utilization and also support dynamic adaptation for changes in network traffic loads. MAC queue is used as the indication of the current network traffic load. When video is transmitted over an IEEE 802.11e WLAN, MPEG4 video packets are placed in AC2 which increase efficiency of accessing channels when the video stream increases in the AC2 queue rapidly filled and problems such as jams and drops occur. Inorder to overcome these problems,the proposed approach arranges most recently received video packets into other available lower priority queues. since UDPlite is used in the transport layer the packets, which are prone to errors due to radio channel variations are also delivered to the receiver by the UDPlite protocol in the transport layer,if the error has occurred in the payload. the corrupted packet will be discarded ,if the error has occurred in the header.

7.1 Simulation Topology

The simulation is performed with 3 types of video sources like YUV QCIF (176 x 144) Foreman, Claire, Akiyo. Each video frame was fragmented into packets before transmission and the maximum packet size over the simulator network is 1000 bytes. Figure 8 presents the simulation topology in the experiment. There are eight ad hoc wireless nodes where one is video server and another is video receiver. The data rate of wireless link is 1Mbps.

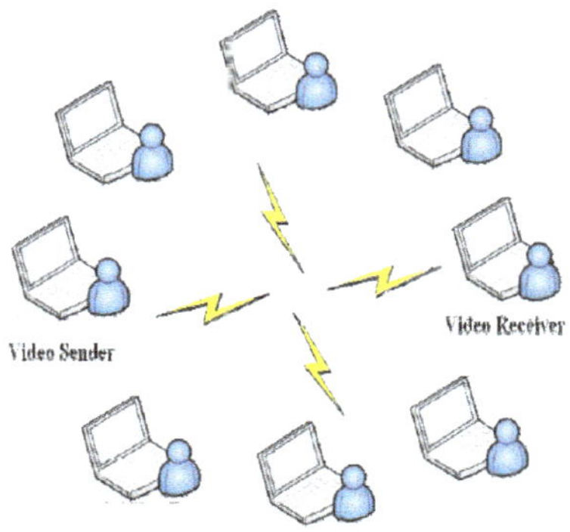

Figure 8: Network topology used in simulation.

7.2 Experiments and Results

In all simulation experiments the performance of 4 cases -MPEG4 UDP EDCA , MPEG4 UDPLite EDCA , cross layer mapping with UDP and cross layer mapping with UDPLite has been compared ,for the video sequences.

7.2.1 Delay

Figure 9 represents the delay produced by MPEG4 UDP EDCA and MPEG4 EDCA CLWP while the video source transmitted is Foreman, Figure 10 represents the Delay produced by cross layer mapping with UDP and cross layer mapping with UDPLite while transmitting Foreman of 400 frames as video source. The delay produced by cross layer mapping with UDP is 0.72 sec because no priority is given for the video packet whereas the delay is 0.6 when adaptive mapping is employed .

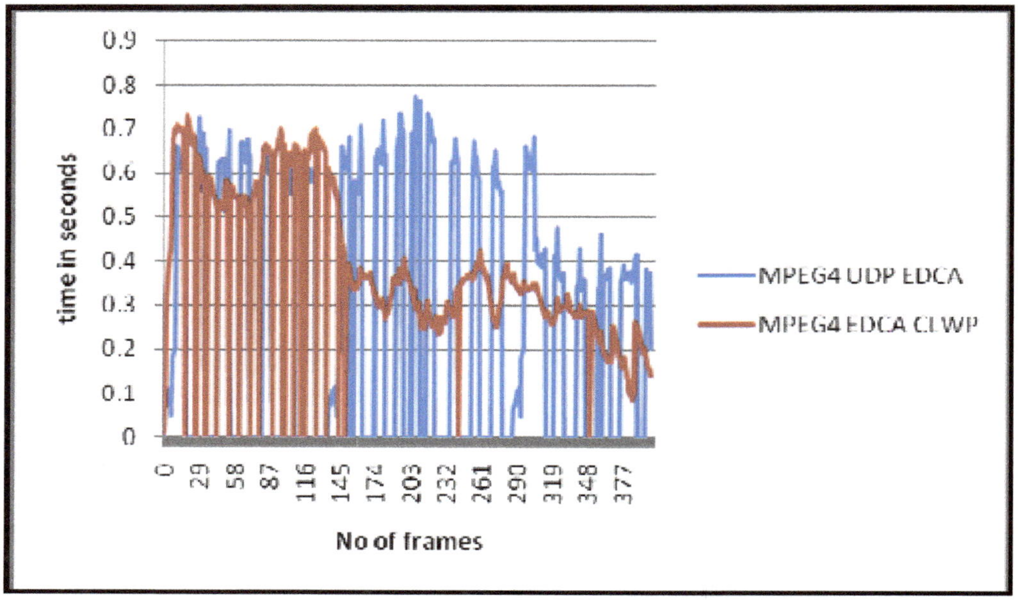

Figure 9: Delay produced by MPEG4 UDP EDCA and MPEG4 EDCA CLWP (Foreman).

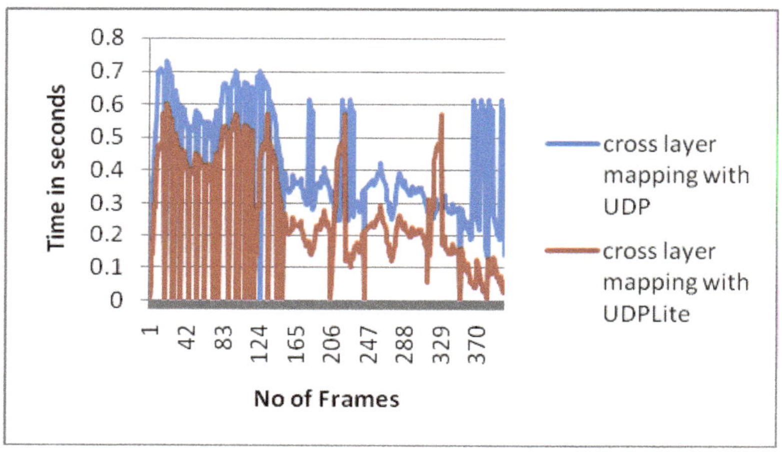

Figure 10: Delay produced by cross layer mapping with UDP and cross layer mapping with UDPLite (Foreman).

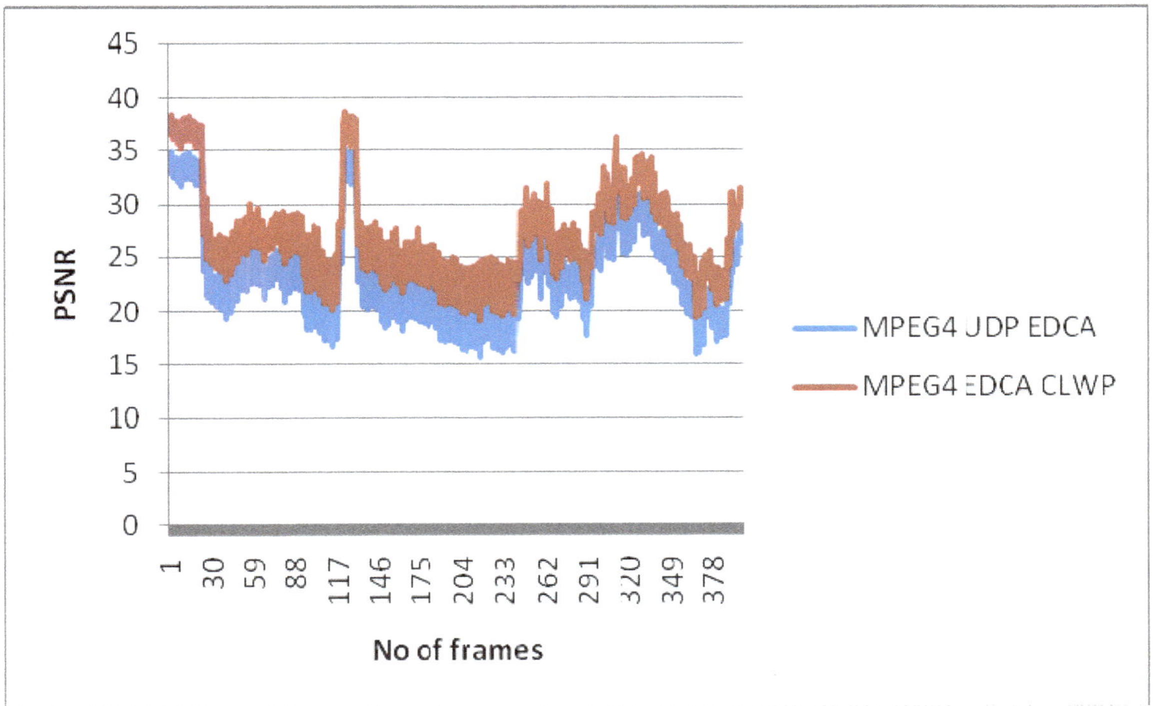

Figure 11: PSNR produced by MPEG4 UDP EDCA and MPEG4 EDCA CLWP (Foreman)

7.2.2 Peak Signal Noise Ratio

Figure 11 represents the PSNR produced by MPEG4 UDP EDCA and MPEG4 EDCA CLWP while the video source transmitted is Foreman, Figure 12 represents the PSNR produced by cross layer mapping with UDP and cross layer mapping with UDPLite while the video source transmitted is Foreman. Without mapping the PSNR is 38 and with mapping the PSNR is 40.

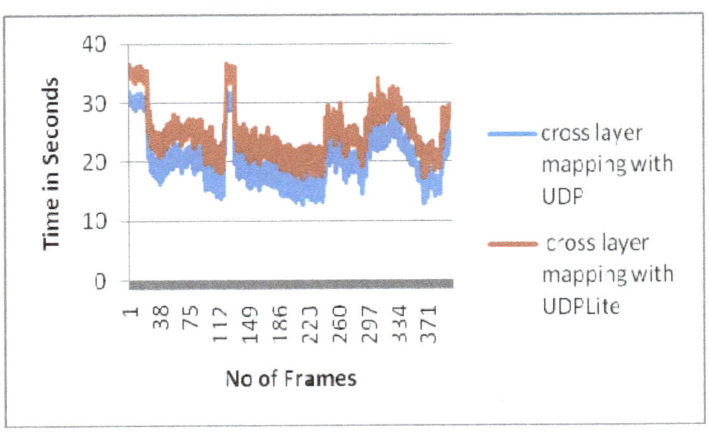

Figure 12: PSNR produced by cross layer mapping with UDP and cross layer mapping with UDPLite (Foreman).

8 CONCLUSION

Although, IEEE 802.11e EDCA has some features for QoS support it is not effective in providing priority to real time traffic such as delay sensitive video. By using dynamic mapping technique, video packets are mapped to the appropriate access category based on the significance of video data and network traffic load. Our proposed approach combines the benefits of UDPLite along with cross layer mapping thereby increasing the PSNR and decreasing the delay to a great extent.

REFERENCE

[1] Lars-Ake Larzon, Mikael Degermark , Stephen Pink: UDP lite for Real Time Multimedia Applications,IEEE International Conference of Communications (ICC) 1999.

[2] http://140.116.72.80/~jhlin5/ns2/crosslayer/NS-2_crosslayer.htm.

[3] C-H. Lin, C-H. Ke, C-K.Shieh, N.K. Chilamkurti, S.Zeadally (2008): A Novel Cross-Layer Architecture for MPEG-4 Video Stream over IEEE 802.11e Wireless Network, In Special Issue of International Journal of Telecommunications System, 2008

[4] Y. Ge, J. C. Hou, and S. Choi: An Analytic Study of Tuning Systems Parameters in IEEE 802.11e Enhanced Distributed Channel Access, *Computer Networks*, Vol. 51, No. 8, pp. 1955-1980, 2007.

[5] C. Hu and J. C. Hou: A Novel Approach to Contention Control in IEEE 802.11e-Operated WLANs, in *Proc. IEEE Infocom'07*, May 2007.

[6] Y. Xiao, H. Li, and S. Choi: Protection and Guarantee for Voice and Video Traffic in IEEE 802.11e Wireless LANs, in *Proc. IEEE Infocom'04*, pp. 2153-2163, Hongkong, Mar. 2004.

[7] A. Ksentini, M. Naimi, and A. Gueroui: Toward an improvement of H.264 video transmission over IEEE 802.11e through a cross-layer architecture, *IEEE Communications Magazine*, pp. 107–114, Jan. 2006.

[8] C. H. Foh, Y. Zhang, Z. Ni, and J. Cai: Scalable Video Transmission over the IEEE 802.11e Networks using Cross-Layer Rate Control, in *Proc. IEEE ICC'07*, Glasgow, UK, Jun. 2007.

[9] Amoolya Sing , Almudena Konrad , Anthony D. Joseph :Performance Evaluation of UDP Lite for Cellular Video, NOSSDAV '01 Proceedings of the 11th international workshop on Network and operating systems support for digital audio and video

[10]. Masala, M. Bottero, J.C. De Martin:. MAC–Level Partial Checksumfor H.264 Video Transmission over802.11 Ad Hoc Wireless Networks, Vehicular Technology Conference, 2005. VTC 2005-Spring. 2005 Vol.5 pp 2864 – 2868 1st June 2005

[11] Yi-Mao Hsiao, Feng-Pin Chang, Jai-Shiarng Chen, Yan-Lin Su, Zhi-Wei Yang, Wei-Shan Chen and Yuan-Sun Chu: High Speed ASIC Design of Multimedia, Network Industrial Electronics and Applications (ICIEA), 2010 the 5thIEEEConference15-17June2010 pp 155 – 159

[12] Xiaoqing Zhu, Bernd Girad and Peter Van Beek: Distributed channel time allocation and rate adaption for multiuser video streaming over wireless home networks, in IEEE International Conference on Image Processing (ICIP 2007), San Antonio, TX, USA, September 2007.

[13] Xiaoqing Zhu, Bernd Girad and Thomas Schiert and Thomas Wiegand: Video multicast over Wireless Mesh Networks with Scalable Video Coding (SVC) in Visual communication and Image processing , 2008, Proc. of SPIE, vol. 6822, pp 682205-682205-8.

[14] N Gomathi, P.seethalakshmi , Dr.A.Govardhan: CROSS LAYER DESIGN TO ENHANCE THROUGHPUT FOR MULTIMEDIA STREAMING OVER MOBILE ADHOC NETWORKS , International Journal on Computer Science and Engineering , vol 3 No.1 Jan 2011,Pg114-126.

10

RESOURCE ALLOCATION ALGORITHMS FOR QoS OPTIMIZATION IN MOBILE WiMAX NETWORKS

L. Nithyanandan[1] and J. Susila Sandhiya[2]

[1, 2] Department of Electronics and Communication Engineering,
Pondicherry Engineering College, Puducherry, India.
nithi@pec.edu
susila@pec.edu

ABSTRACT

WiMAX is based on the standard IEEE 802.16e-2009 for wireless access in Metropolitan Area Networks. It is one of the solutions for 4G IP based wireless technology. WiMAX utilizes Orthogonal Frequency Division Multiple Access which also supports Multicast and Broadcast Service with appropriate Modulation and Coding Scheme. Presently, Scheduling and Resource allocation algorithm in Opportunistic Layered Multicasting provides multicasting of layered video over mobile WiMAX to ensure better QoS. Initially, the knowledge based allocation of subcarriers is used for scheduling. In addition, to reduce the burst overhead, delay and jitter, SWIM (Swapping Min-Max) algorithm is utilized. Another promising technology that can greatly improve the system performance by exploring the broadcasting nature of wireless channels and the cooperation among multiple users is the Cooperative Multicast Scheduling (CMS) technique. The simulation results show, Swapping Min-Max performs better with lesser number of bursts, Zero jitter and with optimal throughput. The results with Cooperative Multicast Scheduling show the enhanced throughput for each member in the Multicasting Scenario.

KEY WORDS

OFDMA, SWIM, CMS,Jitter, QoS, WiMAX.

1. INTRODUCTION

Last few years wireless communication has explored different technologies which include High Speed Packet Access (HSPA), Universal Mobile Telecommunication System (UMTS), Long Term Evolution (LTE), Evolution-Data Optimized (EV-DO), Wireless Interoperability for Microwave access (WiMAX) and Multimedia Broadcast and Multicast Service (MBMS). Out of these EV-DO and WiMAX has gained more popularity among broadband solution. The first acceptable high speed mobile internet access around 5Mbps is EV-DO, however WiMAX approaches to have higher speed as the networks continue to be deployed andcan deliver up to 70Mbps. In recent years, Broadband Wireless Access networks have been rapidly involved to satisfy increasing user scalability and Quality of Service (QoS) [9].

The salient features of Mobile WiMAX are the higher data rate, mobility, scalability and Quality of Service. Certain advanced features are the smart antenna technologies, multicast and broadcast service and fractional frequency reuse.One of the main real time applications of Mobile WiMAX is the Internet Protocol Television. Video, voice and data are all IP data services, but each has its own Quality of Service (QoS) requirements. High availability, sufficient guaranteed bandwidth, low transmission delay and jitter are the QoS requirement for video services. To support QoS for various types of traffic, WiMAX medium access control protocol defines bandwidth request-allocation mechanisms and five types of scheduling classes: extended real time Polling Service (ertPS), Best Effort (BE), Unsolicited Grant Service (UGS), non real time Polling Service (nrtPS) and real time Polling Service (rtPS).Both UGS and rtPS are proposed to support real-time service generating packets periodically. While UGS is based on constant Bit Rate (CBR) traffic such as VoIP, rtPS supports Variable Bit Rate (VBR) traffic such as MPEG video. The core problem in mobile WiMAX network is the scheduling and resource allocation. There are different modes for resource allocation such as partially used Subcarrier (PUSC) and fully used subcarrier [8]. The PUSC mode is recommended for mobile users scenarios. The Mobile WiMAX scheduling is designed to efficiently deliver broadband services such as data, voice and video.

The rest of the paper is organized as follows. Section 2 discusses the literature survey. Section 3 elaborates transmission of the video over Mobile WiMAX. Swapping Min-Max algorithm is discussed in section 4.Section 5describes the system model, Cooperative Multicast Scheme (CMS) and analysis of the CMS scheme. Simulation results are analyzed in section 6 and finally section 7 concludes the paper

2. Literature Survey

Fen Hou, et al [5] developed an analytical model to evaluate the performance of the multicasting scheme, in terms of service probability, power consumption and throughput of each group member and multicast groups. C.W.Huang, et al [4], proposed a resource allocation algorithm that provides enhanced QoS and efficiency for layered video Multicast. S.-M. Huang, et al, [3] proposes a weakly consistent Opportunistic Multicast Scheduling (OMS) scheme for feedback reduction. Chakchai So-In, et al, [2] introduces a technique called Swapping Min-Max (SWIM) for UGS scheduling that not only meets the delay constraint with optimal throughput, but also minimizes the delay jitter and burst overhead. Chih-Wei Huang, et al, [1] proposed Opportunistic Layered Multicasting (OLM), a joint user scheduling and resource allocation algorithm that provides enhanced quality for the layered Video.

3. Wireless Iptv Over Mobile WiMAX

Internet Protocol Television (IPTV) is widespread to deliver the content to users whenever they want and wherever they are. Traditional wired access networks can deliver the contents only to the stable users. Hence, a novel technology which can deliver the content to mobile users when needed and mobile WiMAX is identified as a better choice. For real time application like IPTV and VoIP a guaranteed QoS level is very important because these are CBR (Constant Bit Rate) applications and are delay sensitive.

To support high data rate multicast and broadcast service in WiMAX, the techniques like OFDM and OFDMA are used. Orthogonal Frequency Division Multiplexing (OFDM) is a method which divides the bandwidth into multiple subcarriers at any given time. In mobile WiMAX, Orthogonal Frequency Division Multiple Access (OFDMA) is used to access multiple users simultaneously [6, 7]. The OFDMA symbol structure consists of Data, Pilot and Null sub-carriers. The data and pilot subcarriers are clustered into subsets of subcarriers called sub channels. The WiMAX OFDMA supports sub-channelization in both downlink and uplink. There are two modes in WiMAX for sub-channelization, diversity and contiguous permutation. The contiguous permutation draws the subcarriers as the block to form a sub-channel and the diversity permutation draws sub-carrier pseudo-randomly to form a sub-channel. Due to the frequency diversity and inter-cell interference averaging provided, the diversity permutation is advised for mobile users. IPTV transmission over WiMAX requires coding techniques for high quality video. In order to get high quality video, Scalable Video Coding (SVC) is used. SVC is one of the challenging solutions to the problems caused by the characteristics of modern video transmission systems. The word "scalability" refers to the elimination of parts of the video bit stream in order to adapt it to the user's need. The Scalable Video Coding is an extension of the H.264/AVC standard. The motive of the SVC standardization is to enable the encoding and the decoding of a high-quality video bit stream with the similar quality and the complexity that can be achieved using the H.264/AVC design with the same quantity of data as in the subset bit stream.

4. Swapping Min- Max Algorithm

The scheduler for rtPS service should meet the QoS criteria like Optimal system throughput , maximum latency guarantees, Minimal delay jitter and Minimal number of bursts in order to reduce the MAP overheads[2].The resource is fixed in terms of number of slots per downlink sub frame. This is denoted by variable number of slots. The number of bytes corresponding to a slot depends upon the modulation and the coding. For the rtPS, at the connection setup, MSs declares the total demand (denoted by the subcarriers for the required data size) and period (denoted in terms of frames). For example, connection_1 asks for 540 bytes (1080 frames) every 4 sub carriers. WiMAX profiles specify a size of each frame about 5ms.The complexity of the SWIM in the worst case is in the order of O (n^2log n), where n is the number of active connections.

$$\text{Complexity} = O\ (\text{allocation s} + \text{sorting}) = O\ (n^2 \log n) \qquad (1)$$

$$\text{And the Complexity} = O\ (n^2) \qquad (2)$$

when the information is known about the number of flows.

The procedure for algorithm has two parts, Initialization with the optimal throughput/delay and Resource swapping steps.The swapping procedure is as follows,

1. Determine the min_res connection and the max_res connection.
2. The two connections swap their resources such that min_res gives up its resources in the current frameand it gains an equal amount of the resources in the forecoming frames.
3. The swapping procedure ensures that each burst is at least Minburstsize. (Minburstsizeis considered here as 1).
4. If any connection has the resource equal to Minburstsize, they are excluded from swapping (leaving Minburstsize value as non-zero).

The special cases which may exist are

1. The new max_res connection cannot accept any resources more than it needs and so the min_res connection may not get eliminated.
2. If there is two or more max_res connection in the current frame, the connection which has the resources higher in the next frame will be chosen.
3. If there is two or more max_res connection with same resources in the next frame, the connection with more deadline with be selected.
4. If there are two or more min_res connections, the connection whose deadline is earlier will be chosen.

5. If there are two or more min_res connections with the same deadline, the connection withleast resources in the next will be selected.

4.1 Swapping Min-Max examples

Table 1. Static Flows

	C1	C2	C3	C4	C5
Data Size (bytes)	62	35	500	60	580
Period (frames)	260	160	900	240	1200
No of Subcarriers	4	4	6	6	12

Table 1 shows a simple example of 5 connections (C1-C5) and their demands (Data Size) in bytes, Period in terms of WiMAX frames and the number of subcarriers. The total allocated slots are 395. The throughput is optimal, that is (260*3) + (160*3) + (900*2) + (240*2) + (1200*1) = 4740 frames. Table 2 and 3 show the EQual Allocation (EQA) and Earliest Deadline First (EDF) algorithms respectively. In EQA, the resource is allocated equally in every frame. In EDF, the resource is allocated whose deadline is the first. In the next frame, the scheduler allocates the remaining resources to meet the guaranteed throughput.

Table 2. EQA allocations

Time	C1	C2	C3	C4	C5
0	65	40	150	40	100
1	65	40	150	40	100
2	65	40	150	40	100
3	65	40	150	40	100
4	65	40	150	40	100
5	65	40	150	40	100
6	65	40	150	40	100
7	65	40	150	40	100
8	65	40	150	40	100
9	65	40	150	40	100
10	65	40	150	40	100
11	65	40	150	40	100

Table 3. EDF allocations

Time	C1	C2	C3	C4	C5
0	105	0	290	0	0
1	0	0	215	0	180
2	155	160	0	80	0
3	0	0	295	0	100
4	135	0	100	160	0
5	125	160	0	0	110
6	0	0	295	0	100
7	0	0	0	80	315

8	25	0	210	160	0
9	0	0	395	0	0
10	130	160	0	0	105
11	105	0	0	0	290

Table 4 shows the initial swapping steps. In the first frame, the max_res connection is C3 and the min_res connection is C2. Therefore, C2's allocation in the frame is given to C1 and taken back in the second frame. This results in C3 obtaining 150 + 40 = 190 and C2 obtaining 40 – 40 = 0 in the first frame. C3 obtains 150 – 40 = 110 and C2 obtains 40 + 40 in the second frame. The resulting allocations are shown in Table 4 (a).Thus, the swapping has reduced the number of bursts by one. In the next swapping step, C3 and C4 swap their allocations in the frame 1 and 2 resulting in the Table 4 (b) and swapping proceeds till the last frame. This results in the final allocation shown in Table 5.

Table 4 SWIM initial steps

(a)

Time	C1	C2	C3	C4	C5
0	65	0	190	40	100
1	65	80	110	40	100
2	65	40	150	40	100

(b)

Time	C1	C2	C3	C4	C5
0	65	0	230	0	100
1	65	80	70	80	100
2	65	40	150	40	100

Table 5 SWIM Final Steps

Time	C1	C2	C3	C4	C5
0	0	0	255	0	140
1	195	0	0	40	160
2	0	145	0	0	250
3	65	15	315	0	0
4	0	80	135	0	180
5	0	0	195	200	0
6	0	0	295	0	100
7	260	80	0	55	0
8	0	0	155	0	240
9	0	0	395	0	0
10	195	0	0	145	55
11	65	160	55	40	75

4.2 Need for Cooperative Communication

SWIM performs efficiently, only when newly demanded resources are lesser than the available number of slots. It leads to the complexity of the design when the number of information flow is not known prior to the transmission. To meet all these considerations and support heavy traffic Cooperative Multicast Scheduling Scheme is proposed particularly to cater live multicasting and ensures higher throughput than existing multicast schemes.

5.COOPERATIVEMULTICAST SCHEDULING SCHEME

In Mobile WiMAX network consisting of a BSand multiple subscriber stations (SSs).IEEE 802.16 standards support mesh and Point-to-Multi Point (PMP) mode.

The TDD-OFDM/TDM MAC structure is considered as shown in Fig.1. The time domain is divided into MAC frames withequal duration, each of which is composed of a downlink sub-frame (DL sub-frame), an uplink sub-frame (UL sub-frame),a transmit/receive transition gap (TTG), and a receive/transmittransition gap (RTG).

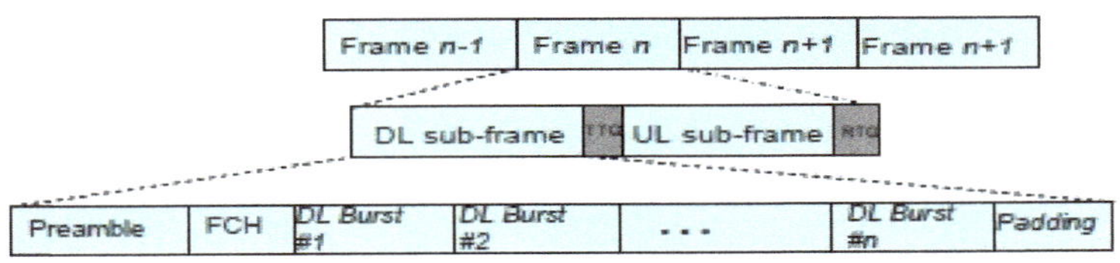

Fig.1 MAC structure for a conventional IEEE 802.16 network

In a conventional IEEE 802.16 network, SSs only receivedata from the BS in DL sub-frames. To achieve cooperativemulticasting, the transmission burst assigned for multicast transmission is divided into two phases as shown in Fig.2

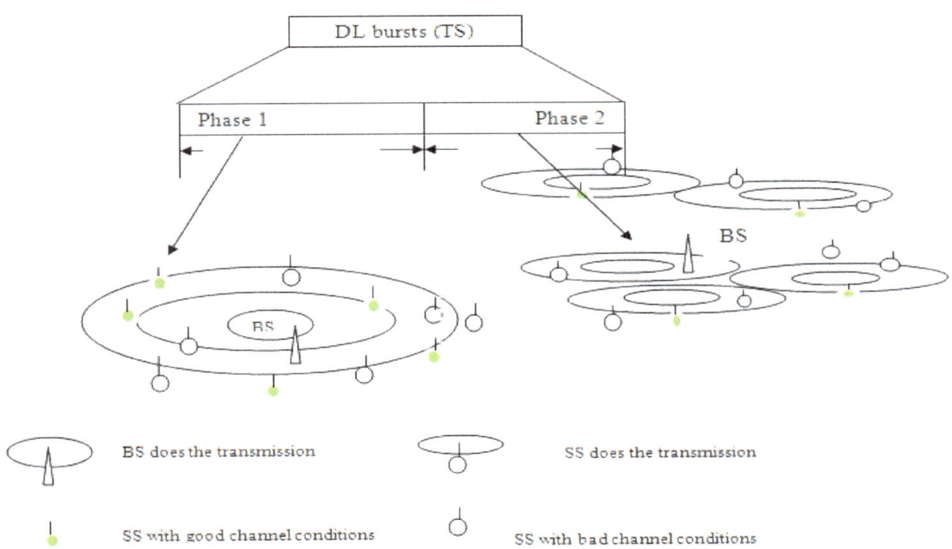

Fig. 2 Cooperative multicast scheduling

In multicast networks, for an IPTV service, an MGroup corresponds to a group of users requesting the same TV channel. As Mobile stations could be a residential house or office building consisting of multiple users accessing different channel simultaneously and thus may belong to several MGroups. To select the appropriate MGroup, the schemes adopted are random MGroup and channel-aware MGroup.

5.1 Multicast Group Selection

1. Random MGroup selection

The Base Station randomly chooses an MGroup for service with a pre-defined probability. Each group is served with the probability of 1/T for achieving a good fairness performance. It is easy to implement. To further improve the performance, the channel-aware MGroup selection is used.

2. Channel-aware MGroup selection

This selection MGroup is considers the channel conditions into account on the group basis, rather than a single group member. By using the multi-group channel diversity and taking into account fairness, a criterion of MGroup selection is given by eqn. (3),

$$i^* = \arg\max_i C_i \quad (3)$$

and the normalized relative channel condition is given by eqn. (4)

$$C_i = \frac{\sum_{j \epsilon G_i} \gamma_{i,j} / \overline{\gamma_{i,J}}}{N_i} \quad (4)$$

Based on (3), the BS selects MGroup which has the maximum value of the normalized relative channel condition.

5.2Co-operative Multicast transmission

After an MGroup is selected, the next step is to multicast data to all group members in the chosenMGroup. To exploit the diversity gain in wireless channels, two phase transmission technique is used to multicast the datain the downlink transmission, where every downlink burst is divided into two phases. In Phase I, the BS multicasts data to all group members of MGroup*i* at a high data rate of R^1suchthat only a particular portion of group members in MGroup*i* can successfully decode the data and is shown in Fig.2. Due to the high data rate, the remaining group members with unfavourable channel conditions may not be able to successfully decode all the data in Phase I. Therefore, in Phase II, cooperative communication is utilized to promise reliable transmission of the remaining group members.To investigate the performance of the Co-operative Multicast Scheduling service probability of each MGroup, throughput of each user, throughput of each MGroup and the whole network throughput is simulated[5].

5.3 Service Probability of MGroup i

Service probability is defined as the probability that the particular Multicast group is chosen to be served when the system is stable. For the channel-aware MGroup selection, the service probability (5) of the MGroup *i* is given by

$$\pi_i = \int_0^\infty \left| \frac{N_i^{N_i}}{(N_i-1)!} x^{N_i-1} e^{N_i x} \prod_{j=1, j\neq i}^{T} \left(1 - e^{-N_j x} \sum_{k=0}^{N_j-1} \frac{(N_j x)^k}{k!} \right) \right| dx \quad (5)$$

5.4 Throughput Analysis

The probability that a group member in MGrcup*i*can successfully receive the data in phase I is

$$P_r\left(E^1 \geq \left(2^{R^1} - 1\right)N_o\right) = e^{-\left(\left(2^{R^1}-1\right)N_o\right)/\bar{E}_B} \quad (6)$$

If the Subscriber Station fails to receive the data in phase I. It is still possible for them to receive the data in phase II and the probability is

$$P_r\left(E^2 \geq \left(2^{R^2} - 1\right)N_o\right) = \sum_{G_i^1 \in C_{i,j}} P_r\left(G_i^1\right) P_r\left(E^2 \geq \left(2^{R^2} - 1\right)N_o | G_i^1\right) \quad (7)$$

The throughput of subscriber's station is given by,

$$\alpha t^2 R^2/(t^1 + t^2) \quad (8)$$

The group throughput of all the group members in MGroup*i*is given by,

$$Th_i^{CMS} = \sum_{j=1}^{N_i} Th_{i,j}^{CMS} \quad (9)$$

In addition to the analysis based on channel capacity, the impact of the promising adaptive modulation and coding (AMC) technique is also investigated. The transmission rate corresponding to different modulation and coding levels is given by

$$r_n = I_n/T_s \quad (10)$$

The throughput achieved by the group member using AMC technique is given by,

$$Th_{i,j}^{CMS} = \pi_i \left[\frac{\alpha t^1 R^1}{t^1+t^2}\right] e^{-\frac{b_n N_o}{E_{i,j}}} + \left[\frac{\alpha t^2 R^2}{t^1+t^2}\right] \left(1 - e^{-\frac{b_n N_o}{E_{i,j}}}\right) .* \sum_{G_i \in C_{i,j}} P_r\left(G_i^1\right) \left[1 - F\left(E^2 \geq \left(2^{R^2} - 1\right)N_o | G_i^1\right)\right] \quad (11)$$

6. Simulation Results

The numerical results in terms of average number of bursts, delay jitter, delay of the EQA, EDF and SWIM algorithms are simulated. The Swapping Min-Max algorithms are analyzed for base and enhancement layers. Further, to effectively utilize the resources and to enhance the throughput performance the multicast scheduling is carried out. The simulation parameters are shown in Table 6 and the list of notations used is quoted in Table 7.

Table 6.performance evaluation parameters

PARAMETERS	VALUE
SWIM	
Frame Length	5ms
System Bandwidth	10MHz
CMS	
Number of MGroups	10
Number of group members in MGroup	20
Frequency Band	3.5GHz
Noise Figure	7 dB
OFDM System duration	23.8μs

Table 7. Table of Notations

Notations	Explanations
T	The total number of MGroups
G_i	The set of all members belonging to MGroup*i*
N_i	The total number of group members in MGroup*i*
G_i^1	A set of members in MGroup*i* successfully receive data
E	The received signal power
R	The rate of the BS
1,2	Phase 1, Phase 2
T	The transmission time
α	Time ratio for multicast transmission
Th_i^{CMS}	The group throughput of the MGroup*i*
I_n	Information bits/ OFDM symbol
b_n	Lower boundaries of SNR
C_i	Normalized relative channel conditions of the MGroup*i*
$\gamma_{i,j}$	Instantaneous channel conditions

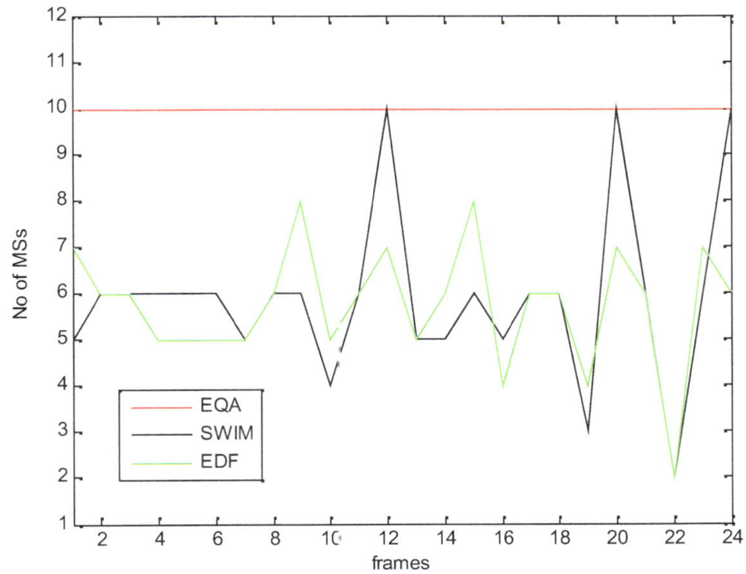

Fig.3 Number of bursts for EQA, EDA and SWIM

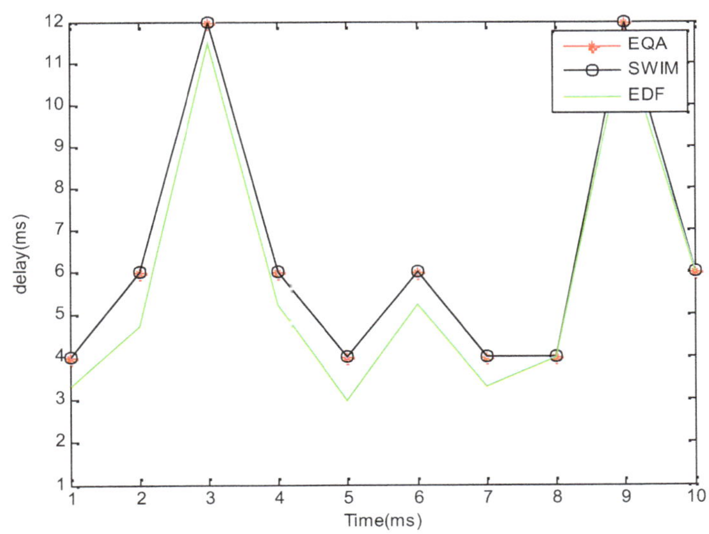

Fig.4 Delay for the EQA, EDF and SWIM

Fig.3 shows the numerical results for the number of bursts in each frame. In equal allocation, there is no reduction in number of bursts as it utilizes all slots in all Mobile Stations. Both the EQA and EDA, minimizes the number of slots to reduce the MAP overheads.

From the Figs. 3, 4 and 5 it is inferred that the SWIM algorithm minimizes the number of bursts with zero jitter and optimizes throughput as shown in Fig.6.

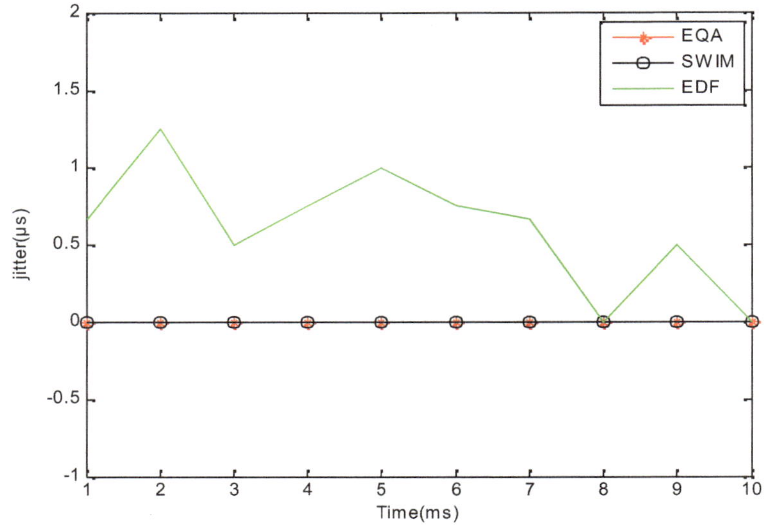

Fig.5 Delay Jitter for EQA, EDF and SWIM

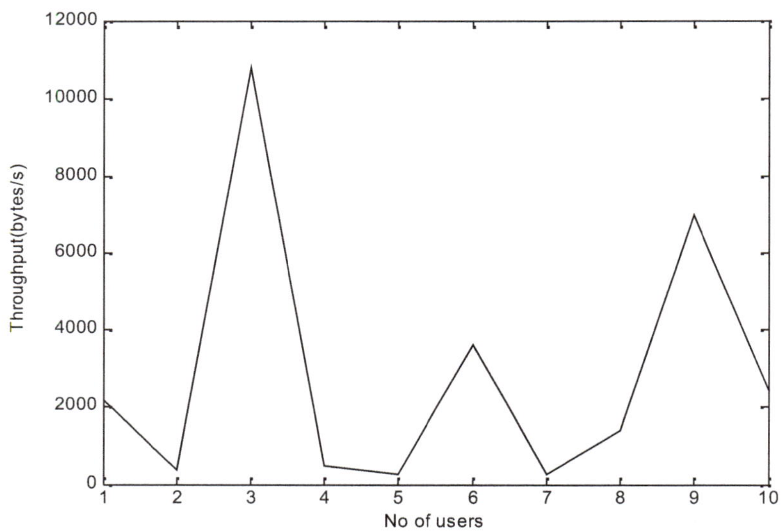

Fig.6 Throughput of the SWIM algorithm for 10 users

Fig.4 shows the delay of EQA, EDF, and SWIM. The delays for EQA and SWIM are the same which are equal to the periods. For EDF, even though delay is reduced when compared to the other two algorithms it will in turn introduce the delay jitter as shown in Fig.5.But the delay jitter is almost zero for SWIM.

Fig.7 shows the steady state probability that an MGroup is selected to be served at an arbitrary frame. It is observed that each MGroup obtains almost the same service probability which achieves the good fairness in terms of channel conditions.

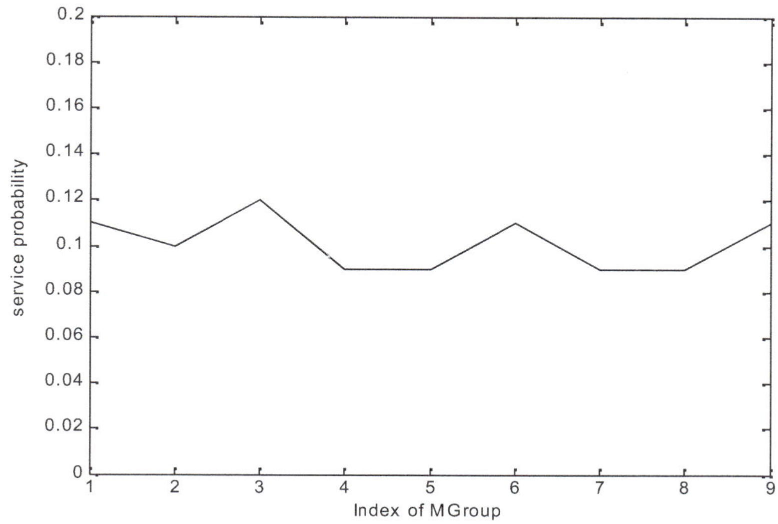

Fig.7 Steady State Service Probability

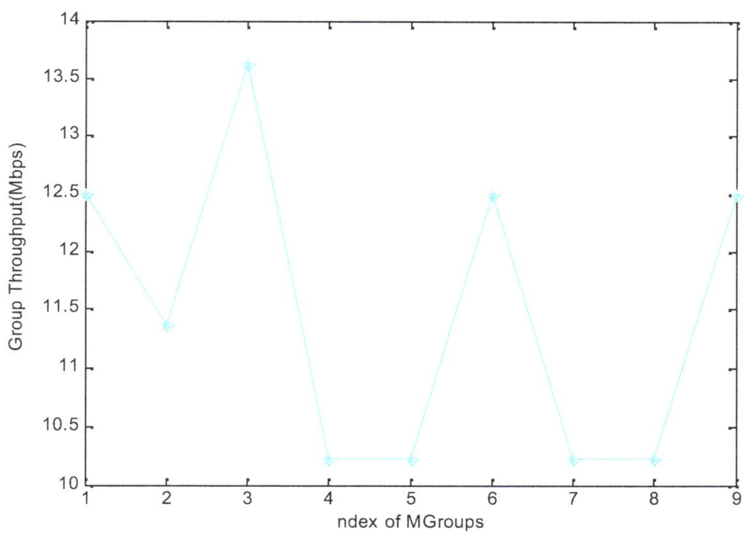

Fig.8 Throughput for each Multicast group

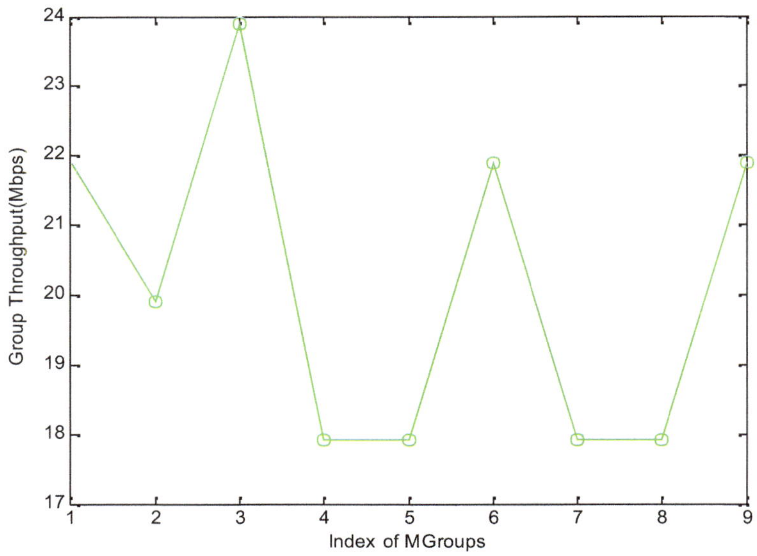

Fig.9 Throughput for each Multicast group using AMC

Fig.8shows the throughput of the each Multicast group for which each MGroupis chosen based on channel aware method. From the Fig.9it is clear that the throughput is still enhanced for the each multicast group due to adaptive modulation and coding scheme used.

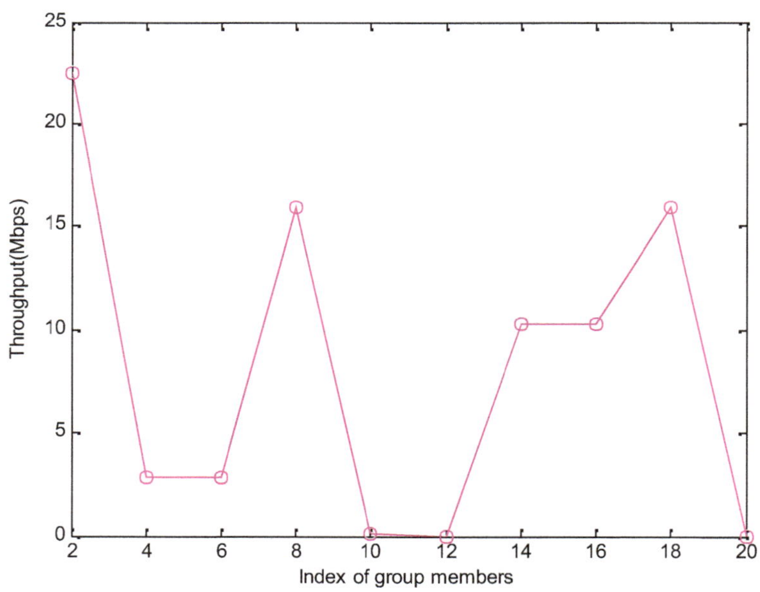

Fig.10 Throughput for each group members

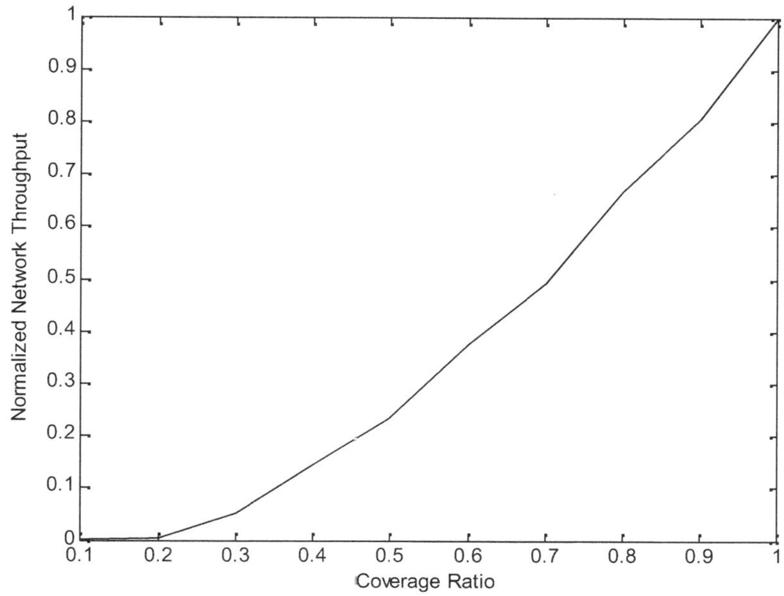

Fig.11 Normalized network throughput Vs Coverage ratio

Fig.10 shows the throughput for each group members and Fig.11 shows the normalized network throughput. From the figure it is seen that the as the coverage area increases the throughput of the network also increases.

7. Conclusion

WiMAX is one of the emerging technologies that can be used for all types of wireless services (voice, data and multimedia, IPTV) which require higher capacity and data rates. In this paper, Wireless IPTV istransmitted over Mobile WiMAX. To support various bandwidth requirements, video streams are coded into base and enhancement layers by Scalable Video Coding (SVC) technology. Efficient algorithms are carried cut for the effective utilization of the resources. The performance of the Swapping Min-Max algorithm minimizes the number of bursts and achieves approximately zero jitter. Further, Cooperative Multicasting Scheduling is realized which still optimizes throughputand exploits multi-channel diversity among the group members which are involved in multicasting.

References

[1] Chih-Wei Huang,Shiang-Ming Huang, Po-Han Wu, Shiang-Jiun Lin and Jenq-Neng Hwang, "OLM: Opportunistic Layered Multicasting for scalable IPTV over Mobile WiMAX," *IEEE Trans. Mobile Computing*,vol.11,no.3,453-463,Mar. 2012

[2] Chakchai So-In, Raj Jain and Abdel-karim Al Tamini, "A Scheduler for Unsolicited Grant Service (UGS) in IEEE 802.16e Mobile WiMAX Networks," revised September, 2010.

[3] S.-M. Huang, J.-N.Hwang, and Y.-C. Chen, "Reducing Feedback Load of Opportunistic Multicast Scheduling over Wireless Systems," *IEEE Comm. Letters*, vol. 14, no. 12, pp. 1179-1181, Dec. 2010.

[4] Chih-Wei Huang,Shiang-Ming Huang, Po-Han Wu, Shiang-Jiun Lin, and Jenq-Neng Hwang, "Layered Video Resource Allocation in Mobile WiMAX using Opportunistic Multicasting," *Proc. IEEE Wireless Comm. And Networking Conf., Apr.* 2009

[5] Fen Hou, Lin X Cai, Pin-Han Ho, XueminShen and Junshan Zhan, "A Cooperative Multicast Scheduling Scheme for Multimedia Services in IEEE 802.16 Networks," *IEEE Transactions on Wireless Communications*, Vol. 8, no. 3, March 2009

[6] S. Deb, S. Jaiswa and K. Nagaraj, "Real-Time Video Multicast in WiMAX Networks," *Proc. IEEE INFOCOM*, pp. 1579-1587, 2008.

[7] Q. Qu and U. Kozat, "On the Opportunistic Multicasting in OFDM-Based Cellular Networks," *Proc. IEEE Int'l Conf. Comm. (ICC '08)*, pp. 3708-3714, May 2008

[8] WiMAX Forum, "Mobile WiMAX- Part I: A Technical Overview and Performance Evaluation," Aug.2006.

[9] J. Susila@Sandhiya and L.Nithyanandan, "Resource and QoS Optimization for multicasting in Mobile WiMAX Networks," *Proceedings of IEEE International Conference on Communication & Signal Processing – 2013,* pp. 1536-1539, April 2013.

11

A FAIRED QUALITY OF SERVICE ASSURED MAC PROTOCOL FOR MOBILE ADHOC NETWORK & ITS PERFORMANCE EVALUATION

D.D.Seth[1], Srikanta Patnaik[2] and Srikanta Pal[3]

[1]School of Electronics, KIIT University, Bhubaneswar, Odisha, India
ddgayatri@yahoo.co.in

[2]Interscience Institute of Technology & Management, Bhubaneswar, Odisha, India
patnaik_srikanta@yahoo.co.in

[3]Department of Electronics and Communication Engineering, Birla Institute of Technology, Mesra, Ranchi, India
pal_srikanta@yahoo.co.uk

ABSTRACT

The medium access control (MAC) protocol for Mobile Adhoc Network (MANET) need to be distributed, QoS assured, fair to all flows and should work in a multihop environment. Although there are numerous MAC protocol proposed for MANET, very few of them possesses all the above properties. This paper, which proposes a MAC protocol for adhoc network have all the above mentioned properties. The protocol supports quality of service and exhibit flow based fairness over multihop adhoc network. In accordance with the protocol, high prioritized traffic flows are allowed to access the channel earlier than the low prioritized traffic flows. If multiple numbers of flows of the same priority level are active in the network then the fairness factor comes into account .The less served flow, and then gets the priority over the highly served flow in accessing the channel. The performance of the protocol is evaluated analytically. The result shows that the high prioritized traffic flow experience less delay and have more throughputs as compared to others.

KEYWORDS

MANET, QoS, DCF,Fairness,Saturation Throughput

1. INTRODUCTION

In the past decade wireless networks have experienced unprecedented development. Mobile Adhoc Network (MANET) is one of the most rapidly developing areas. Briefly adhoc networks are stand-alone wireless networks that lack the services of a backbone infrastructure .They consist only of a collection of mobile stations voluntarily act as forwarders or routers for other mobile stations in the network. Such networks are initially designed for use in military and emergency-relief application. Lately, by virtue of better spatial reuse characteristics in comparison to the conventional wireless network model [1] they are used in regular wireless network application, sensor network, and personal area networks.

Most of the research works on routing and transport layer protocols in adhoc networks assume the use of CSMA/CA as the Medium Access Control (MAC) protocol in their protocol stacks. The very reasons for the assumption are (i) the infrastructure-less mode of operation that the

IEEE 802.11 DCF supports (ii) the popularity of the IEEE standard itself that requires no additional hardwires for the adhoc mode of operation.

The explosive growth of multimedia application has arisen the problem of the required Quality of Service (QoS) of these applications such as guaranteed delay, jitter and bandwidth. Although IEEE 802.11 DCF has gained the momentum and popularity in recent times and has become the de-facto standard for adhoc networks, it has some limitations.(a)It does not support Quality of Service (QoS).That means it does not differentiate between the bandwidth and delay sensitive real time traffics (RT) and non sensitive non-real time (n-RT) traffics. (b) It is also poor with regard to fairness. A channel is said to be fair if it is able to provide each and every individual nodes without giving preference to one node over the others when there is no service differentiation. As in IEEE 802.11 DCF, the successful node (the node which successfully transmitted the packet) sets its contention window to CW_{min} where as collided node (the node which experiences collision) sets its contention window to $2*CW_{old}$. Hence the successful nodes remain in a better position to transmit the next packet in comparison to the collided nodes. This leads to unfairness. (c) This IEEE 802.11 DCF performs poorly in a multihop adhoc network. In a multihop adhoc networks the intermediate nodes not only have to transmit their own generated packets but also have to forward the packets of other nodes. Hence in absence of proper queue management at the MAC level overall throughput decreases.

A new Medium Access Control (MAC) protocol has been proposed in this work by extending the mandatory Distributed coordinated Function (DCF) of the MAC sub layer. The proposed protocol has the following features.(i) Distributed in nature (ii) Supports Quality of Service (QoS) (iii)Exhibits fairness (iv) Applicable to Multihop adhoc network (v) Avoids hidden node problem. The performance evaluation model for the protocol has also been introduced.

The rest of the paper is organized as follows. Section II discusses the related work in this field. Section III give the detailed description of the proposed protocol, In Section IV Analytical model and performance evaluation described in Section V. Section VI discusses the results. Concluding remark and suggestion for future work is included in Section VII. This document describes, and is written to conform to, author guidelines for the journals of AIRCC series. It is prepared in Microsoft Word as a .doc document. Although other means of preparation are acceptable, final, camera-ready versions must conform to this layout. Microsoft Word terminology is used where appropriate in this document. Although formatting instructions may often appear daunting, the simplest approach is to use this template and insert headings and text into it as appropriate.

2. RELATED WORK

In recent times a number of medium access control (MAC) protocol supporting quality of service has been proposed. Some of the protocol ensures quality of service but very poor with respect to fairness. Others although assures QoS for one hop performs poorly when it is applied to a multihop network. Some of the protocols proposed in this field are described in the following section. As the medium access control (MAC) protocol proposed in this thesis is based on IEEE 802.11 DCF [2], it is discussed in the following subsection.

2.1. DCF (Distributed Coordination Function)[2]

IEEE 802.11 DCF is an implementation of CSMA /CA protocol which follows four way handshaking (RTS/CTS/DATA/ACK) for data transmission. Whenever a station is ready to transmit data, it senses the channel for DIFS period and when it observe that the channel is idle for this period ,it generates a random back off timer chosen uniformly from a (0,CW-1), where CW = contention window. After the back off timer expires, it sends a RTS packet to the destination node. If the destination node is ready to receive the data, it replies with a CTS packet. Both RTS and CTS messages carry the duration information for which the channel is going to be occupied by the proposed data transmission. All other nodes in the vicinity of the sender and destination update their NAV (Network Allocation Vector) with the information for which channel is going to be busy. Hence possibility of collision is reduced as those stations defer their transmission and reception. The CTS message is followed by the DATA transmission which is acknowledged by the receiver by sending an ACK message if the DATA is received successfully. The data is repeatedly retransmitted in the absence of ACKs till a threshold number of retransmissions are carried out. Once the retransmissions exceed the threshold, the transmission is assumed to be unsuccessful. After an unsuccessful transmission attempt, the sender follows a binary exponential back off (BEB) and doubles its contention window size. This is done in order to reduce the channel contention between nodes. The contention window is not incremented further if it already equals CWmax. The contention is reset back to CWmin, after every successful transmission. RTS, CTS, DATA, and ACK are separated by a time spacing of Short Inter Frame Space (SIFS).

The IEEE 802.11 DCF in its original form does not support service differentiation. It give similar treatments to all types of packet and all stations contend for the channel with same argument by which the CW is set and BEB(binary exponential back off) algorithm is run. Hence the delay sensitive traffics may experience intolerable delay and required Quality of Service (QoS) is not maintained by using the legacy IEEE 802.11 DCF.

Further, CSMA/CA protocol suffers from fairness problem. Initially all the stations have equal opportunity to access the channel. But after successful transmission the station sets its CW size to CWmin whereas the collided station doubles its CW size. Hence a successful node remains always in a better position to access the channel as compared to unsuccessful one, which leads to unfairness. The unfairness of MAC has a far reaching impact on the behavior of higher layer protocols and the applications using the network. Application like audio/video streaming are sensitive to packet delays and jitters. When the underlying link behavior is unfair, some applications may be starved of bandwidth just because their share is unfairly distributed somewhere else.

Although IEEE 802.11 MAC support some kind of adhoc network architecture it is not intended to support the wireless mobile adhoc network, in which multi-hop connectivity is one of the most prominent features[17] .It is seen that 802.11 based multihop wireless network suffer from serious exposed node problem and collisions. This MAC layer problem may cause the failure of routing protocol.From the above discussion, it is clear that the legacy 802.11 DCF need to be modified to work in a multihop adhoc network.

2.2.Related works on QoS

EDCF is the distributed medium access control protocol [9] is the enhanced version of DCF. It supports service differentiation by employing eight priority classes. Although it supports Quality of Service, it does not address fairness and multihop issue.

Black Burst [4] protocol provides QoS by jamming the channel through pulses of energy for the node with time sensitive traffics. It continues to access the channel until all the packets are transmitted. Again it does not fully consider the fairness as well as the multihop issues.

DBASE [3] protocol supports both data and multimedia traffic over the IEEE 802.11 adhoc network. Nodes with real time traffics (RT-node) are prioritized higher than the non-real time node by choosing a shorter observation period. In this protocol bandwidth is allocated among the nodes according to their requirement in the next cycle. But the allocation does not exceed the average bandwidth requirement. The bandwidth requirement in the next cycle, if, lower than the present bandwidth, the residual bandwidth can be shared with the needy nodes. The bandwidth so shared can be withdrawn when own requirement exceeds. Although this protocol is a noble one in providing QoS , it also does not consider the cases of multihop adhoc network. Fairness issue also not addressed.

TCMA [7] is a QoS based distributed MAC protocol. It is a CSMA/CA protocol. It provides prioritization to different traffics in terms of arbitration time, persistence factor and MAC layer dwell-time limit.TCMA provides QoS in single hop and does not discusses the fairness issue at all. It has no explanation for multihop scenario. Hidden node problem and exposed node problems are also not addressed here.

The above protocols [9,4,3,7] and other protocols such as [5,6,8] supports service differentiation and provide special treatment to bandwidth–delay sensitive real time traffic. But these protocols are not fully faired. Further they are suitable for single hop adhoc network and do not address the issues involved in a multihop scenario.

2.3. Distributed Fair Scheduling (DFS) [10]

Distributed Fair Scheduling is a protocol which allows the stations to contend for the channel with the contention window in proportion to the weight assigned to them. The stations with higher weight set the contention window of less size compared to the other stations. Although the DFS protocol deals with fairness issue to a large extent it is however, not clear how the channel shares Φ_i is assigned to each node. It provides node based fairness. That means all node should be fairly served. Node based fairness may cause problem when it is to be applied to a multihop adhoc network. Due to possibility of some nodes (such as intermediate node) servicing more flows than the other, a per – node fairness will cause the degradation of performance. Hence in such situation, flow based fairness is the best choice. Our proposed protocol has been designed taking this point into consideration. The other protocols such as Dugar and Vaidya [12] also discusses the fairness issue.

2.4. QCMA (Queue-driven Cut-through Multiple Access) [16]

Legacy IEEE 802.11 DCF is for single hop adhoc network . The performance of DCF in multihop adhoc network has been studied in [17]. It performs very poorly with respect to throughput as well as delay in such Multihop case. QCMA is a recent algorithm proposed by D. Raguin & others to overcome the inefficiency of IEEE 802.11 DCF protocol for multihop adhoc network. In a multihop adhoc network, the intermediate node may require frequent channel access to forward the packets of the other node. This QCMA protocol has presented a noble idea for access procedure in multihop adhoc network. However it does not address the fairness issue as well as QoS issue.

The related work on the performance analysis may be presented as follows. The performance of the IEEE 802.11 DCF [20] has been investigated based on the Discrete Time Markov Chain (DTMC) model. The saturation throughput analysis employed the Markov chain model and it considered several assumptions namely: (i) saturation condition, (ii) no hidden terminals, (iii) no capture effect, (iv) constant packet arrival rate for all the stations, (v) same radio conditions for all the stations and (vi) no transmission error. These assumptions are not suitable for real wireless environment. A lot of works has been carried out to evaluate the performance of IEEE 802.11 DCF over ideal error-free channel. Earlier studies [25-27] have presented the influence of the physical layer on the MAC layer performance.

3. PROPOSED PROTOCOL

In this section, the detailed description of the protocol is given. This protocol employs the IEEE 802.11 as a subroutine for channel contention. The proposed protocol which is designed for the MAC layer of adhoc network and has the following key features: (1) It supports quality of service (2) It provides flow based fairness. That means it exhibits fairness to different flows but of the same priority class. (3) It also works in a multihop adhoc network. The detailed working of the protocol is given through the flow chart in Figure: 2and 3. In order to provide service differentiation to different types of traffic different priority level is assigned. Hence if the traffic is of higher priority level, it will contend for the channel with a lower CWmin and CWmax. In this work, three types of traffics are involved: voice traffic (constant bit rate), video traffic (variable bit rate) and datagram traffic. The priority level assigned to voice, video and data are 2, 1 & 0 respectively. Hence voice traffic will contend the channel with less contention window size and data traffic with larger window size. Typically, CWmin and CWmax has been taken as (7, 15), (15, 31) and (31, 1023) for priority class 2,1and 0 respectively. When the MAC protocol is to be applied in a multihop enviroment, it becomes a new challenge. Because the intermediate nodes not only have to transmit own generated traffic but also have to forward the packets received from its previous nodes.

3.1.Enque operation

Each node maintains three queues q2, q1, q0 to enque the packets of different priority class 2, 1 and 0.After receiving the packet and knowing the type of traffic they are enqued in their respective queues. As per the priority level, the queues are dequed.In our work the queue maintained for voice traffic that is q2 is attended first .then q2 and lastly q1.The operations are explained through flow chart in Figure 2. Initially set the serv_flow =1 for all flows on seeing the packets of those flow for the first time, where serv_flow is the number of service obtained by the flow in the network. It is simply the number of packets of the particular flow has reached the destination successfully. So when the packets pass to the intermediate node, they also make the record of the serv_flow along with the ID of the flow in the locally maintained table of serv_flow. When the intermediate stations receive the packets of a new flow they set the serv_flow=1 with flow ID locally. When the packet finally reaches to the destination the destination node will reply with a ACK. On receiving the ACK, the intermediate node increment serv_flow of the concerned flow by 1 and forward the ACK to the downstream node upto source node of the particular flow. Hence on sending the 1st packet of a particular flow successfully to the destination, all the nodes passed by the flow have set the serv_flow=2. So each node keep record of the services obtained by each flow generated and passing through it. When the packets of two different flows but of same priority has been enqued in IFQ(interface queue) , the packet belonging to the minimum served flow is dequed . The node contends for the access of the channel with a contention window as follows.

$$CW = scaling\_factor * CW^i$$

Where , scaling_factor = serv_PF/ serv_BSF

serv_PF = number of service obtained by the current flow in question & serv_BOF = number of service obtained by the better of the other flows of same priority class & CW^i = contention window corresponding to the priority class 'i'. Hence station will reset the contention window for accessing the channel for the packets of a given flow with

$$CWmin = scaling\_factor * CW^i_{min}$$

$$CWmax = scaling\_factor * CW^i_{max}$$

From the above expression it may be observed that scaling factor will be less than 1, when service obtained by the present flow is less than the other flows. Hence it will contend for the channel with still less contention windows.

The working of the proposed protocol is described in flow charts (Figure 2 &3)

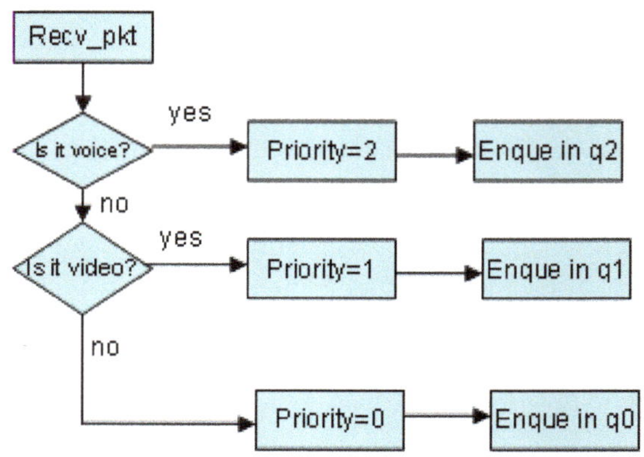

Figure:2: Flow Chart for Enque Operation

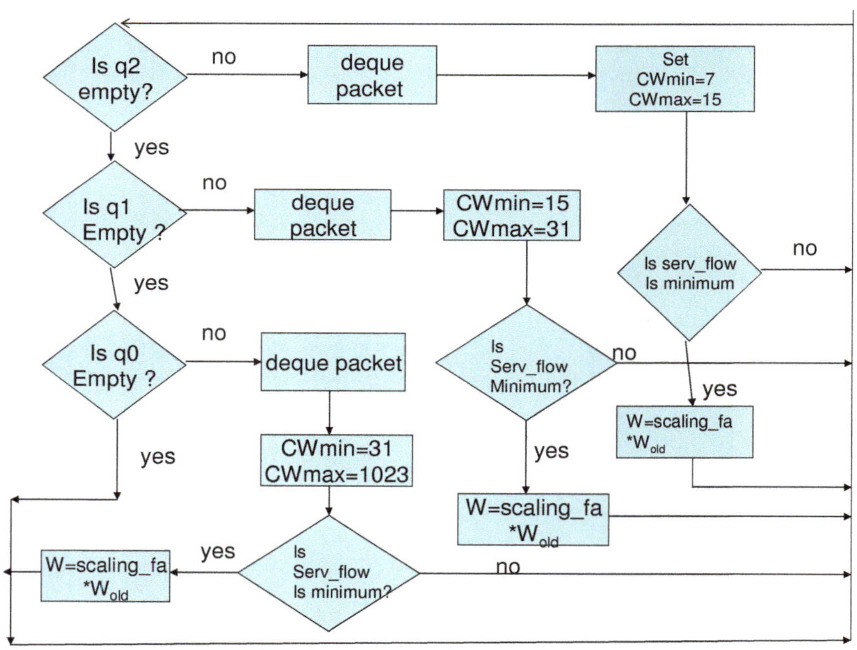

Figure: 3 Flow chart for Deque operation

3.2. Deque Operation:

Figure:3 explains the details of how the packets are dequed and processed.q2 the highest priority queue is checked first and if it found to be non –empty, the packet is dequed.The contention window is set for that prioritized traffic , which is typically taken as 7,15 in this work. Then CWsize is multiplied by the scaling factor. If the number of service obtained by the flow to which the packet belongs to is less compared to the minimum service among other flows then the CWsize will be reduced and the node will contend for the channel with a less window size for the packet belonging to less served flow. If the service obtained by the concerned flow is more than the minimum served flow the contention window size will be set more. The operations are repeated until all the queues are empty.

Illustration with an example: The protocol is explained by taking an example of a simple multihop adhoc network with different flows as shown in the Figure: 4

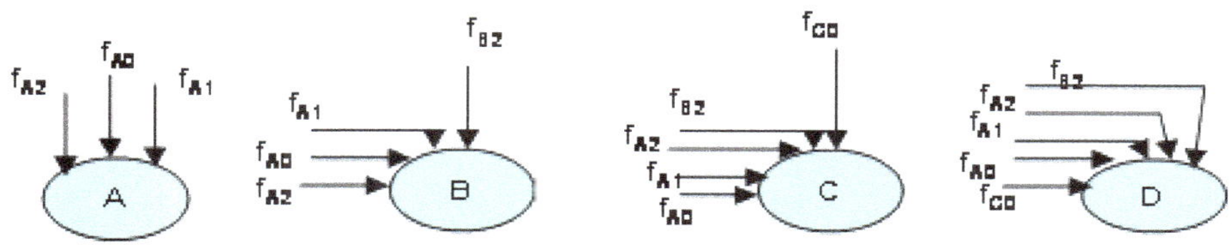

Figure: 4 : Simple multihop scenario with different flows.

Referring to above Figure:4, suppose a node A generates voice, video and data packets which belongs to the priority level 2, 1 and 0 in accordance with the proposed protocol. Let the individual flows be named as f_{A2} ,f_{A1} and f_{A0}.The packets are enqued in its own prioritized queue. Node 'B' generates voice packets belonging to the priority level 2 and let it be designated by flow ID f_{B2}. This node 'B' not only have to transmit from its own f_{B2}, but also forward the packets of f_{A2} ,f_{A1} and f_{A0} .Node 'C' generates data packets which belongs to the priority level '0' and let the flow ID be set as f_{C0}. The node 'C' forwards the packets of flows f_{B2} , f_{A2} ,f_{A1} and f_{A0} also transmits the packets from its own f_{C0}. Initially all the flows f_{B2} , f_{A2} ,f_{A1} , f_{A0} and f_{C0} service obtained by the flow (serv_flow) equals to 1.That means, serv_ f_{B2} = 1, serv_ f_{A2} = 1, serv_ f_{A1} = 1 , serv_ f_{A0} = 1 and serv_ f_{C0} = 1. Let node 'A' simultaneously have voice, video and data packets to transmit. According to the rule the packets belonging to the priority 2 will be dequed and node 'A' will contend to access the channel with CWmin, CWmax belonging to the priority 2. If the node becomes successful in accessing the channel and send the packets which consequently reach its intended destination 'D' then 'D' increment serv_f_{A2} by 1 that is equal to 1. This information is piggybacked in the ACK packet and send to the downstream nodes (towards source).When node 'C' receives the ACK, it set serv_f_{A2} =2 ; The node B and hence node A also set their serv_f_{A2} =2 on receiving the ACK .

Now node 'B' suppose want to transmit its own voice packets belonging to the flow f_{B2} at the same time node 'A' wants to transmit the packet belonging to f_{A2} . As serv_f_{A2} =2 and serv_f_{B2} = 1 at B, the packets belonging to the less served flow that is f_{B2} will be dequed.Node 'B' will contend for accessing the channel with contention window size with the scaling factor 0.5 where as node A will contend with the scaling factor 1. That means the node B will contend for the channel with less sized window as compared to node A .Hence node B will be in a better position to access the channel. Hence the packets belonging to a less served flow such as serv_f_{B2} will be transmitted thereby showing the fairness.

4. ANALYTICAL MODEL

In this paper, we have assumed that there are 'n' active stations in the network. Each station has implemented 3 queues of priority level 2, 1 and 0 .Hence there are total 'n*3' queue entities. This is equivalent to '3*n' stations in the legacy IEEE 802.11 DCF.

Further let it be assumed that any queue in any station has packets ready to send. Table:1 may be referred for the symbol and expression used here.

Table: 1: Symbols used in the analysis

Sl. No	Parameters	Description	Sl. No	Parameters	Description
1	P_c	Collision probabilty	16	T_{DIFS}	DIFS period
2.	P_b	Busy Probability	17	£	Propagation delay
3.	ζ_i	Transmission probability	18	$N_{c,i}$	Random variable representing the number of collisions before transmitting a frame
4	$P_{s,i}$	Success Probability	19	X_i	Random variable representing the time interval during which the counter reaches zero without considering the case when the counter freezes
5	S_i	Normalized throughput	20	F_i	Time duration that the backoff counter of a station freezes
6	E[P]	Average packet payload size	21	N_{Fi}	The number of time that the backoff counter freezes
7	n_i	No. of station	22	B_i	Backoff delay of a station for the priority class 'i'
8	σ	Slot duration	23	D_i	Random variable representing the frame delay
9	$T_{s,i}$	Successful transmission time	24	T_0	Time duration after collision that a station has to wait before sensing the channel again
10	$T_{c,i}$	Average collision	25	T_{timeou}	Duration of the ACK
11	T_H	Header tx time (both PHY and MAC)			
12	$T_{E[P]}$	Payload tx Time			
13	$T_{E[P^*]}$	$T_{E[P]}$ for longest frame			
14	T_{ACK}	Time to transmit the ACK			
15	T_{SIFS}	SIFS period			

Assuming P_c and P_b are both constant and independent , then in accordance with Bianchi's model [20] the bidirectional random process {s(i,t),b(i,t)} is discrete–time Markov chain and is depicted in the following Figure 5 .The state of each station in the priority class 'i' is described by {i,j,k}.

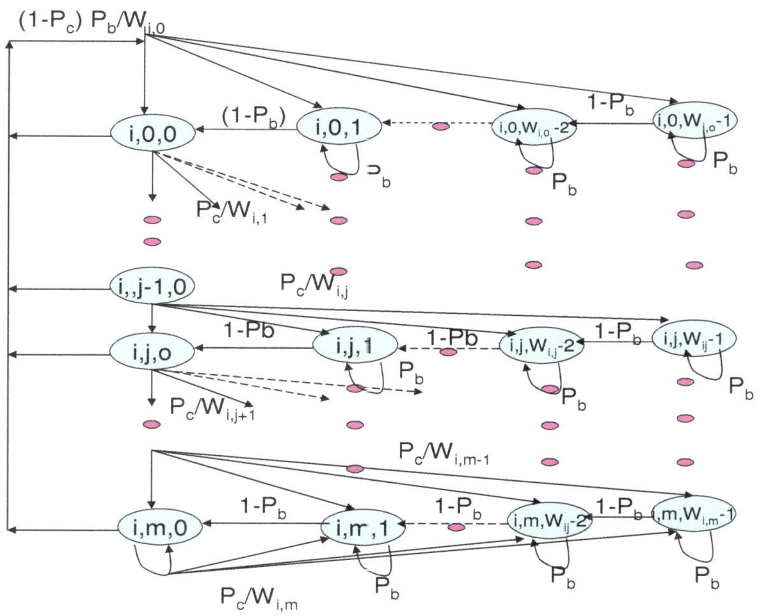

Fig 4: The state transition diagram for the priority class "i"

In accordance with the above Markov Model, the non –null one step transition probabilities are

$P\{(i,j,k) | (i,j,k)\} = P_b$ for $1 \le k \le W_{i,j} - 1$ & $0 \le j \le m$ ---------- (1)

The above expression indicates that the channel is busy and hence there is no decrement in the backoff counter.

$P\{(i,j,k) | (i,j,k+1)\} = 1 - P_b$ for $0 \le k \le W_{i,j} - 2$ & and $0 \le j \le m$ ------ (2)

This indicates that the channel is free and hence backoff counter is decremented by 1.

$P\{(i,0,k) | (i,j,0\} = (1 - P_c) * P_b / W_{i,0}$ for $0 \le k \le W_{i,0} - 1$ and $0 \le j \le m$ ------ (3)

The above expression account for the fact that a new packet following a successful packet transmission starts with backoff stage '0' for the priority class 'i' and the backoff is initially chosen in the range (0 , $W_{i,0} - 1$).

$P\{(i,j,k) | (i,j-1,0\} = P_c / W_{i,j}$ for $0 \le k \le W_{i,j} - 1$ and $1 \le j \le m$ ----- (4)

The above expression models the system after unsuccessful transmission.

$P\{(i,m,k) | (i,m,0\} = P_c / W_{i,m}$ for $0 \le k \le W_{i,m} - 1$ ----------- (5)

Let $b_{i,j,k} = \lim_{t\to\infty} P\{ s(i,t) = j, b(i,t) = k\}$;
where $0 \le j \le m$ and $0 \le k \le W_{i,j} - 1$ --- (6)
be the stationary distribution of the Markov Chain .
In steady state the following relations can be derived through chain regularities.

$b_{i,j,0} = P_c^j * b_{i,0,0}$ where $0 \le j \le m-1$ -----(7)

$b_{i,m,0} = P_c^{m} * b_{i,0,0} / (1- P_c)$ where $j=m$ ----- ----------------(8)

$b_{i,j,k} = \{(W_{i,j} - k)/ W_{i,j} \} *\{ 1/ (1- P_b)\} * b_{i,j,0}$; ------------(9)
Where $0 \leq j \leq m$ and $1 \leq k \leq W_{i,j} - 1$

By applying normalized condition $b_{i,0,0}$ can be determined as follows:

$$1 = \sum_{j=0}^{m} \sum_{k=0}^{W_{i,j}-1} b_{i,j,k} \quad \text{------- (10)}$$

Now plugging (7) to (9) in (10) , We have,

$$1 = \sum_{j=0}^{m} \sum_{k=0}^{W_{i,j}-1} b_{i,j,k}$$

$$= \sum_{j=0}^{m} b_{i,j,0} \sum_{k=0}^{W_{i,j}-1} \{(W_{i,j} - k)/W_{i,j}\} * \{1/(1 - P_b)\}$$
$$= \{1/(1 - P_b)\} \sum_{j=0}^{m} \{ b_{i,j,0} (W_{i,j} + 1)/2\}$$
$$= \left\{\frac{1}{1-P_b}\right\} * b_{i,0,0} /2[W_i (\textstyle\sum_{j=0}^{m-1}(2P_c)^j + \frac{(2P_c)^m}{1-P_c} + \frac{1}{1-P_c}] \quad \text{-----}$$

(11)

From the above equation we can get the value of $b_{i,0,0}$ as follows.

$b_{i,0,0} = 2(1- P_b) (1- 2P_c) (1- P_c)/\{(1-2P_c) (W_i+1)+ P_cW_i(1-(2P_c)^m\}$ ----- (12)

Now,

$$\zeta_i = \sum_{i=0}^{m} b_{i,j,0} = b_{i,0,0} /(1 - P_c)$$

$= 2(1- P_b) (1- 2P_c)/ \{ (1- 2P_c) (W_i + 1) +P_cW_i (1- (2P_c)^m\}$ ------------- (13)

$$P_b = 1 - \prod_{h=0}^{3}(1 - \zeta_i)^{nh} \quad \text{---------- (14)}$$

$$P_c = 1 - [\prod_{h=0}^{i-1}(1 - \zeta_h)^{nh}](1 - \zeta_i)^{ni-1}[\prod_{i+3}^{3}(1 - \zeta_h)^{nh}] \quad \text{--------.(15)}$$

<u>Saturation Throughput</u>

Saturation throughput is the maximum limit of the throughput that the system can carry in stable condition. In several access mechanisms the throughput increases with the offered load .It increases to a maximum value called maximum throughput and then starts decreasing as the offered load increases further. This causes in the practical impossibility to operate the random access scheme at its maximum throughput for a long period time. Hence it is meaningless to take the maximum throughput as the performance figure and hence *saturation throughput* is taken for performance measurement [22].

The probability with which a packet belonging to priority class 'i' is transmitted is given by

$$P_{s,i} - n_i\zeta_i(1 - \zeta_i)^{n-1} \prod_{h=0,h=i}^{3}(1 - \zeta_h)^{nh}$$

$= n_i \zeta_i / (1- \zeta_i)(1- P_b)$ ---------------- (16)

In accordance with Bianchi [19] the normalized throughput for priority class 'i' is given by

S_i = E [payload information transmitted in a slot time for the priority i class] / E [length of a slot time]

$= P_{s,i} E[P] / (1- P_b) \sigma + P_{s,i} T_{s,i} + [P_b - P_{s,i}] T_{c,i}$ --------- (17)

where,

$T_{s,i} = T_H + T_{E[P]} + T_{SIFS} + £ + T_{ACK} + T_{DIFS} + £$ -----------(18)

$$T_{c,i} = T_H + T_{E[P^*]} + T_{DIFS} + £ \quad \text{-------------(19)}$$

SATURATION DELAY

Saturation delay is the delay calculated in stable condition.

Average delay experienced by the frame ($E(D_i)$) = average time period for which the channel was busy + average backoff period+average time consumed while it experienced number of collision ----------- (20)

Average backoff period:($E(B_i)$)

Average backoff period , $E(B_i)$, can calculated as the sum of average time period taken by the counter to be decremented to zero without considering the case when the counter freezes($E(X_i)$) and average time duration for which the station freezes before transmitting the frame ($E(F_i)$)

Mathematically

$$E(B_i) = E(X_i) + E(F_i) \quad \text{---------- (21)}$$

Where , $E(X_i) = \sum_{j=0}^{m} \sum_{k=1}^{W_{i,j}-1} k.\, b_{i,j,k}$

$= \{b_{i,0,0} / 6(1 - P_b)\}\{(W^2_{i,0} (1- P_c - 3 P_c (4 P_c)^m) + 4P_c - 1)/ (1 - 4 P_c)(1 - P_c)\}$ ---- (22)

And $E(F_i) = E(N_{Fi})[(P_{s,i} / P_b) * T_{s,i} +(1- (P_{s,i}/ P_b) * T_{c,i}]$ -- ---------- (23)

Hence , Saturation delay and hence the equation (20) can be expressed by

$$E(D_i) = E(N_{c,i}) [E(B_i) + T_{c,i} +T_0] + E(B_i) + T_{s,i} \quad \text{------------- (24)}$$

Where $T_0 = T_{SIFS} + T_{timeout}$.

5. PERFORMANCE EVALUATION

In order to evaluate the protocol three different type of traffic is used. The voice traffic is assigned with highest priority 2.The video traffic is assigned the priority level 1 and the least priority level 'o' is assigned to the datagram traffic. Voice traffic usually considered as a service with CBR traffic. The data bit rate of the voice traffic is taken to be 64kbps.Video traffic is usually modeled as variable bit rate traffic. Continuous bit stream is generated for a certain holding period. The bit rate of different states is obtained from a truncated exponential distribution with minimum and maximum bit rate values. The holding times of the state are assumed to be statistically independent and exponentially distributed. The following table gives the numerical values for the video traffic which were taken for the analysis.

Table: 2 : Numerical values for video traffic model

PARAMETER	VALUE
Peak Bit Rate(PBR)	420 kbps
Average Bit Rate (ABR)	240 kbps
Minimum Bit Rate(MBR)	120 kbps
Maximum Packet Delay	75 ms

It is assumed that the data traffic arrive at each station following a poisson process with the mean value of λ. According to the traffic model defined above, the traffic parameters used for evaluating our protocol is listed in the following Table.3.

Table:3: The parameters are for the 2Mbps channel rate.

Sl. No.	Parameter	Symbol	Value (in μs)
1.	Slot time	T_{slot}	20
2.	Propagation delay	£	1
3.	Transmission time for physical pramble	$T_{preamble}$	72
4.	DIFS period	T_{DIFS}	50
5.	SIFS period	T_{SIFS}	10
6.	Time to transmit physical header	T_{PHH}	24
7.	Transmission time for MAC header	T_{MACH}	136
8.	Transmission time for ACK	T_{ACK}	152

Performance Metrics:

_A. Transmission Delays: The time elapsed between the instant the source transmits the packet & the instant the destination node receives it.

B. Saturation Throughput

Saturation throughput is the maximum limit of the throughput that the system can carry in stable condition. In several access mechanisms the throughput increases with the offered load .It increases to a maximum value called maximum throughput and then starts decreasing as the offered load increases further. This causes in the practical impossibility to operate the random access scheme at its maximum throughput for a long period time. Hence it is meaningless to take the maximum throughput as the performance figure and hence *saturation throughput* is taken for performance measurement. In this section the saturation throughput and delay obtained through analysis are given. In order to simplify the analysis it is assumed that the channel is error free and become corrupted when collision occurs.

6. RESULTS

Analysis is carried out here for three scenarios. Scenario_1 have 6 nodes, where number of nodes having voice, video and data traffic is 3, 2 and 1 respectively. Scenario_2 have 14 nodes out of which 7,5,2 are voice, video and data nodes. In Scenario_3, there are 15,10,5 nodes are with voice, video and data traffic respectively. The channel rate was taken to be 2 Mbps. Table I shows $T_{\Xi[P]}$ values.

Table.3: Transmission time of payloads under 2 Mbps

$T_{E[P2]}$: Time for pay load transmission(voice: cbr)	800 μs
$T_{E[P1]PBR}$: Time for pay load transmission (video: PBR)	15750 μs
$T_{E[P1]MBR}$:Time for pay load transmission (video:MBR)	4500 μs
$T_{E[P1]ABR}$: Time for pay load transmission (video: ABR)	9000 μs
$T_{E[P0]}$:Time for pay load transmission (data)	4092 μs

Table:4: Saturated Throughput for Voice Traffic(S2).

Scenarios	Values of S2
Scenario _1	0.6875
Scenario_2	0.2067
Scenario_3	0.052

Table:5: Saturation throughput for Video Traffic (S1)

	Peak bit rate(PBR)	Minimum bit rate(MBR)	Average bit rate(ABR)
Scenario_1	0.304	0.091	0.1785
Scenario_2	0.096	0.027823	0.055549
Scenario_3	0.0023	0.0006641	0.0013282

Table:6:Saturation Throughput for Datagram Traffic(S0)

Scenarios	Values of S0
Scenario_1	0.071
Scenario_2	0.0182
Scenario_3	0.00055

Table 4, 5 and 6 shows the calculated value of saturation throughput for different traffic categories & scenarios. It may be observed from the table that the throughput decreases with number of nodes & with less prioritized traffics. It is shown in Figure 6 & 7.

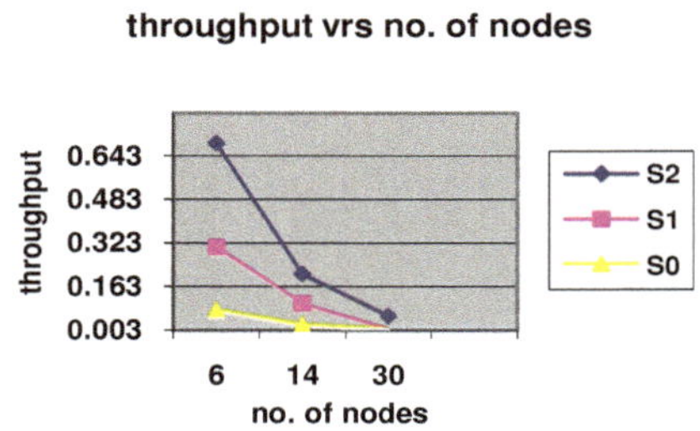

Figure:6: Throuput Vrs Number of nodes

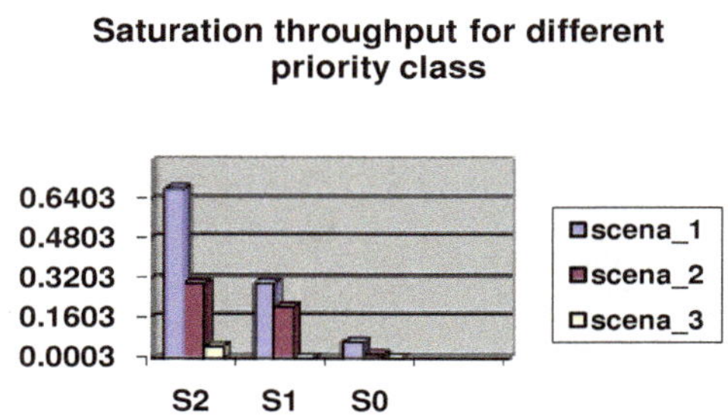

Figure:7: Saturation Throughput for different Priority Class.

The calculated values of transmission probabilities for different priority classes (ζ_0, ζ_1, ζ_2) are tabulated in Table 7. It may be observed that the values decrease with traffic categories and also with number of nodes. The variation of saturation throughput with respect to transmission probabilities shown in Fig.7,8 and 9.

Table: 7: Transmission Probabilities for different Traffic categories.

SCENARIOS	ζ_2	ζ_1	ζ_0
Scenario_1	0.0955	0.04775	0.024
Scenario_2	0.026	0.013	0.0065
Scenario_3	0.002	0.001	0.0005

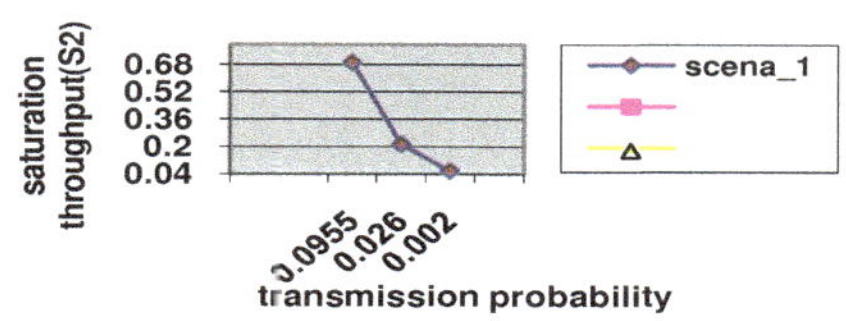

Figure:7: Transmission Probability Vrs Saturation Throughput for Voice Traffic

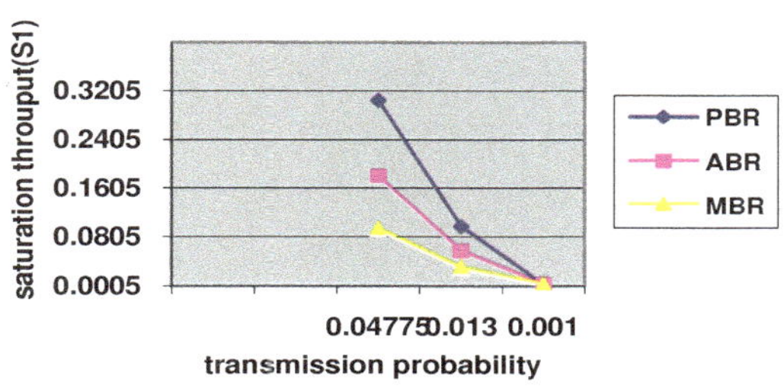

Figure:8:Transmission Probability Vrs Saturation Throughput for Video Traffic

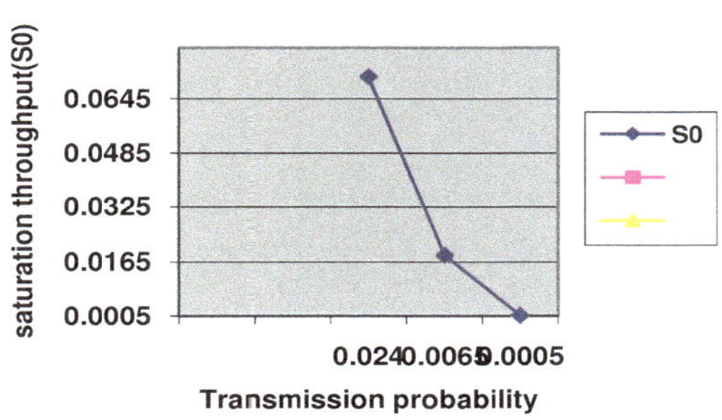

Figure:9:Transmission Probability Vrs Saturation Throughput for Data Traffic.

The delay experienced by various traffic categories voice, video and data is shown in the Table: 8. Here D2, D1 & D0 are the delays in ms experienced by Voice, Video & Data traffics respectively. The high priority traffic which is time sensitive is experiencing fewer delays. Delay increases for low priority traffic. The high priority traffic also experiences much delay when the number of nodes in the network increases.Figure:10: shows the delay variation for different priority classes with respect to node density. Delay increases at a slower rate with increase in node density in case of a high priority traffic such as voice where as it increases sharply for low priority traffic such as data.

Table:8: Delays experienced by different priority traffics.

Scenarios	D2	D1	D0
Scenario_1	7.89	12.566	59.720
Scenario_2	21.041	42.036	219.671
Scenario_3	60.491	106.016	583.196

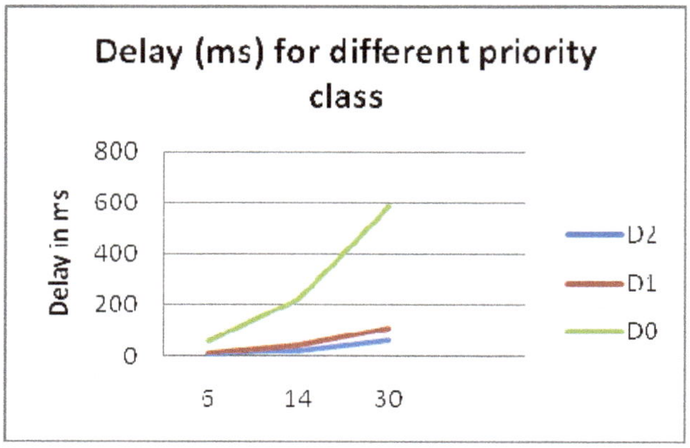

Figure:10: Delay experienced by different prioritized Traffic

7. CONCLUSION

In this paper, a medium access control protocol has been proposed for adhoc network. The protocol supports quality of service, exhibits fairness over multihop adhoc network. Throughput and delay experienced by different type of traffics were also analyzed. It was found that the high prioritized traffics have high throughput and experienced less delay as compared to the low prioritized traffics. Hence time sensitive traffics may be assigned high priority and its delay and bandwidth requirement may be served qualitatively. Although, Quality of Service assured protocol has been proposed in this paper, there are some works still to be addressed. In this paper, channel is assumed to be ideal. Hence the working of the protocol under noisy condition may need to be studied. As in Mobile Adhoc Network (MANET), any node may join or leave at any instant of time. Hence the topology of the network changes dynamically. The working of the protocol under such condition still remains to be explored.

REFERENCES

[1]. H.Y. Hsieh, R.sivakumar, performances comparison of cellular and multihop wireless networks: A quantitive study .- in proceedings of ACM SIGMETRICS, Boston , MA,USA June 2001.

[2]. IEEE, "Wireless LAN medium access control (MAC) and physical layer (PHY) specification," IEEE Std. 802.11, June 1999.

[3]. Shiann-Tsong Sheu and Tzu-Fang Sheu, "DBASE:A Distributed Bandwidth Allocation/Sharing / Extension Protocol for Multimedia over IEEE 802.11 Ad Hoc Wireless LAN," IEEE INFOCOM 2001.

[4]. J. L. Sobrinho and A. S. Krishnakumar, "Quality-of-Service in Ad Hoc Carrier Sense Multiple Access Networks," *IEEE Journal on Selected Areas in Communications,* 17(8):1353-1368, Aug. 1999.

[5]. A. Veres, A. T. Campbell, M. Barry, and L. Sun, "Supporting Service Differentiation in Wireless Packet Networks Using Distributed Control," *IEEE Journal on Selected Area in Communication,* 19(10):2081-2093,Oct. 2001.

[6]. Zhou Ying, A.L. Anand and L.jacob " A QoS Enabled MAC protocol for Multihop Adhoc Wireless Networks"IEEE international conference on performance , computing and communication`` 9-11 April 2003 pp 149-156.

[7]. Benveniste , M., 'Tiered contention multiple access' (TCMA), a QoS-based distributed MAC protocol 'The 13th IEEE International Symposium on Personal, Indoor and Mobile Radio Communications , Volume: 2 , 15-18 Sept. 2002 Pages:598 - 604 vol.2.

[8]. J. Sobrinho and A. Krishnakumar, "Real-time traffic over the ieee802.11 medium access control layer," in *Bell Labs Technical Journal*, 1996, pp. 172–187.

[9]. S. Mangold, S. Choi, P. May, O. Klein, G. Hiertz, and L. Stibor,"IEEE 802.11e wireless lan for quality of service (invited paper)," in *Proceedings of the European Wireless*, vol. 1, Florence, Italy,
Feb. 2002,pp. 32–39.

[10]. N. H. Vaidya, P. Bahl, and S. Gupta, "Distributed Fair Scheduling in a Wireless LAN," MobiCom 2000, pp. 167-178

[11]. T.Nandagopal, T.E.Kim, X.Gao, and V. Bharghavan, "Achieving MAC Layer Fairness in Wireless Packet Networks," MobiCom 2000,pp. 87-98.

[12]. A.Dugar & N.Vaidya "Priority and fair scheduling in WLAN" IEEE - MILCOM 2001 page 993-997 vol.2.

[13]. P. Gupta and P. R. Kumar, The Capacity of Wireless Networks, *IEEE Transaactions on Information Theory*, Vol 46, pp 388-404, March 2000.

[14]. S. Xu and T. Saadawi, "Does the IEEE 802.11 MAC Protocol Work Well in Multihop Wirelles Ad Hoc Networks?" *IEEE Communications Magazine*, June 2001.

[15]. G.Holland, N.Vaidya and P.Bahl " A Rate –adaptive MAC protocol for Multihop wireless Networks" ACM/IEEE Conference on Mobile Computing and Networking (MOBICOM'01) , Rome, Italy, July 2001.

[16]. D. Raguin, M. Kubisch, H. Karl, A. Wolisz " Queue-driven cut-through medium access in wireless adhoc network In proceeding of IEEE Wireless communications and Networking conference(WCNC) , Atlanta, Georgia, USA March 2004.

[17]. Shugong Xu, Tarek Saadawi "Revealing the problems with 802.11 medium access control protocol in multi-hop wireless adhoc networks" in www.elsivier.com computer networks - 2002

[18]. C-K Toh , V. Vassiliou etc " MARCH : A medium access control protocol for multihop adhoc network" in the proceedings of IEEE – MILCOM'2000 22-25 Oct. 2000 Pages:512 - 516 vol.1

[19]. F. Borgonovo, A. Capone, M. Cesana, L. Fratta "ADHOC MAC: a new, flexible and reliable MAC architecture for ad-hoc networks in the proceedings of IEEE WCNC `2003".

[20]. G. Bianchi, "Performance Analysis of the IEEE 802.11 Distributed Coordination Function," *IEEE J-SAC.* Vol. 18, Mar. 2000, pp. 535-547.

[21]. T. S. Ho and K. C. Chen, "Performance evaluation and enhancement of the CSMA/CA MAC protocol for 802.11 wireless LAN's", in *Proc. IEEE PIMRC*, Taipei, Taiwan, Oct. 1996, pp. 392-396

[22]. B. Bing, "Measured Performance of the IEEE 802.11 Wireless LAN," IEEE LCN'99

[23]. Y chen , Q.A.Zeng and D.P.Agrawal " Performance Analysis and Enhancement for IEEE 802.11 MAC Protocol" IEEE'2003

[24]. V. Bharghavan, A. Demers, S. Shenker, and L. Zhang. "MACAW: A Media-Access Protocol for Packet Radio". In *Proceedings of ACM SIGCOMM 1994.*

[25] Latkoski, P., Hadi-velkov, Z., & Popovski, B. (2006). Extended model for performance analysis of non-saturated IEEE 802.11 DCF in erroneous channel. In *Proceedings of International Conference on mobile Adhoc and Sensor Systems*, 783-788.

[26] Daneshgaran, F., Laddomada, M., Mesiti, F., Mondin, M., & Zanolo, M. (2008). Saturation throughput analysis of IEEE 802.11 in presence of non ideal transmission channel and capture effects. *IEEE transactions on Communications, 56*, 1178-1188.

[27] .Daneshgaran, F., Laddomada, M., Mesiti, F., & Mondin, M. (2008). Unsaturated throughput analysis of IEEE 802.11 in presence of non ideal transmission channel and capture effects. *IEEE transactions on Wireless Communications, 7*, 1276-1286.

A COMPARATIVE STUDY IN WIRELESS SENSOR NETWORKS

1 Hasan Al-Refai 2 Ali Alawneh 3Khaldoun batiha

1 Department of Computer Information Systems, Philadelphia University, Amman Jordan
2Department of Management Information Systems, Philadelphia University, Amman Jordan
3Dean of IT, Philadelphia University, Amman Jordan

ABSTRACT

Sensor networks consist of a large number of small, low-powered wireless nodes with limited computation, communication, and sensing abilities, in a battery-powered sensor network, energy and communication bandwidth are a precious resources. Thus, there is a need to adapt the networking process to match the application in order to minimize the resources consumed and extend the life of the network. In this paper, we introduce a comparative study in different routing algorithms that propose vital solutions to the most important issues that should be taken into account when designing wireless network which are reliability, lifetime, communication bandwidth, transmission rand, and finally the limited energy issue, so we will introduce their algorithms and discuss how did they propose to solve such of these challenges and finally we will do some evaluation to each approach.

KEYWORDS

Wireless sensor networks; Geographic routing; Localized routing; Power-aware routing, reliability, timeliness, query processing, redundancy, energy conservation, QoS.

1. INTRODUCTION

Recent advances in technology have made low-cost, low-power wireless sensors a reality. Sensor networks formed from such sensors can be deployed in an ad hoc fashion and cooperate to sense and process a physical phenomenon. As each sensor has a finite battery source, an important feature of sensor network is energy efficiency to extend the network's lifetime.

Thus, the unique characteristics and constraints present many new challenges for the development and application of WSNs. Due to the severe energy constraints of large number of densely deployed sensor nodes, it requires a suite of network protocols to implement various network control and management functions such as synchronization, node localization, and network security.

There are a common routing protocols that have several shortcomings when applied to WSNs, which are mainly due to the energy- constrained nature of such networks [3]. In most cases it is very difficult and even impossible to change or recharge batteries for the sensor nodes. WSNs are characterized with denser levels of sensor nodes deployment, higher unreliability of sensor nodes, and sever power, computation, and memory constraints.

In [1] the authors have proposed an algorithm that investigate how to make on-demand routing protocols adapt quickly to topology changes, this problem is important because such protocols use route caches to make routing decisions; it is challenging because topology changes are frequent. They propose proactively disseminating the broken link information to the nodes that have that link in their caches. Proactive cache updating is key in making route caches adapt quickly to topology changes, it is also important to inform only the nodes that have cached a broken link to avoid unnecessary overhead. Thus, when a link failure is detected, the goal is to notify all reachable nodes that have cached the link about the link failure. They define a new cache structure called a cache table to maintain the information necessary for cache updates. A cache table has no capacity limit; its size increases as new routes are discovered and decreases as stale routes are removed.

The algorithm is fully-distributed, adaptive to the topology changes and fault-tolerant, since there is always a way to handle broken link or node failure.

Moreover, this algorithm eliminates the overhead of flooding route request into the network and can construct the multipaths on the fly. Thus, this algorithm is scalable and efficient and it has less communication and storage overhead than other algorithms while providing greater resilience to node failures.

In [3, 9] the authors have proposed a fast, randomized, distributed algorithm for organizing the sensors in a wireless sensor network in a hierarchy of clusters with an objective of minimizing the energy spent in communicating the information to the information processing center, but in the other hand, they assume that the communication environment is contention and error-free, so sensors do not have to retransmit any data, in the other words the overall reliability of the system is on a stake, since there is no redundancy mechanism to cope with network failures.

So the main concentration in this approach is how to reduce the overall power consumption in the system to cope with limited battery power or battery constraints.

In [4, 7] the authors have proposed an algorithm that handle the fault tolerant and reliability requirements for data information in sensor networks, so in this approach each node has a maximum transmission range to avoid link failure due to the power limitations of sensors(small battery), also it uses Total Time to Live (TTL) to adjust the level of data reliability, so that the more TTL the more possible paths used by the data the more robust it would be in the face of link or node failures, and TTL is always chosen to be slightly higher than the diameter of a network for the sake of necessary redundant packet.

In [5, 8] the authors have proposed an algorithm that is very based on query-based sensor systems, in which a user would issue a query with QoS requirements in term of reliability and timeliness and expect a response within a specific time which is called the deadline, where the fault-tolerant for the overall system can be achieved by redundancy of path and source and the optimal path and source redundancy is chosen depends on the routing algorithm they mentioned.

They also determine the optimal redundancy level that could satisfy the QoS requirements while prolonging the lifetime of the wireless sensor network. Since sensor networks are constrained with resources, they develop a hop-by-hop data dissemination paradigm to dynamically form multiple paths for data delivery, instead of incurring extra overhead to first formulate multiple paths before data delivery.

2. THE PROPOSED ROUTING ALGORITHM

In [13] the authors have proposed a dynamic source routing protocol to address the cache staleness issue which limits the mobility of the overall network because of the delay that is, a node is not notified when a cached route breaks until it uses the route to send packets.

When a node detects a link failure, the goal is to notify all reachable nodes that have cached that link to update their caches. To achieve this goal, the node detecting a link failure needs to know which nodes have cached the broken link and needs to notify such nodes efficiently. This goal is very challenging because of mobility and the fast propagation of routing information.

The solution is to keep track of topology propagation state in a distribution manner. Topology propagation state means which node has cached which link. In a cache table, a node not only stores routes but also maintains two types of information for each route which are how well routing information is synchronized among nodes on a route and which neighbor has learned which links through a route reply. Each node gathers such information during route discoveries and data transmission, without introducing additional overhead.

This approach enables dynamic source routing (DSR) to quickly remove stale routes, thus reducing packet losses, delivery latency, and replies.

In [14,17] the authors have proposed the geographical power efficient routing (GPER) protocol where each node knows whereabouts of the nodes within its power range and each intermediate node locally identifies the most power efficient path to the sub destination within its reach and forwards the packet to the next node on this locally identified path. The packet however, may not follow this path exactly since the sub destination changes from node to node. Two mechanisms, force routing and perimeter routing, are applied in GPER to prevent infinite loops.

So, each sensor node makes local decisions as to how far to transmit; therefore, the protocol is power-efficient, localized, highly distributed, and scalable, in GPER, given a final destination, each node first establishes a sub destination within its maximum radio range, the node, however, may decide to relay the packet to this sub destination through an intermediary node or alter the sub destination if this will preserve power.

Three geographic multipath routing algorithms have been introduced (MGPSR, MGPERsub, and MGPERhop), which generate alternative paths probabilistically based only on local information. These algorithms identify and use different routes for different messages between a source and a destination, they take away from the load on the nodes that are on the preferred path between the end-points and distribute this load on the other nodes that are in the neighborhood. Of course, as these paths are less perfect than the preferred path, they are expected to introduce a degree of overhead, but experiments show that multipath routing improves the lifetime of the network, especially in cases where the communication is non-uniformly distributed in the network

In [1,20] the authors have proposed a distributed, randomized, clustering algorithm to organize the sensors in a wireless sensor network into clusters. Then they extend this algorithm to generate a hierarchy of cluster heads and observe that the energy savings increase with the number of levels in the hierarchy, they assume that sensors transmit at the same power level and hence they all have the same radio range, moreover they suppose that each sensor uses 1 unit of energy to transmit or receive I unit of data and finally, they assume that the network is fault-tolerant and the communication environment is contention and error-free and sensor do not have to retransmit any data.

A routing infrastructure has been used in this approach, hence when a sensor communicates data to another sensor, only sensors on the routing path forward the data, which means the route should be established first and then data can be transmitted which adds time overhead to the system at all.

The algorithm states that each sensor in network becomes a cluster head and advertises itself to the sensors within its radio range, and it is called the volunteer cluster head, this advertisement is forwarded to all the sensors that are no more than k hops away from the cluster head where k is chosen by the algorithm and any sensor that receives such advertisements and is not itself a cluster head joins the cluster of the closest cluster head, any sensor that is neither a cluster head no has joined any cluster itself becomes a cluster head and then it is called the forced cluster head and since this algorithm is a distributed one, there is no need for clock synchronization between the sensors at all even if they were in the same cluster. Since this algorithm is a hierarchy based then each level does the same and elects the cluster head in the same manner.

In [15,18] the authors have proposed an algorithm based on majority voting in sensor network, under the assumptions that the topology of the network is fixed, each node has a maximum transmission range to overcome the power limitations of sensors, the propagation delay on each functioning link is equivalent, and the processing delay in each node is non zero.

In this approach, when nodes are first deployed, the sink node sends a control packet informing each node of the value of TTL (total time to live) where TTL is chosen under a certain criterion that is, the more TTL the more possible paths used by the data the more robust to cope with link or node failures, on the other hand, TTL Is always taken to be equal to the diameter of the network to eliminate redundant traffic while broadcasting, but for the sake of necessary redundant packet to perform majority voting, TTL should be set to be slightly higher the diameter. The majority voting could be trigger whenever 3 packets with the same sequence number arrive.

In [16,19] the authors have proposed an algorithm based on a hop-by-hop data dissemination paradigm to dynamically form multiple paths for data delivery, instead of incurring extra overhead to first formulate multiple paths before data delivery. They assume that sensors have the same initial energy level and sensors group themselves into distinct clusters in the system for energy conservation purpose, each cluster has a cluster head (CH) which is responsible for data reading and aggregating from sensor within same cluster and the role of cluster head is rotated fairly among sensors in same cluster using a specific algorithm such as HEED or LEACH which will help to conserve energy and prolong the lifetime of the network so that sensors will not quickly deplete their energy, not only that but also the transmission power of a sensor even when it is a cluster head is reduced to a minimum level to enable it to communicate within one-hop radio range, thus a sensor node should use a multi-hop route protocol to enable it to communicate with another sensor distant away.

Two forms of redundancy are used in this approach to help fault tolerance and lifetime reasons which are path and source redundancy, that is, instead of having a single path to connect a source cluster to the processing center, m disjoint paths may be used, on the other hand, they use m sensor nodes to return readings to cope with sensor hardware and software faults.

In cluster network a source cluster head must send data information to the processing center and this will consume energy depends on the length of the path between CH and the processing cluster, if the overall system energy falls below a threshold value say threshold energy then the system is considered as having depleted its energy. That is, the consumed energy can be computed in terms of the energy to run the transmitter and receiver circuitry, the amplification energy that is used to achieve acceptable signal to noise ratio, the loss energy due to channel transmission.

Thus, the system lifetime can be defined here as the total number of queries the system can answer correctly until it fails due to either channel or sensor faults or when the system energy approaches the threshold level energy.

In [23] their work based on [21] to conduct a measurement study on a large-scale operating sensor network system, GreenOrbs, with up to 330 nodes deployed in the wild. they aim to understand how the sensor network performs when it scales to contain hundreds or even thousands of nodes.

3. EVALUATION

The approach in [13] achieves fault-tolerance to the wireless network against link or node failures by using caching technique in dynamic source routing which quickly removes stale routes, thus it reduce packet losses, delivery latency, and replies.

The approach in [14,17] achieves power efficient routing method which enables the nodes to adaptively chooses the transmission range to minimize the power consumption of the sensor as well as of the non-participating listeners, and by using three families of probabilistic geographical multipath routing protocols, they improve the network lifetime of wireless network, in addition, all of these protocols are highly power efficient, distributed and scalable, on the other hand, the reliability of the network still not handled in this approach.

The approach in [1,19] achieves conservation to the total energy by minimizing the total energy spent in the system to communicate the information gathered by these sensors to the information processing center and even this approach has a significant improvement on a time complexity over many clustering algorithms and this makes this approach suitable for networks of large number of nodes, it is still has a problem with reliability since it states that the communication environment is contention and error free, moreover it neglects the fact that the cluster heads in this architecture spend relatively more energy than other sensors because they have to receive information from all the sensors within their clusters, aggregate this information and then communicate to the higher level cluster heads or the information processing center.

The approach in [15,18] achieves reliability and fault tolerance for a simple network with fixed topology which is free of crash failures but it is still unable to cover dynamic and sophisticated networks reliabilities which are changing very quickly, on the other hand it, didn't handle the problem of power limitations of sensors or manage energy usage.

The approach in [16,19] achieves quality of service requirements in terms of reliability and timeliness while maximizing the lifetime of query-based sensor networks by using hop-by-hop packet broadcasting protocol, in the other hand this approach is based on the average behavior without considering dynamics such as more energy may be consumed by some sensors over others or some sensors may fail earlier than others. It doesn't handle multiple QoS levels from multiple queries and the bandwidth here is a bottle-neck, since much redundancy is achieved by hop-by-hope packet broadcasting protocol.

The approach in [23] achieves two contributions, first, they argue that they are the first to conduct a long term and large-scale measurement study on an operating sensor network in the wild. Where they present observations across a variety of layers in the network that provide research community empirical experiences on how practical problems affect when the sensor network scales.

Second, based on their basic findings from the system measurement, they propose and initially attempt to validate three conjectures that provide guidelines for future algorithm and protocol designs with larger scale sensor networks. That is, 1- they thought it possible that some of the intermediate nodes bottleneck the entire network, and most of currently used indicators may not accurately capture them; 2- most of the wireless links in large scale sensor networks are physically stable. But due to the dynamic features mainly come from the inherent concurrency of network operations that need to be further investigated and considered in designing scalable network protocols. 3- the environment, while with insignificant dynamics, has an unpredictable impact on thesensor network under it. They suggest that an event based routing structure can be trained optimized and thus better adapt to the wild environment when building a large-scale sensor network.

REFERENCES

[1] Seema Bandyopadhyay and Edward J, Coyle, An Energy Efficient Hierarchical Clustering Algorithm for Wireless Sensor Networks.

[2] D.B. Johnson, D.A. Maltz, Dynamic source routing in ad hoc wireless networks, in: Mobile Computing, 1996, pp. 153–181.

[3] S. Murthy, J.J. Garcia-Luna-Aceves, An efficient routing protocol for wireless networks, Mobile Networks and Applications 1 (2) (1996) 183–197.

[4] V.D. Park, M.S. Corson, A highly adaptive distributed routing algorithm for mobile wireless networks, in: INFOCOM, 1997, pp. 1405–1413.

[5] C. Perkins, Ad hoc on-demand distance vector routing, in: MILCOM '97 Panel on Ad Hoc Networks, 1997.

[6] C. Perkins, P. Bhagwat, Highly dynamic destinationsequenced distance-vector routing (DSDV) for mobile computers, in: SIGCOMM, 1994.

[7] A. Rao, S. Ratnasamy, C. Papadimitriou, S. Shenker, I. Stoica, Geographic routing without location information, in: MobiCom, 2003, pp. 96–108.

[8] V. Rodoplu, T. Meng, Minimum energy mobile wireless networks, in: ICC, 1998, pp. 1633–1639.

[9] L. Li, J. Halpern, Minimum energy mobile wireless networks revisited, in: ICC, 2001.

[10] Q. Li, J. Aslam, D. Rus, Distributed energy-conserving routing protocols for sensor networks, in: IEEE HICSS, 2002

[11] J.H. Chang, L. Tassiulas, Energy conserving routing in wireless ad hoc networks, in: INFOCOM, 2000, pp. 22–31.

[12] Y.B. Ko, N.H. Vaidya, Location-aided routing (LAR) in mobile ad hoc networks, in: MobiCom, 1998, pp. 66–75.

[13] Xin Yn, Distributed cache updating for the dynamic source routing protocol, IEEE TRANSACTIONS ON MOBILE COMPUTING, VOL. 5, NO. 6, JUNE 2006

[14] Shibo Wu, K. Selc, uk Candan, Power-aware single and multipath geographic routing in sensor networks. Elsevier, Ad Hoc Networks 5 (2007) 974–997

[15] Jen-Yeu Chen, Yen-Shiang Shue, Hakeem Ogunleye, Saurabh Bagchi, A comparative study on data fault tolerant requirements for data propagation in sensor networks.

[16] Anh Phan Speer Ing-Ray Chen Virginia Tech, On Optimal Path and Source Redundancy for Achieving QoS AND Maximizing Lifetime of Query-Based Wireless Sensor Networks

[17] Haojun Huang*, Guangmin Hu, Fucai Yu. Energy-aware multipath geographic routing for detouring mode in wireless sensor networks. European Transactions on Telecommunications. Volume 22, Issue 7, pages 375–387, November 2011

[18] Ehsan, S. A Survey on Energy-Efficient Routing Techniques with QoS Assurances for Wireless Multimedia Sensor Networks. Communications Surveys & Tutorials, IEEE. Second Quarter 2012

[19] Debnath Bhattacharyya , Tai-hoon Kim , and Subhajit Pal. A Comparative Study of Wireless Sensor Networks and Their Routing Protocols. Journal of MDPI Sensors. 2010.

[20] Guangjie Han, Lei Shu, Jianhua Ma, Jong Hyuk Park, Jianjun Ni. Power-Aware and Reliable Sensor Selection Based on Trust for Wireless Sensor Networks . Journal of Communications, Vol 5, No 1 (2010)

[21] T. He, P. Vicaire, T. Yan, Q. Cao, G. Zhou, L. Gu, L. Luo, R. Stoleru, J. A. Stankovic, and T. F. Abdelzaher, "Achieving Long-Term Surveillance in VigilNet," TOSN, vol. 5, no. 1, pp. 1–39, 2009.

[22] Shio Kumar Singh , M P Singh , and D K Singh, "Routing Protocols in Wireless Sensor Networks –A Survey", International Journal of Computer Science & Engineering Survey (IJCSES), November 2010.

[23] Does Wireless Sensor Network Scale? A Measurement Study on GreenOrbs Yunhao Liu, Yuan He1, Mo Li, Jiliang Wang1, Kebin Liu, Lufeng Mo, Wei Dong, Zheng Yang, Min Xi, Jizhong Zhao, Xiang-Yang Li. 2013 Zhejiang University. Singapore.

13

DCT & DWT IMAGES COMPRESSION ALGORITHMS IN WIRELESS SENSORS NETWORKS: COMPARATIVE STUDY AND PERFORMANCE ANALYSIS

Oussama Ghorbel[1] Walid Ayedi[1] Mohamed Wasim Jmal[1] and Mohamed Abid[1]

[1]National Engineers School of Sfax, Sfax University, Tunisia. CES Research Unit

oussama.ghorbel@gmail.com , ayediwalid@ieee.org ,
wassim.jmal@gmail.com , mohamed.abid@enis.rnu.tn

ABSTRACT

The recent availability of inexpensive hardware has enabled the new research field of wireless sensor networks. This is a network of interconnected devices, capable of retrieving images from the environment. The nodes, in this type of network, have very limited resources, in terms of processing unit, bandwidth and energy. Efficient coding of the image content is therefore important. In Wireless Sensors Network, many image compression algorithms are used. Indeed, the most popular methods such as JPEG or JPEG2000 can yield higher energy consumption than when transmitting uncompressed images. However, the most of the compression algorithms are inapplicable on sensor nodes because the limitation in terms of memory as well as processor speed.

To solve this problem, we use DWT and DCT image compression algorithm that allows an efficient trade-off between energy consumption and image distortion. These experimental results demonstrated by the performance evaluation are in terms of image quality, execution time and memory space, as shown in this paper.

KEYWORDS

Wireless Sensors Networks (WSNs), Image compression, Discrete Wavelet Transform (DWT), Discrete Cosine Transform (DCT), Computational complexity, Energy consumption, Experimentation.

1. INTRODUCTION

Uncompressed data (graphics, audio and video) requires considerable storage capacity and transmission bandwidth. The demand for data storage capacity and data transmission bandwidth continues to outstrip the capabilities of available technologies, yet the rapid change in terms of the mass storage density, processor speed and system performance digital communication.
Previously, to capture data from the natural or built environment are not possible, but actually, with Wireless Sensor Networks (WSN), which is a new tool to sense all type of data, all is possible. In WSN, a tiny embedded device has the capability for sensing, processing and wireless communication.
Thanks to recent advances in microelectronics and wireless communications, it is predicted that wireless sensor networks (WSN) will become ubiquitous in our daily life and they have already been a hot research area for the past couple of years [1]. A wide range of emerging WSN applications, like object detection, surveillance, recognition, localization, and tracking, require vision capabilities [2].
Although, the hardware prerequisites are met, application aware and energy efficient algorithms for both the processing and communication of image have to be developed to make vision

sensor applications feasible [3]. Most of the work in literature is devoted to image processing (data extraction, compression and analysis) (Tang and Raghavendra (2004); Magli et al. (2003); Song et al. (2006); Wagner et al. (2003)) while the case of image transmission over WSN is still in an earlier stage of investigation.
We propose in this paper, two image transmission scenario which both are based on discrete wavelet transform (DWT) and discrete cosine transforms (DCT) [4, 5] driven by energy efficiency considerations suitable for WSN.

In many applications based on image compression and transmission, intermediate nodes consuming much more energy to forwarding data toward the sink than nodes collecting and forwarding data such as temperature value. As the radio transceiver is one of the most power greedy components of sensor nodes, compression seems a natural answer to the image transmission problem over a WCSN. Unfortunately, most of the compression algorithms are inapplicable on sensor a node which is due to resource limitation in terms of memory or processor speed. For instance, Ferrigno et al. have shown in [6] that the energy consumption involved by popular algorithms such as JPEG, JPEG2000 or SPIHT could be even greater than for the uncompressed image case [7,8,9]. As a consequence, special challenges for energy-efficient image compression in WSNs must be addressed to maintain a long network lifetime. These nodes are equipped with a processing unit with limited memory and energies resources and a communication unit, usually a radio transceiver. Nodes are powered by small batteries which generally cannot be changed. The application of multimedia (image, video, etc.) on wireless sensor networks is being, these days, a great requirement for the research and industrial community [10]. The current researches deal with image processing like data extraction, image processing and analysis [11]. Thus, the case of image compression and image transmission over WSN is not a new concept because there are many researchers who worked on it [12].

Our work is to experimentally test two techniques for image compression to ensure that they are effective or not for image transmission through wireless sensor network. The two selected techniques are from two different families which are: discrete wavelet transforms (DWT) and discrete cosine transforms (DCT). DWT is an operation that is applied on the entire image and allows image decomposition into separable subbands for multi-resolution representation purposes but DCT is applied on each block with size 8x8 and not on the entire image which allows a high compression ratio. The implementations of these methods on a real wireless sensor platform are allowed to reconcile with the real problems related to image processing applications. Wireless Sensor Network (WSN) [13] has been considered as an active area of research which many researcher worked on it and a wide range of applications have been developed. Sensor nodes are mainly characterized by their scarce resources and limited energy.

The rest of this paper is organized as follows: A state of the art, which deals with the various researches related to wireless sensors network, was presented in section 2. Section 3 and 4, gives an idea about compression methods and their advantages in general and in wireless sensor network. Section 5, it's a performance evaluation that describes the real implementation of DWT and DCT on TelosB sensor nodes. Section 6 and Section 7 present the applied scenario for the two compression technique DWT and DCT. A summary table between DWT and DCT is deals in Section 8. Section 9 shows quantity of consumed energy by the TelosB mote and how long time the network can be survived in section 10. Conclusion is drawn in section 11.

2. Related Work

Recently, WSN (Akylidiz et al. (2002)) has been an active area of research and a wide range of applications have been developed. Sensor nodes are mainly characterized by their scarce resources and limited energy. As a result, a considerable effort has been given to energy

efficient data transmission schemes ranging from the hop by hop medium access control level (Ye and Heidemann (2003); Langendoen and Halkes (2005)) to that of sensor to sink data delivery (Heinzelman et al. (2000); Tian and Georganas (2003);Kim et al. (2004); Liu et al. (2004)).
Out of concern for energy efficiency, both image data processing and transmission must be considered. A lot of work has been done in data processing research area. Recent results on data compression and issues in their deployment in wireless sensor networks have been exposed by Tang and Raghavendra (2004) and Mishra et al. (2007). Chiasserini and Magli (2002); Gerla and Xu (2003); Magli et al. (2003) adopted JPEG with change/difference compression. This latter has the advantage of low processing power needs, however, it does not support error control.
In order to save energy, mainly in a densely deployed WSN, many proposals (Song et al. (2006); Wagner et al. (2003); Gerla and Xu (2003); Wu and Abouzeid (2004)) adopt a distributed coding approach where neighboring sensor nodes cooperate in performing image coding. However, a significant exchange of data, which would waste energy, may be necessary. Maniezzo et al. (2002) addressed the tradeoff between computing and communication and show that there are an optimal number of nodes involved in a distributed coding process which minimizes the total energy consumption. Wu and Abouzeid (2004) adopted a multi-layer coding using JPEG2000 based on wavelet compression. Two methods of parallelizing the compression process are proposed and compared in terms of consumed energy and image quality. Yu et al. (2004) used DWT in order to obtain different layers with different quality. There is a focus on point to point image transmission with energy saving considerations. The optimal number of layers to be transmitted and the optimal strategies for each layer are determined thanks to an algorithm that minimizes the overall processing and transmission energy consumption given the expected end-to-end distortion constraint. This algorithm has the following inputs: the estimated channel condition, the characteristics of image content, and the set of available coding and transmission strategies. Other approaches to save energy use buffering techniques (Magli et al. (2003); Wanghong and Nahrstedt (2003)). Another work proposed by Wanghong and Nahrstedt (2003)), they realized a scenario in which a source node buffers encoded frames and transmits them in bursts in order to better exploit idle intervals of the processor and network card. The most current research related to image compression in wireless sensor networks [14] are limited to the evaluation by simulation. The authors have used several methods in the compression CWHN as LBT, SPIHT [15], ISEC [16] tested by simulation and tested on a real platform on Mica2. For instance, ISEC makes a compression method at the source, which uses a coding block of 2x2 pixels and removes one pixel from the 4 to minimize the compression rate and then finds the missing pixel using three present pixels. In [17], there is a comparison between two models of selection zonal coefficient of the DCT, one using a square shaped area and the other a triangular area.

In our work, a considerable effort has been given to compare two models for image compression to ensure they are effective or not for image transmission through wireless sensor network called DWT and DCT. In our work, and compared to other projects in WSN, we focuses more on low-bitrate image transmission, image quality and execution time over long-range indoor and outdoor sensor networks.

The aim of our paper is as follow: Firstly, we compare two methods of compressing images which are the Discrete Cosine Transform (DCT) and Discrete Wavelet Transform (DWT) and test their capability on wireless sensor networks (WSNs). Secondly, implement these methods on a real WSNs platform with TelosB sensor type. Thirdly, we execute performance evaluation and we compare the two methods in terms of image quality, execution and transmission time, packet lost and memory usage and finally energy consumption.

3. IMAGE COMPRESSION

3.1 Preview

To solve the problem of reducing the amount of data required to represent a digital image, we use Image compression method to remove the redundant data. Based on mathematical theory, we transforming a 2-D pixel array into a statistically uncorrelated data set which is applied prior to storage or to transmit the image. After some time, the compressed image is decompressed to reconstruct the original image. Image compression is recognized as a famous technology used until now to solve the compression problem. It plays a major role in many diverse applications, including remote sensing (weather and earth applications), medical imaging, facsimile transmission [18], video-conferencing, military and space application.

3.2 Fundamentals of image compression

To minimise a given quantity of information, we use a compression method which is refers to the process of reducing the amount of data. The neighboring pixels are correlated and therefore contain redundant information. So, the solution is to find less correlated representation of the image. There are two fundamental components of compression which are the redundancy and irrelevancy reduction: first, aims to removing duplication from the signal source and second, remove parts of the signal that will not be noticed by the signal receiver.

We show now a mathematically quantifiable entity but not an abstract concept. If n1 and n2 denote the number of information-carrying units in the two data sets that represent the same information, the relative data redundancy RD of the first data set (the one characterized by n1) can be defined as: RD = 1 - 1/ CR, where CR called the compression ratio, so CR = n1/n2.

For the case when n2 = n1 and RD = 0, the first representation of the information contains no redundant data. Then, if n2 << n1, so CR -> ∞ and RD ->1, implying significant compression and highly redundant data. Finally, in the case of n2 >> n1, CR ->0 and RD -> $-\infty$, we conclude that the second data set contains much more data than the original representation. Generally CR =10 (10:1) defines that the first data set has 10 information carrying units for every 1 unit in the second or compressed data set. Thus, the redundancy of the data in the first data set is redundant with respect to the second one [19].

In order to be useful, a compression algorithm has a corresponding decompression algorithm that reproduces the original file once the compressed file is given. There are many types of compression algorithms developed.

3.3 Image Compression and Reconstruction

Three basic data redundancies can be categorized in the image compression standard.

- Spatial redundancy which is due to the correlation between neighboring pixels.
- Spectral redundancy due to correlation between the color components.
- Psycho-visual redundancy which is due to properties of the human visual system.

The spatial and spectral redundancies are present because certain spatial and spectral patterns between the pixels and the color components are common to each other, whereas the psycho-visual redundancy originates from the fact that the human eye is insensitive to certain spatial frequencies. The principle of image compression algorithms are (i) reducing the redundancy in the image data and (or) (ii) producing a reconstructed image from the original image with the introduction of error that is insignificant to the intended applications. The aim here is to obtain an acceptable representation of digital image while preserving the essential information contained in that particular data set [20].

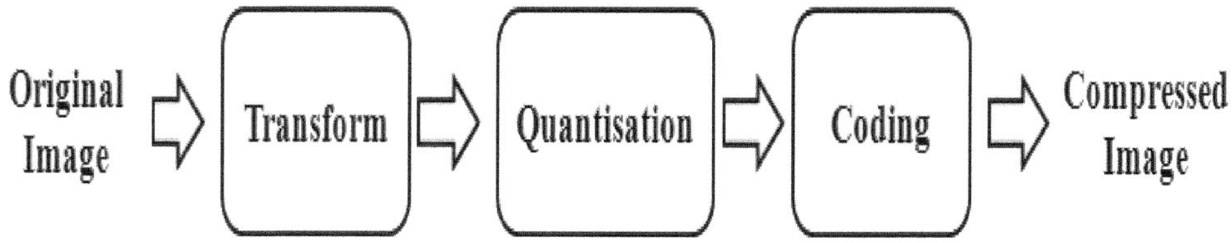

Figure 1. Image Compression System

The problem faced by image compression is very easy to define, as demonstrated in figure 1. First the original digital image is usually transformed into another domain, where it is highly de-correlated by using some transform. This decorrelation concentrates the important image information into a more compact form.

The compressor then removes the redundancy in the transformed image and stores it into a compressed file or data stream. In the second stage, the quantisation block reduces the accuracy of the transformed output in accordance with some pre- established fidelity criterion. Also this stage reduces the psycho-visual redundancy of the input image. Quantisation operation is a reversible process and thus may be omitted when there is a need of error free or lossless compression. In last step of the data compression model the symbol coder creates a fixed or variable-length code to represent the quantizer output and maps the output in accordance with the code.

A variable length code is commonly used to represent the mapped and quantised data set. It assigns the shortest code words to the most frequently occurring output values and thus reduces coding redundancy. The operation in fact is a reversible one. The decompression reverses the compression process to produce the recovered image as shown in figure 2. The recovered image may have lost some information due to the compression, and may have an error or distortion compared to the original image.

Figure 2. Image Decompression System

4. Compression Method for Sensor Networks

Compression method reduces the amount of data that the source will be transmitted. The data processing at the source is also necessary to anticipate the possible loss of information during the transmission of packets to the desired node, or to control the amount of data to send (the quality of the image) under the conditions of network such as the distribution of nodes and the distance separating them. There are many compression method used in literature for digital signal processing applications. We use in our work (in section IV) the two DWT and DCT transforms, and outline their relevance to the implementation of compression algorithm.

Data compression has several advantages such as:

- Extending the life of the source node. Unless the source has data to transmit, and consumes less energy at the radio transceiver. This is true as the complexity of the compression algorithm adopted is reliable enough to be profitable. The compression process should not cost more in terms of energy gain that it brings about communication, if the presence of a compression process could reduce the lifetime of the node.

- Extending the life of the intermediate nodes. For the same reasons, reducing the amount of data at the source will be necessarily beneficial to the nodes responsible for relaying packets between the source and the destination node. They receive fewer packets of data, so they have fewer packets and acknowledgments to transmit.
- Reduced risk of congestion. A decrease in the amount of data transmitted over the network will lead to reduced risk of network congestion, thus a decrease in packet loss and transmission delays.
- Tolerance to losses. Some reinforcement on tolerance to packet loss can be achieved by applying some processing mechanisms at the source, such as mixing or tattoo image.

5. Performance Evaluation

We evaluate our work on the real platform named TelosB nodes. The comparison will be very detailed depending on the results obtained through the PSNR values taking into account studies on image quality. In our experiment, we evaluate two compression techniques in image processing applications. We focus on the following question: Are these techniques valid for the context of WSNs? For there, we follow the steps of scenarios, measurement parameters and materials used that allow us to do our assessment.

The performance of the two compression algorithm has been evaluated considering a JPEG compression chain that is including the quantization and Huffman tables. In the following, we discuss about performance of the two approaches regarding quality of the reconstructed images and energy consumption.

Experimental results in terms of quality are given in Figure 5 which represents image Lena by DCT [21] and DWT [22] for different values of k. Obviously, even if the overall visual quality is good, image distortion increases as k decreases for both proposed methods. Furthermore, DWT and DCT provide different level of quality (visually speaking and with respect to the PSNR) for large values of k (i.e., 2 and 8 here). On the other hand, for small values of k the DWT method offers a better performance, which is normal since more coefficients are processed.

To discuss energy consumption, we have considered the TelosB mote [23], an ultra-low power sensor plateform developed by the University of California, Berkeley. TelosB mote consists of a MSP430 16- bit microcontroller and a 802.15.4-compliant CC2420 radio transceiver. The microcontroller operates down to 1.8V and consumes 3 mW in active mode (clock speed of 8 MHz). The radio transceiver has a transmit power of 35 mW at 0 dBm and provides a data rate of 250 kbps. We used also the TinyOS operating system to validate and measure the performance of our proposal. The TelosB motes have the following characteristics.

Tableau 1. Characteristics of Sensors Nodes

Manufacturer	**Processor**	**Program Memory**	**RAM**
Crossbow	IT MSP430	48 KB	10 KB
Clock	**Radio Unit**	**Band/Data rate**	**Max Consumption**
8MHz	Chipcon CC2420	2.4 GHz / 250 kbps	1.8 mA

5.1. The Process of the Algorithm Implementation using DCT

Image compression at the sensor node includes several steps. First, image is transformed into a format suitable for image compression. Then, the image is split into 8x8 blocks. The next step involves transformation of the block into the frequency plane. This is done by using a fast discrete cosine transform (FDCT). To exploit spatial correlation between pixels, we use this transform. Most of information is concentrated to a few low-frequency components after the transformation.

To reduce the number of bits needed to represent the image, these components are then quantized. This step will lower the quality of the image by reducing the precision of the components. The trade-off between quality and produced bits can be controlled by the quantization matrix, which will define the step size for each of the frequency component. We use ZigZag method to put the most likely non-zero components first and the zero components in the bit-stream. We use then entropy coding which is a combination of variable length coding and Huffman encoding. Finally, we create a suitable data packet for transmission over the wireless sensor network.

5.2. The Process of the Algorithm Implementation using DWT

For the second algorithm, the image is transformed into a suitable format to an image compression. On the first level, the filters divide the input image into four non-overlapping multi-resolution sub-bands LL, LH, HL and HH. The sub-band LL represents the coarse-scale DWT coefficients while the sub-bands LH, HL and HH represent the fine-scale of DWT coefficients. We show that each subband contains the low-pass information and the others contain high-pass information of horizontal, vertical and diagonal orientation. The next step is quantification and coding of subbands used to reduce the number of bits needed to represent the image. This step will lower the quality of the image but not like DCT. The quality will depend on the value of the quantization used. The bit plane coding and subbands provides various coding modes, the compressed image can indeed be represented by increasing resolution or by increasing quality. The next step is arithmetic coding. It is a variable length coding. Unlike other encodings, it encodes the source message fully and represents a single number. Finally, we create a suitable data packet for transmission over the wireless sensor network.

6. Applied Scenario for the Two Compression Techniques DWT and DCT

6.1. First Scenario

In this scenario, we have used 2 TelosB motes. Compression and decompression are performed respectively at the source and destination nodes. For radio communication, our implementation is based on 802.15.4 PHY and MAC layers which is a standard protocol for low-rate wireless personal area networks (LR-WPANs).

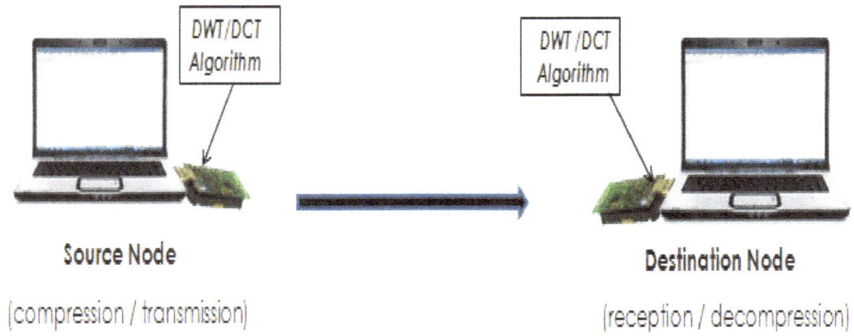

Figure 3. From the source to destination nodes using radio communication

There are different images uses for experimentation, we choose for example Lena to do the different scenario.

Figure 4. Original Image

6.2. Peak Signal to Noise Ratio (PSNR)

As a measure of the quality of image, the peak signal to noise ratio (PSNR) is typically used. This PSNR ratio expresses the difference in quality among the original Lena image and the decomposed one, while the higher the PSNR is, the better the quality of the decomposed image is. The expression of PSNR in decibels (dB) is given below in (1).

$$PSNR = 10 \log_{10}\left(\frac{255^2}{MSE}\right) \quad (1)$$

$$where: \quad MSE = \frac{1}{m\,n}\sum_{i=0}^{m-1}\sum_{j=0}^{n-1}\left\|I_0(i,j) - I_r(i,j)\right\|^2$$

$$m, n = image\ size > 0,$$

$$I_0 = original\ value,$$

$$I_r = compressed\ value$$

The following figure illustrates the obtained PSNR by varying nodes distance for 64x64 images. According to these figures, images quality is better using DWT regardless of the distance and resolution. For distances higher than 30m, a higher gap of PSNR values is obtained. Quality is reduced when increasing distances between source and destination nodes.

The reconstructed images (Lena) for different distance (d) between sensor nodes are given by figure 5.

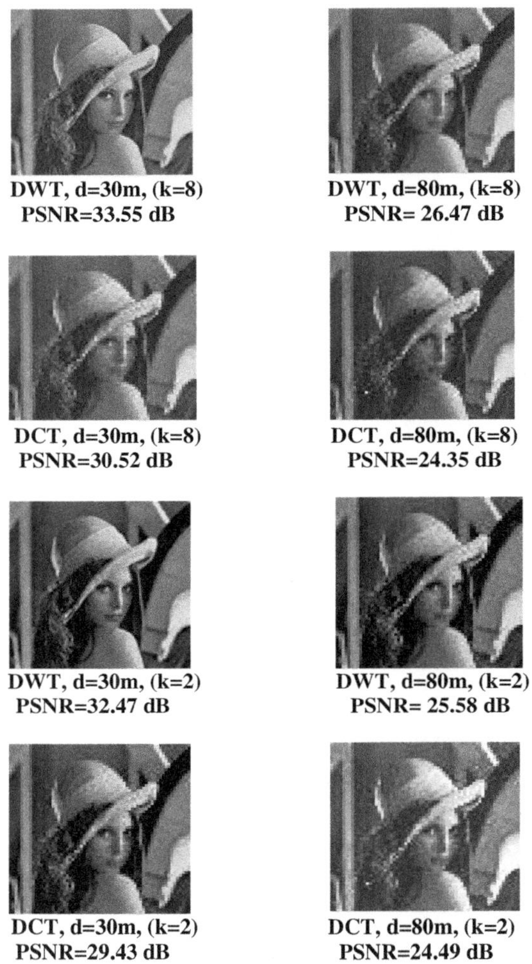

Figure 5. Obtained images with different distance for Lena 64x64

We conclude that the image quality decreases when the distance between the nodes increases. We show that at some distances we get blurred images that are due to packet loss. In addition, the image size has a great influence. A question can be asked that why we used the Lena image with only two sizes 32x32 and 64x64 and not other sizes? The answer is simple because it depend by the memory available by the sensors and consequently it affects the transmission time because the coding phase depend of the microcontroller used (hardware). We can of course use images size 128x128 and more, but the coding phase will be processed by the PC which it won us in terms of transmission time which is almost equal to 50%.

7. Transmission With an Intermediate Node

7.1. Second Scenario

We are interested in this section to a topology with larger number of nodes. Our experiments were performed on TelosB wireless sensor nodes.

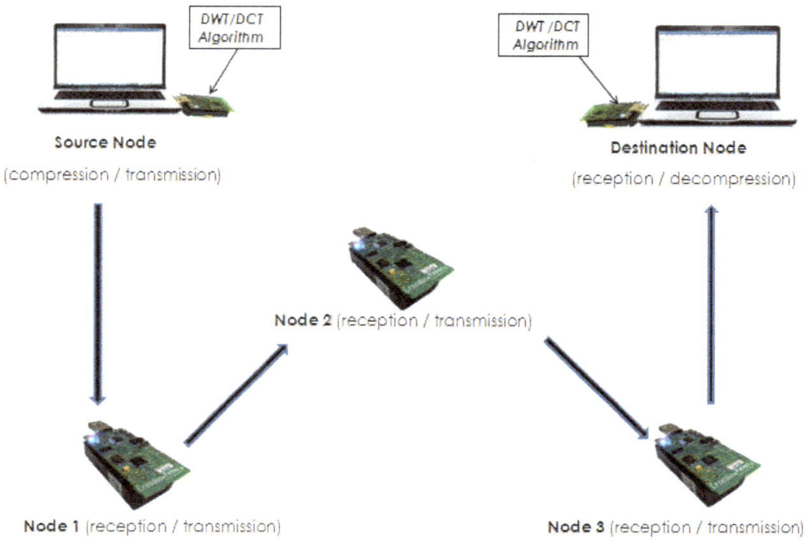

Figure 6. Transmission scenario with intermediate nodes.

This figure represents an image transmission scenario from source to destination node through intermediate node 1, node 2 and node 3. Compression and decompression are performed respectively at the source and destination nodes. Intermediate nodes only transmit compressed images. We used different image types and different resolution. We varied distance between source and destination nodes. For each configuration, tests are performed with DCT and DWT. Measured parameters are the quality of reconstructed images (using PSNR).

7.2. Obtained Images with an Intermediate Node

This figure shows the reconstructed images obtained with the use of intermediate sensors that are used across different distances. The first two images represent the images using DCT and the last two images are obtained by using the DWT technology.

DCT, d=30m, (k=2)
PSNR=28.54 dB

DCT, d=80m, (k=2)
PSNR=23.36 dB

DWT, d=30m, (k=2)
PSNR=31.23 dB

DWT, d=80m, (k=2)
PSNR=26.12 dB

Figure 7. Obtained images with different distance for Lena 64x64

From this figure, we can conclude that the more we increase the distance, the more there is degradation. Thus, it is clear that there is a difference between the images using DCT and DWT. In fact, the compression based on DWT gave better results in terms of image quality and energy consumption.

8. Summary Table: DWT vs DCT

Table 2 shows a comparison between DWT and DCT with the evaluation criteria terms: Compression Performance, Execution Time and used Memory. Character (+) refers the quality and performance task. When the number of (+) increases, it means that the operation is successfuly used by this algorithm and gives good results.

Tableau 2. Comparative criterion between DWT and DCT

Efficiency Criterion	***DCT***	***DWT***
Compression performance	+++	+++++
Execution Time	+	+++
Memory Used	+++	+

We finally note from this table that the DWT is better than the DCT in terms of compression performance and execution Time in the context of Wireless Sensors Networks using TelosB motes.

9. Energy Consumption Measurement

The experiment of measuring energy consumption in the WSN show that the average voltage corresponding to the maximum energy consumption is approximately: V Havg = 212.9 mV. This value assumes that the microcontroller and radio transceiver are you running (ie

processing and radio transmission in progress). The intensity when the radio transceiver of the sensor Telosb we used in our experience, is off is not zero, the consumption of the microcontroller must be considered and it is: I Lavg = 1800 uA.
Considering both values and the measured value of the resistance is 10 ohms connected in series, it is possible to calculate the batteries lifetime.

10. Batteries Lifetime

The most suitable power source for the motes is batteries. Each mote needs 3V to work, or, considering batteries, two units of AA batteries.
Nowadays, market is full of different brands that offer many different kinds of batteries, but, considering the low environmental impact of the network, the most suitable are rechargeable batteries. The reason to choose this kind of batteries is they can be recharged several times, which involves saving money in batteries and the possible contamination that batteries can cause if they are not recycled in the correct way.
Rechargeable batteries can be based on different technologies. The first one where based on NiCd, but due to memory effect, after each recharge the life cycle was reduced. Now the principal technology is NiMH; they have no memory effect and higher charge and discharge capacity which makes them more useful than the previous ones.
One of the possible rechargeable batteries to be used is Alkaline AA-NiMH 2850 mAh.

Tableau 3. Battery characteristics

Maximum charge voltage	1.5 V
Nominal voltage	1.2 V
Nominal capacity	2850 mAh
Standard charge (current charge /charge time)	270 mA/16 h
Fast Charge (current charge /charge time)	2700 mA/1.1 h

We then calculated the duration of battery life (estimated theoretical). According to the experiment, the intensity of the average load of the battery is given by equation (2) follows:

$$I_{Pavg} = (V_{Havg} / R * TH + I_{Lavg} * TL) / (TH + TL) \tag{2}$$

V Havg was measured over R = 10 Ω and I Pavg = 11,623 mA Max.

Considering the nominal capacity shown in previous table, it is obtained:

$$\frac{Capacity}{I_{Pavg}} = \frac{2850 \ mAh}{11.623 \ mAh} = 245.20h$$

Total DCT time is 245.20 h = 10 days, 5 hours, 20 minutes

Total DWT time is 281.34 h = 11 days, 17 hours, 34 minutes

Those values represent the total of energy consumption and battery life time. So, there is no big difference between de duration of time of DWT and DCT, but the method using DWT is more adopted in our experimentation because it gives a perfect performance in terms of image quality and energy consumption.

11. Conclusions

In this paper we have proposed and evaluated two image compression methods dedicated to transmission over Wireless Sensor Networks (WSNs) in order to decrease the energy consumption of sensors and thus to maintain a long network lifetime.

Despite discrete cosine transforms is considered as a good compression algorithm that gives a high compression ratio, the discrete wavelet transform have been widely recognized to be more prevalent than others. Due to the wavelets' excellent spatial localization, frequency spread, and multi-resolution characteristics, this algorithm is considered similar to the theoretical models of the human visual system. In future, our works will be concerned by transposing these results to other fast implementation methods of the literature and to other real sensor platforms. In addition, we aim to implement our proposal on dedicated platform for multimedia wireless sensor networks (MWSNs).

References

[1] Naumowicz, Freeman, Heil, Calsyn, Hellmich, Brandle, Guilford & Schiller (2008). "Autonomous monitoring of vulnerable habitats using a wireless sensor network". In : Proceedings of the Workshop on Real-World Wireless Sensor Networks, REALWSN'08. Glasgow, Scotland.

[2] Marcelloni & Vecchio (2009). "An Efficient Lossless Compression Algorithm for Tiny Nodes of Monitoring Wireless Sensor Networks". The Computer Journal 52(8), pp 969-987.

[3] Akyildiz, Melodia, Kaushik & Chowdhury (2007). "A survey on wireless multimedia sensor networks". Journal Computer Networks: The International Journal of Computer and Telecommunications Networking, Volume 51, Issue 4, Inc . New York, NY, USA, United State.

[4] Liang and Tran (2001). Fast Multiplierless Approximations of the DCT with the Lifting Scheme. IEEE Trans. on Signal Processing, 49(12): 3032–3044.

[5] Jeong, Kim and Cho (2004). Low-Power Multiplierless DCT Architecture Using Image Correlation. IEEE Trans. on Consumer Electronics, 50(1): 262–267.

[6] Ferrigno, Marano, Paciello & Pietrosanto (2005). "Balancing computational and transmission power consumption in wireless image sensor networks". In: IEEE International Conference on Virtual Environments, Human-Computer Interfaces, and Measures Systems (VECIMS 2005). Giardini Naxos, Italy.

[7] Mammeri, Khoumsi, Ziou and Hadjou (2008). "Modeling and Adapting JPEG to the Energy Requirements of VSN". In International Conference on Computer Communications and Networks (ICCCN 2008), Proceedings, 806–811, St. Thomas, Virgin Islands-USA.

[8] Feig and Winograd. Fast algorithms for the discrete cosine transform. IEEE Trans. on Signal Processing, 40(9): 2174–2193, 1992.

[9] C. N. Taylor and S. Dey. Adaptive Image Compression for Wireless Multimedia Communication. In IEEE International Conference on Communications (ICC 2001), Proceedings, 6: 1925–1929, Helsinki-Finland, June 2001.

[10] Enesi, Zanaj, Kamo, Kolici, Shurd (2010). "Image Compression for Wireless Outdoor Sensor Networks". BALWOIS 2010 - Ohrid, Republic of Macedonia.

[11] Soro & Heinzelman (2009). "A survey of visual sensor networks. Advances in Multimedia": Article ID 640386, pp 21.

[12] Tang & Raghavendra (2004). "Compression techniques for wireless sensor networks". Wireless Sensor Networks, chapter Kluwer Academic Publishers 2004.

[13] Akyildiz, Melodia & Chowdhury (2007). "Wireless multimedia sensor networks: a survey". IEEE Wireless Communications Magazine, 32-39.

[14] Chew, Ang, & Seng (2008), "Survey of Image Compression Algorithms in Wireless Sensor Networks", in Proc. of Int. Symp. on Information Technology (ITSim 2008).

[15] Li, Peng Pang & Liu (2010). "SPIHT algorithm combined with Huffman encoding". Third International Symposium on Intelligent Information Technology and Security Informatics.

[16] Duran-Faundez (2009). " Transmission d'images sur les réseaux de capteurs sans fils sous la contrainte de l'énergie" . Thesis, Nancy University.

[17] Makkaoui, Lecuire, & .Moureaux (2010). "Fast Zonal DCT-based image compression for Wireless Camera Sensor Networks", in "2nd International Conference on Image Processing Theory, Tools and Applications, IPTA 2010, Paris : France (2010)" DOI : 10.1109/IPTA.2010.5586798.

[18] Digital Compression and coding of Continuous-tone still images, part 1, requirements and Guidelines. ISO/IEC JTC1 Draft International Standard 10918-1, Nov. 1991.

[19] Gonzalez and Woods, "Digital Image Processing", Reading. MA: Addison Wesley. 2004.

[20] David Salomon, "Data Compression". The Complete Reference, 2nd Edition Springer-Verlag 1998.

[21] Caben & Gent (2008). "Image Compression and the Discret Cosine Transform, in International Conference on Machine Learning and Cybernetics, pp. 407- 410.

[22] Antonini, Barlaud, Mathieu & Daubechies (1992) . "Image coding using wavelet transform". IEEE Transactions on Image Processing 1, 2, 205-220.

[23] Crossbow, Telosb datasheet, www.xbow.com, Document part Number: 6020-0094-01 Rev B.

14

A STUDY OF VOICE TRAFFIC BLOCKING IN A MODEL CELLULAR NETWORK

Oliver Mitch Maguitte[1], Mohammad Sameer Sunhaloo[1], Ben Oodit and Vinaye Armoogum[1]

[1]School of Innovative Technologies and Engineering, University of Technology, Mauritius

molivierm@gmail.com, sameer.s@utm.intnet.mu, ben.oodit@gmail.com and varmoogum@utm.intnet.mu

ABSTRACT

In this paper, we present results obtained from a statistical analysis of a model cellular network which is similar to the one used in Mauritius. We have tested system samples with different cell cell configurations with two to four transceiver units, which serve a delimited region of several square kilometres. The Erlang B, Erlang C and Poisson models have been considered in this study. The experimental work has showed that the Erlang C Model provided the closest match for the studied environment.

KEYWORDS

Cellular Network, Voice Traffic, Erlang B, Erlang C

1. INTRODUCTION

GSM (Groupe Spécial Mobile) technology is used worldwide and is the basis of public mobile telecommunication [6, 12, 13]. Mobile networks have adopted the cellular concept for reasons related to capacity, flexibility, scalability and cost effectiveness. Currently there are 3 GSM operators in Mauritius and mobile penetration is reported to be 92% by the ICTA (Information and Communication Technologies Authority of Mauritius), the Mauritian regulator [17, 18].

Mobile networks are traffic sensitive due to the rapid increase of subscribers throughout the world. It is thus crucial to save on operating cost and to consume bandwidth effectively with respect to traffic load as spectrum is limited. There exist many traffic models and not all of them may be appropriate for a specific cellular network system. Since, different traffic models behave differently due to differences in underlying assumptions, one model may prove to be more useful and provide more realistic feedback of the performance of a given network than others. These traffic models help in traffic engineering by highlighting bottlenecks which can be proactively handled by the mobile operator through the optimization of cell capacity.

Based on the ability of traffic models to simulate real-life conditions, appropriate models have been developed so as to approximate blocking probabilities in order to help network engineers in dimensioning [3, 9, 10, 11]. Some common factors which affect blocking probabilities are call arrival patterns, number of servers (or channels), channel holding time and traffic intensity [15, 16]. Most operators throughout the world favour the use of the Erlang B model for blocking probability but that model was initially developed for fixed telephone networks and mainly trunks. The Erlang B model has been ported to mobile telephone network for dimensioning the air interface.

Although traffic distribution has been an active research, only a few researchers have been questioning the reliability of different traffic models, including the famous Erlang B for mobile networks at cell level. Tunnicliffe et al. [16] and Smith et al. [15] cleared doubts about the accuracy of using a single model, like Erlang B model, to evaluate call blocking. Investigations of several parameters like arrival rate, mean holding time, with varying sample size and network resources have led to the conclusion that Erlang B is not appropriate for cells with less than 12 traffic channels as blocking tends to be over estimated. However, Erlang B does provide a good model for cells with number of channels greater than 12 and with blocking rate above 1%.

In this paper, we investigate the appropriateness of various models, including the Erlang B model, for the estimation of blocking at the air interface of a model mobile network. The focus is on the voice call class of service.

2. Related Work

The distribution of channels holding and inter-arrival times were analyzed using real-life data by Jedrzycki and Leung [8]. They observed that calls which had been handoff from previous cells had low holding times. Weak signals were another cause of calls handoff. The holding time followed a lognormal distribution and not a negative exponential distribution. However, call inter-arrival times followed approximately a negative exponential distribution.

Rappaport and Lon-Rong Hu [14] studied the concept of layered cells, in which different cells help with the traffic load at various levels. Different parameters like blocking, carried traffic, forced terminations and handoff failures were observed. The aim was to find improvements on coverage which could be achieved with more capacity. It was noted that frequency reuse was more efficient with smaller cells. They also considered handoff interruption. Through simulations with different cell layers, it was noted that both the higher and the lower layers could be helpful to each other in reducing congestion under low and moderate offered traffics.

Land to mobile communications and mobile to mobile calls based on Erlang and Engset were studied by Baky and Ackroyd in 1981 [1]. Calls in the same cell were considered. We note that in reality calls are between cells of different base stations. They derived formulae for measuring congestion in a single cell. It was noted that average traffics gave a better picture of blocking probabilities as in real-life different types of calls are mixed. In 1982, Baky and Ackroyd studied a multi cell mobile radio telephone system in view of deriving formulae related to congestion [2]. They considered calls towards other networks, to mobile phones in other locations and calls to mobiles served by the same cell. The main concern of the study was to minimize congestion within a radio mobile telephone system. The study was made on 30 cells and it was found that dynamic channel allocation resulted in more congestion compared to fixed allocation.

In [7], Jabbari studied a new generation of networks by considering various parameters including voice service. He investigated about the strength of signal and propagation variation for a micro cell located in a region. He also presented an overview of different factors concerning mobility and layers. A model based on fluid flow assumptions was used to have a better system capacity based on parameters like traffic carried and different blocking probabilities.

Guerin performed simulations with some assumptions, like mobility of mobile phones towards a fixed direction, to investigate about channel occupancy, the distribution of holding time and to derive a model for channel occupancy [5]. It has been noted that channel occupancy is exponential even when mobile phones move across the cells. This study did help to make proper dimensioning and better estimation of handoff traffic.

In this paper, we concentrate on the voice traffic and the performance of network for standard cells using less than 5 transceiver units.

3. MODEL CELLULAR NETWORK

We consider a model cellular network, where generated traffic measurement is close to real life conditions. The measurements in real life are typically obtained from the Base Station Subsystem Performance Management System, part of the OSS (Operation and Support System), following the parsing of scanner files containing cell level counters. Voice traffic in different clutter categories as described in Table 1 was analyzed.

Table1. Number if cells in different clutters

	Rural	Urban
Commercial	4	3
Residential	11	1

The model network consisted of standard cells with the number of transceiver units ranging from two to four. For any standard cell to work it is mandatory to have one broadcast and at least one signalling channel configured and the others configured as traffic channels. For instance, 2 Transmitter/Receiver (Transceiver) TRX consists of 14 Traffic Channels (TCH) whilst 4 TRX consists of 28 TCH. Each channel is mapped onto a time slot. The cells are located in different regions and thus are able to sample varying traffic patterns.

3.1. Physical Layout of Scenario

The layout configuration of our model network is listed in Tables 2 and 3, where BCCH means Broadcasting Control Channel (Control Info), PDCH represents Packet Data Channel (Data Traffic) and SDCCH stands for Slow Dedicated Control Channel (Signalling).

Table2. Cells configuration for control channels and respective clutter serving with one BCCH.

Location	TRX	SDCCH	Physical TCH	Logical TCH
C RR 0	3	1	22	33
C RR 1	3	1	22	33
C RR 2	3	1	22	33
L RR 0	2	1	14	14
L RR 1	2	1	14	24
L RR 2	2	1	14	24
Z RR 0	2	1	14	14
Z RR 1	2	1	14	14
Z RR 2	2	1	14	14
W RC 0	2	1	14	14
W RC 1	2	1	14	24
W RC 2	4	3	28	53
D RC 0	2	1	14	14
D RR 1	2	1	14	14
D RR 2	2	1	14	14
A UR 0	2	1	14	14
B UC 0	2	1	14	14
B UC 1	2	1	14	14
B UC 2	2	1	14	14

Table3. Logical traffic channel configuration per cell considering half rate activation with respective clutter.

Location	Clutter	Area	Logical TCH
C RR 0	Rural	Residential	22
C RR 1	Rural	Residential	22
C RR 2	Rural	Residential	22
L RR 0	Rural	Residential	14
L RR 1	Rural	Residential	14
L RR 2	Rural	Residential	14
Z RR 0	Rural	Residential	14
Z RR 1	Rural	Residential	14
Z RR 2	Rural	Residential	14
W RC 0	Rural	Commercial	14
W RC 1	Rural	Commercial	14
W RC 2	Rural	Commercial	28
D RC 0	Rural	Commercial	14
D RR 1	Rural	Residential	14
D RR 2	Rural	Residential	14
A UR 0	Urban	Residential	14
B UC 0	Urban	Commercial	14
B UC 1	Urban	Commercial	14
B UC 2	Urban	Commercial	14

3.2. Physical Topology of Selected Network

Our model network simulates a mountainous topology. Target threshold for this segment is 5% Grade of Service which is acceptable taking in consideration the landscape and access facility to the sites. In Figures 1 and 2, the topology and coverage zone of the cells used for the experiments are shown.

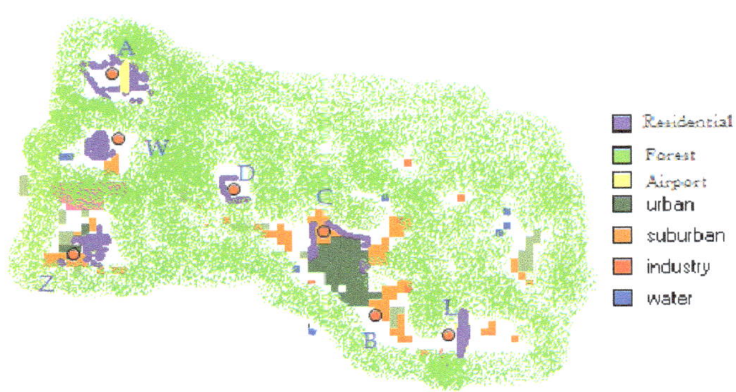

Figure 1. Topology of model network

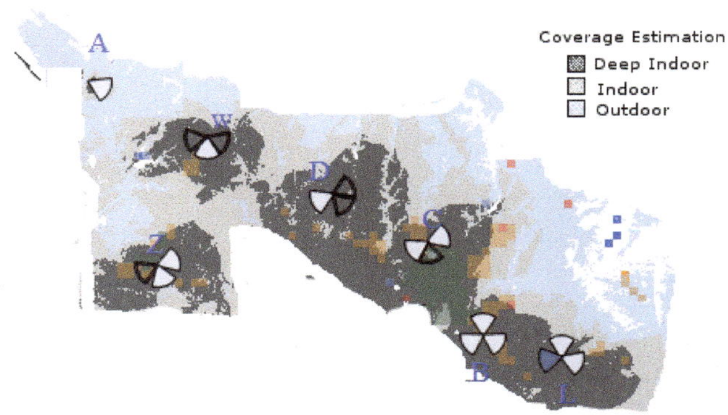

Figure 2. Coverage zone of model network

We note that the residential regions C, D, L, W and Z include rocky valleys with steep slopes and forest. There is also a nature park surrounded by forests. The residential zones A and B are more developed, with location B being mainly a commercial zone with many shops and major banks in the surroundings. An airport is situated near Location A.

3.3. Measurement

Traffic measurement files for the model network had to be captured over several days on an hourly period from the BSS (Base station subsystem) and transferred to the OSS (Operation and Support System) performance monitoring tool. Individual cell KPI (Key Performance Indicator) were obtained and processed from the individual hourly measurement traffic files. To ensure the model network performance is not altered, only routine maintenance was performed with no modification to antenna settings and cell configuration, which led to constant and standard sampling with the only variation of subscribers which perform calls in a random and uncontrolled manner.

4. TRAFFIC MODELLING

Among the various types of traffic models which exist, only some are commonly employed in real life environment by operators based on arrival rate within the network. The models help operators to know about the blockings faced by users within their networks [4, 15, 16]. The different types of traffic model are discussed below.

Traffic Intensity is related to average call duration multiply by the average number of call requests and is measured in a dimensionless unit which is the Erlang. When one radio resource channel is utilized for one hour, it is said that 1 Erlang of traffic has flowed through. Clearly,

$$A_C = \lambda\, h,$$

where

- A_C is the traffic intensity in Erlang generated by each user,
- h is the average duration of a call measured in hour and
- λ is the average number of call requests per hour.

4.1. Erlang B

Erlang B is a distribution originally used by Public Switched Telephone Network to dimension trunks. However, it is also applicable to mobile telephone network dimensioning of air interface. This model is employed when callers do not reattempt when they face a block call as the system reroute the call so as it does not return to the original cell where call is initiated. In the Erlang B model, it is assumed that the number of sources is infinite, calls arrive randomly and are served in order of arrival, blocked calls are lost and holding times are exponentially distributed. The Erlang B loss probability B_B is given by

$$B_B = \left(\frac{A^N}{N!}\right) \Bigg/ \sum_{i=0}^{N} \frac{A^i}{i!},$$

where N and A denote the number of trunks available in resource group and traffic offered to group in Erlangs, respectively.

4.2. Erlang C

The Erlang C model is popular in networks where queuing is applied. Calls are not rerouted as in the case of the Erlang B model, but are put in a waiting queue until resources are free for call to be processed or queue time expires. Thus, in the Erlang C model, blocked calls are delayed. The relationship between B_B and the Erlang C probability of blocking B_C is given by

$$B_C = \frac{NE_B}{N - A(1 - B_B)}.$$

It should be noted that when resources start to be unavailable, for the Erlang C model the blocking rate is higher as it take in consideration the number of instances in the queue and the resources utilized.

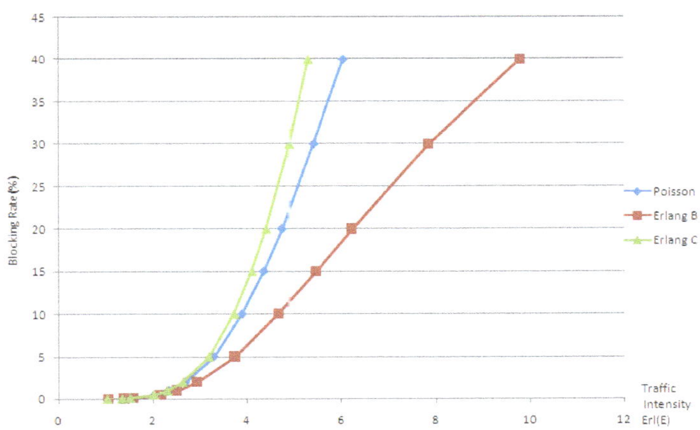

Figure 3. Traffic model comparison with limited servers (7 TCH).

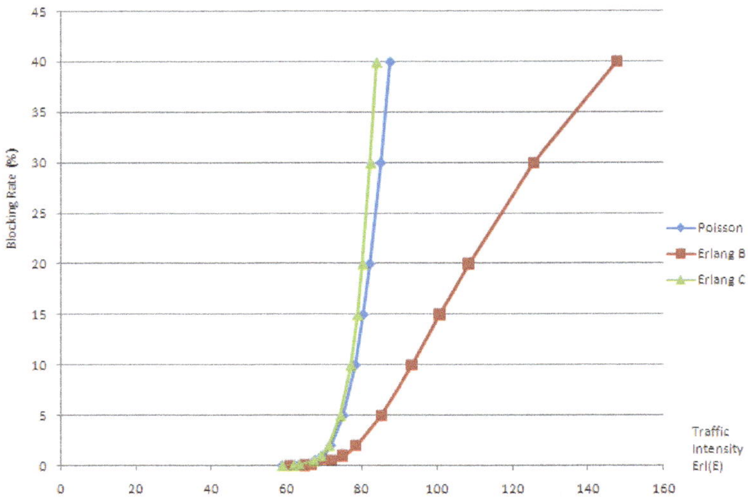

Figure 4. Traffic model comparison with huge number of servers (90 TCH).

Figures 3 and 4 show that Erlang C for any cell with more than 14 channels installed behaves nearly like the Poisson traffic model. Erlang C behaves as Erlang B as long as the channel in busy state does not reach number of channel installed. When there is no blocking, Erlang C behaves similarly to Erlang B as seen in Figure 3 and 4.

5. Network behaviour and Analysis of Data

In this section, measured blockings are compared with theoretical values based on the traffic models, Erlang B and Erlang C depending on system being investigated. The sample size of the measured values has been made over a period for the same cell with no alteration to both the physical equipments and logical parameters, ensuring that the test bed being in the same condition for the sampled period. Figure 5 illustrates the arrival rate over a 24 hour period. In practice, the study of the arrival instances at a finer granularity is hard to obtain as it involves a huge load on the Operation Support Subsystem and network equipment storing the scanner files. Thus, we considered values for a day only.

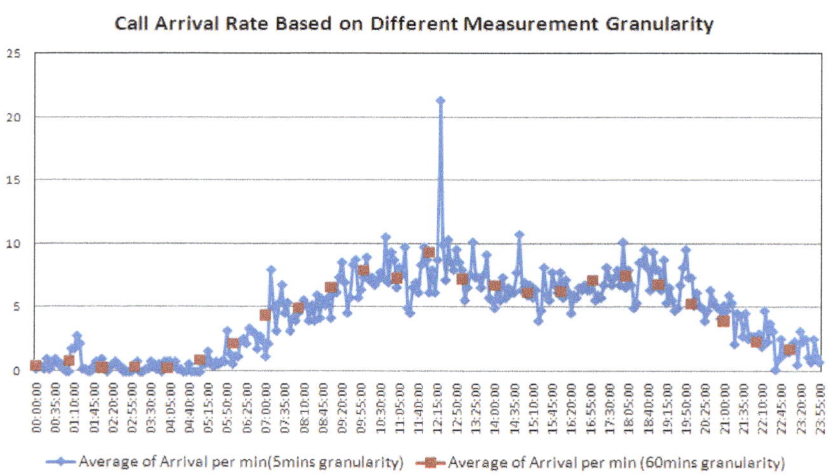

Figure 5. Call arrivals per minute and per hour.

In Figures 6 and 7, the plots of blocking probability against traffic intensity for cells with 14 and 33 voice channels, respectively, are given. Due to the dynamic nature of the network and features implemented, it can be noticed that more traffic is carried with less blocking for some instances, but most of the time adhere to the assumptions made.

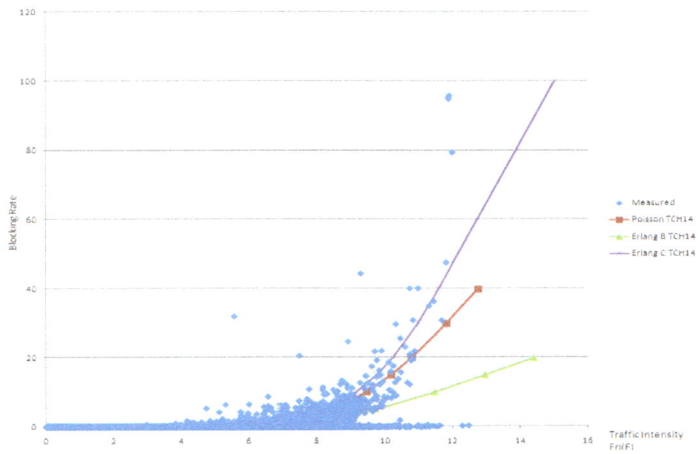

Figure 6. Graph of blocking rate against traffic intensity of collected data, Poisson, Erlang B, Erlang C for 14 TCH cells.

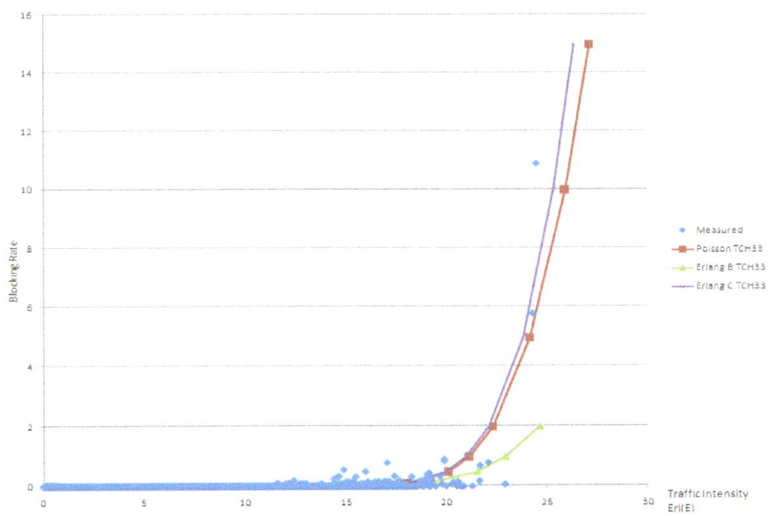

Figure 7. Graph of blocking rate against traffic intensity of collected data, Poisson, Erlang B, Erlang C for 33 TCH cells.

5.1. Data Analysis

In this section, the data captured from the Operation Support Subsystem is analysed in details and selected graphs are plotted to illustrate the trend and comparison of measured against theoretical values. The network segment has different cells with 2, 3 and 4 TRX.

First we consider 14 TCH cells. As shown in Figure 6, the Erlang C theoretical value lies midway in most cases. There are some values which do not follow the Erlang C model but this can be explained by the use of GSM features to manage load in cells. The Erlang B Model is not proper as measured blocking is higher than the hypothetical values. Each point referred as measured represent a measured value captured during the sampling duration and is usually archived by the Key Performance Indicator (KPI) system.

Figure 8 shows a 95% confidence interval calculated based on the measured information. It can be seen clearly that the Erlang C model for this class of cells with 14 TCH fit within the confidence limits.

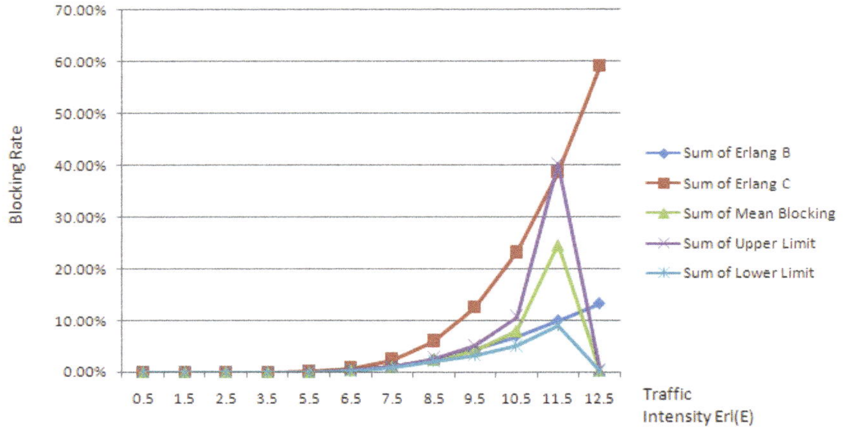

Figure 8. Confidence interval for all cells with 14 TCH .

Next, we consider 53 TCH cells. The graph of blocking probability against traffic intensity is given in Figure 9.

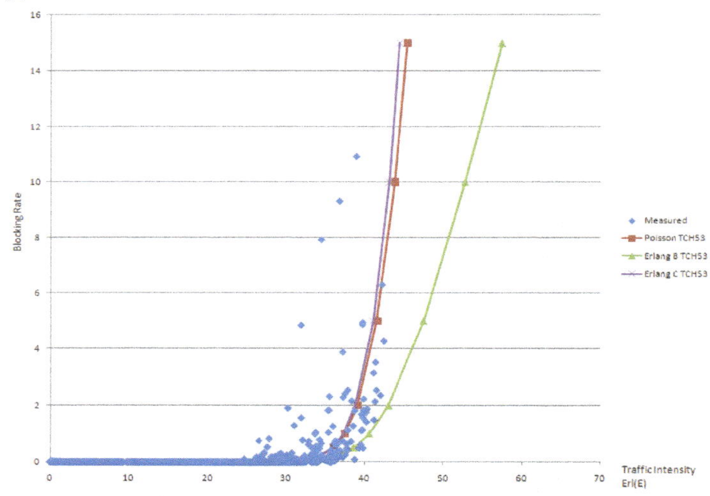

Figure 9. Graph of blocking rate against traffic intensity of collected data, Poisson, Erlang B, Erlang C for 53 TCH cells.

Figure 10 shows the confidence interval calculated based on the measured information. It can be seen clearly that the Erlang C model for cells with 53 TCH also fit within the confidence limits.

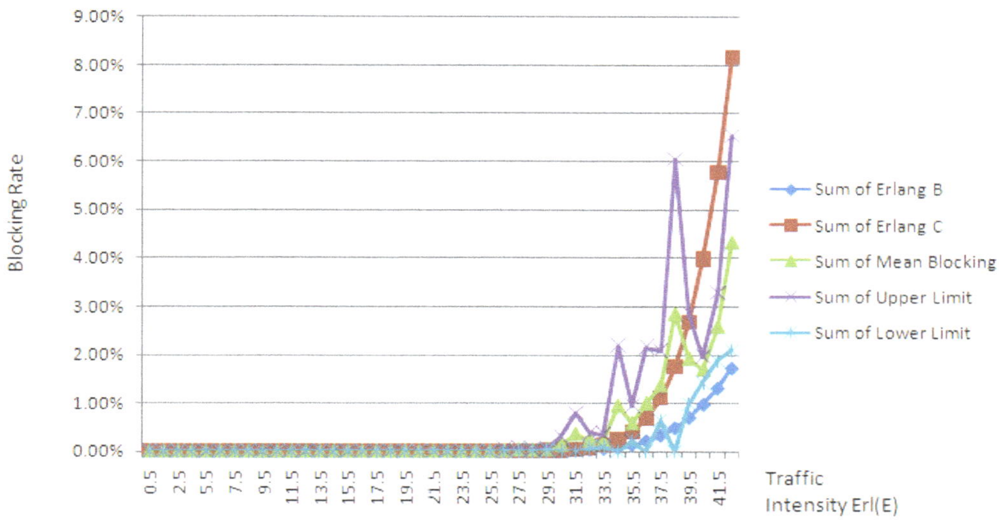

Figure 10. Confidence interval for all cells with 53 TCH.

The distribution of samples with respect to the traffic bin each at an interval of 1 Erlang is shown in the form of a histogram in Figure 11. The measurement window is based on a 24-hour range sampled at interval of 60 minutes as a general rule. It can be seen that a normal bell shaped distribution is obtained.

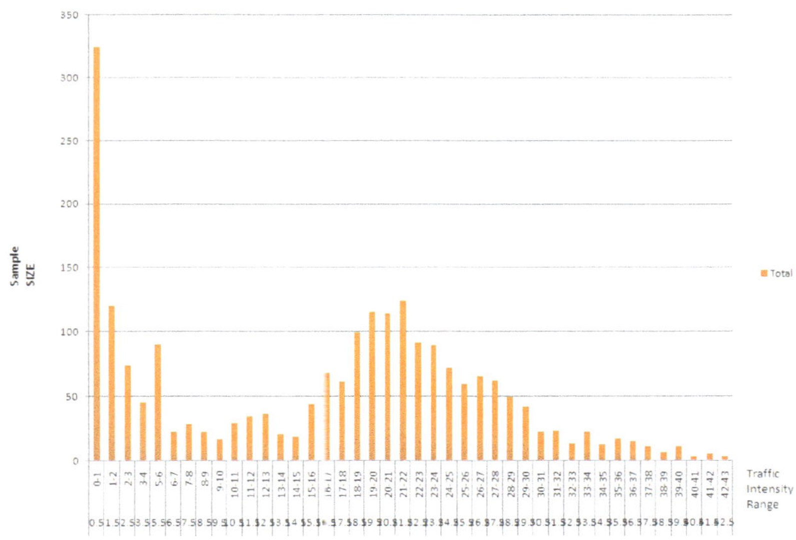

Figure 11. Sample distribution for all cells with 53 TCH.

Based on the above, it can be said that the Erlang C model is appropriate for the studied network. The samples are evenly distributed on the network, and as 24-hour sampling is used, there are several samples with low traffic intensity. We also note that in some cases, Erlang C did not provide the best fit because of GSM features change the configuration of the cell dynamically and therefore it may happen that less traffic channels are available instantaneously thus causing more blocking. In Figure 12, we show the graph of traffic intensity against time for 53 TCH cell in a rural commercial area from Monday to Sunday.

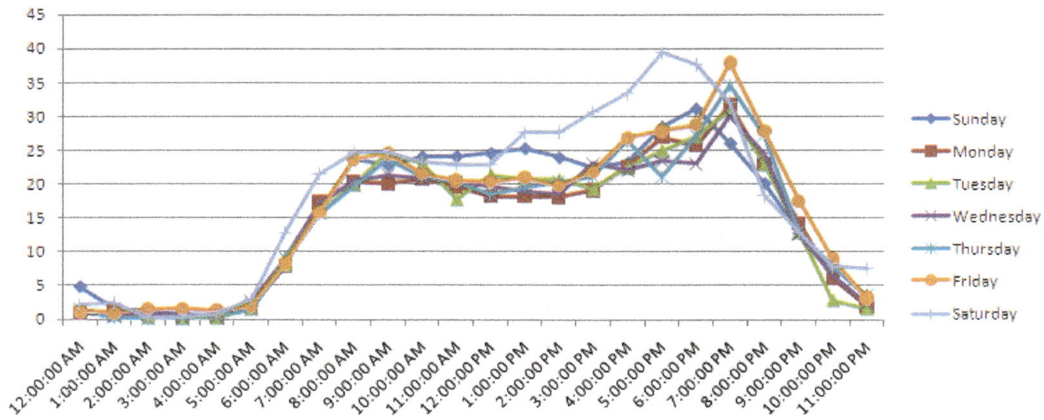

Figure 12. Traffic intensity against time for 53 TCH cell in a rural commercial area from Monday to Sunday.

5.2. Half Rate Traffic

In Figures 13 to 16, the shaded boundaries indicate where half rate traffic is considered on the traffic plots. We observe that mixed traffic is triggered from solely normal full rate traffic to half rate traffic.

Figures 14, 15 and 16 indicate that the measured samples fit within the boundary for most cells. However for Figure 13 which considers cells with lower amount of channels, it is seen that some samples are still outside the boundary which is due to the limited amount of channels and GSM feature of dynamic channel allocation.

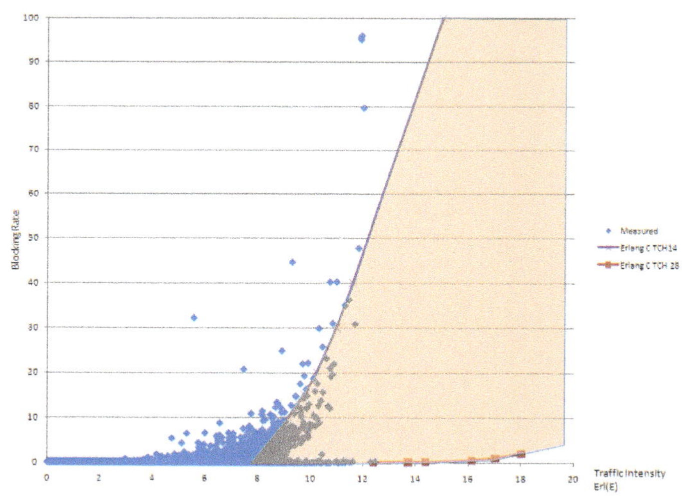

Figure 13. Graph of blocking rate against traffic intensity for 14 TCH with performance information and the acceptable range.

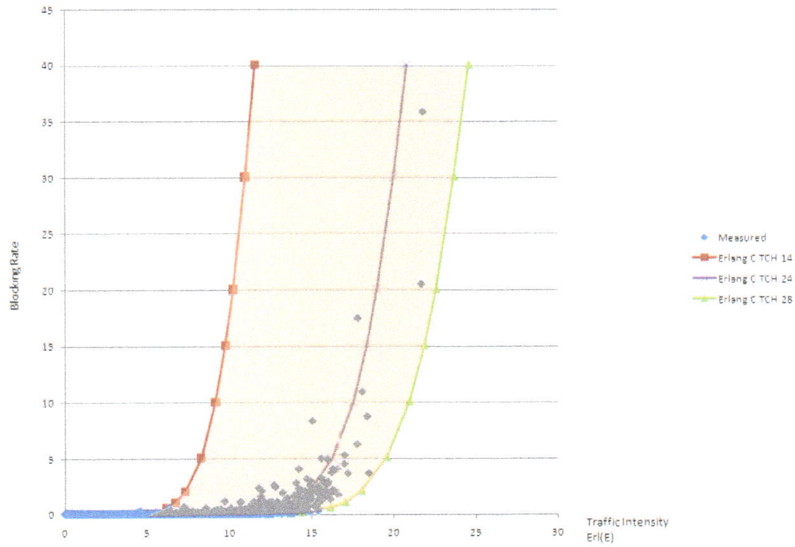

Figure 14. Graph of blocking rate against traffic intensity for 24 TCH with performance information and the acceptable range.

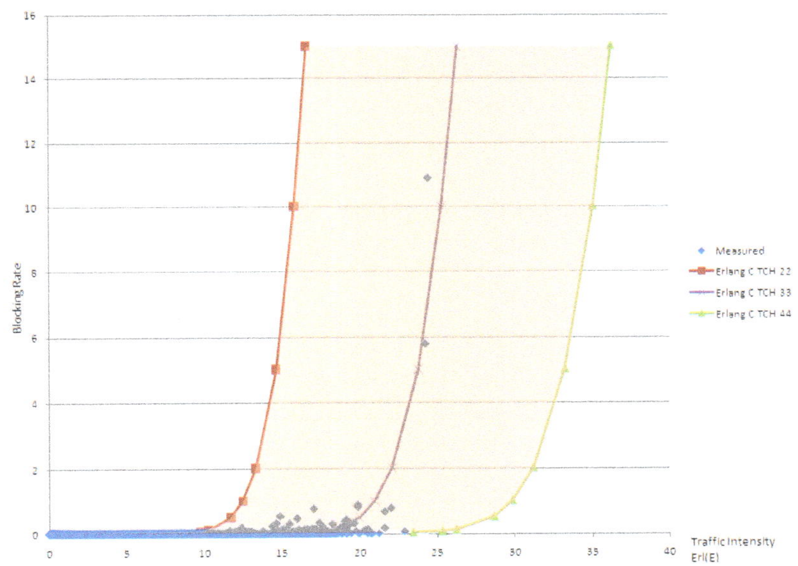

Figure 15. Graph of blocking rate against traffic intensity for 33 TCH with performance information and the acceptable range.

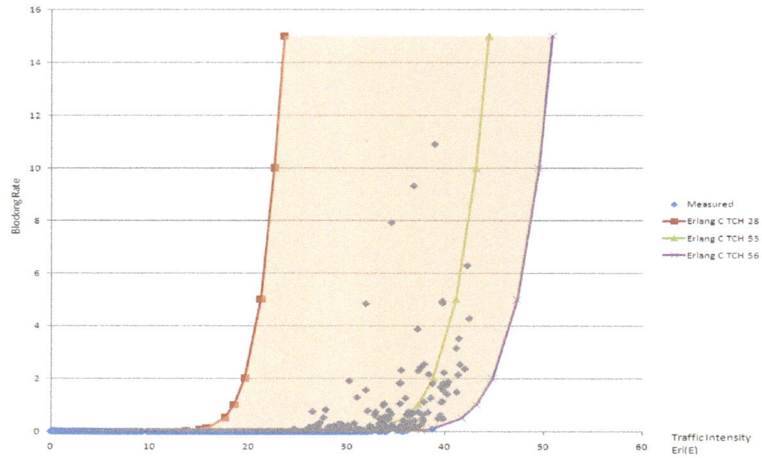

Figure 16. Graph of blocking rate against traffic intensity for 53 TCH with performance information and the acceptable range.

6. Conclusion

We have observed that cells with more than 14 channels have a behaviour which show a good fit to Erlang C theoretical values. We also note that activation of GSM enhanced features such as GSM Half Rate speech codec does help the network by reducing blocking probabilities.

From the analysis, it can be concluded that the studied network is properly sized and the probability of high blocking is quite low and occurs only during the busiest hour, which is normal. The network operates within industry standards for Grade of Service which should lead to user satisfaction with respect to network accessibility.

We note that new GSM features, which are traffic control measures, do provide a lot of benefits to cellular networks in the sense that they help to reduce blocking. When dynamic channel allocation is enabled, cell configuration is dynamic with respect to time and a dimensioning exercise becomes more difficult due to the number of channels being variable and subject to traffic load.

Smith et al. [15] and Tunnicliffe et al. [16] have mentioned that the real blocking faced on network with cells with low number of channel is less than the blocking predicted by the traffic model which they have considered. This is in line with the results which we have obtained. However, blocking depends on the number of subscribers and the arrival rate. Based on the observations obtained from our experimental work, we have found that both Erlang C and Poisson converge towards the same values as the number of resources increases for our model network. Furthermore, using Erlang C is proposed as it will give higher blocking values which would lead the operator to have extra capacity margins when relevant cell expansions are made.

References

[1] Baky, S. and Ackroyd, M. (1981) "Teletraffic Analysis for Single-Cell Mobile Radio Telephone Systems" *IEEE Transactions on Communications*, vol. 29, no.3, pp298–304

[2] Baky, S. and Ackroyd, M. (1982) "Teletraffic Analysis for Multicell Mobile Radio Telephone Systems" *IEEE Transactions on Communications*, vol. 30, no. 8, pp1905–1909

[3] Everit, D. E. (1994) "Traffic Engineering of the Radio Interface for Cellular Mobile Networks" *Proceedings of the IEEE*, vol. 82, No. 9, pp1371-1382

[4] Freeman, R. (2004) *Telecommunications System Engineering,* John Wiley & Sons

[5] Guerin, R. A. (1987) "Channel Occupancy Time Distribution in a Cellular Radio System" *IEEE Transactions on Vehicular Technology*, vol. 36, no. 3, pp89–99

[6] Hillebrand, F. (2001) *GSM and UMTS: The Creation of Global Mobile Communication,* John Wiley & Sons

[7] Jabbari, B. (1996) "Teletraffic Aspects of Evolving and Next Generation Wireless Communication Networks', *IEEE Personal Communications*, vol.3, no. 6, pp4-9

[8] Jedrzycki, C. and V. C. M. Leung (1996) "Probability Distribution of Channel Holding Time in Cellular Telephony Systems", *Proceedings of the IEEE Vehicular Technology Conference*, vol. 1, 247 – 251

[9] Karray, M. K. (2010) "Evaluation of the blocking probability and the throughput in the uplink of wireless cellular networks", *Proceedings of the International Conference on Communications and Networking*, pp1-8

[10] Kolaczyk, E. D. (2010) *Statistical Analysis of Network Data: Methods and Models,* Springer

[11] Nasser, N. (2006) "An acceptable trade-off between new call blocking and handoff call dropping probabilities in multimedia cellular networks", *Proceedings of the 4th Communication Networks and Services Research Conference*, pp69-75

[12] Oancea, C. D. (2011) "GSM infrastructure used for data transmission", *Proceedings of the 7th International Symposium on Advanced Topics in Electrical Engineering*, pp1-4

[13] Perkov, L., Klisura, A. and Pavkovic, N. (2011) "Recent Advances in GSM Insecurities" *Proceedings of the MIPRO 34th International Convention*, pp1502-1506

[14] Rappaport, S. S. and Lon-Rong Hu (1994) "Microcellular Communication Systems with Hierarchical Macrocell Overlays: Traffic Performance Models and Analysis", *Proceedings of the IEEE*, vol. 82, no. 9, pp1383-1397

[15] Smith, P. J., Sathyendran, A. and Murch, A. R., (1999) "Analysis of Traffic Distribution in Cellular Networks", *Houston, TX, USA: 49th IEEE Vehicular Technology Conference*, pp2075-2079

[16] Tunnicliffe, G. W., Murch, A. R., Sathyendran, A. and Smith, P. J., (1998) "Analysis of Traffic Distribution in Cellular Networks", *48th IEEE Vehicular Technology Conference IEEE*, pp1984-1988

[17] Information and Communication Technologies Authority of Mauritius ICTA., (2011) "Mobile Telephony Services and mobile penetration Rate", *http://www.icta.mu/market/mobile_telephony.htm, Last Accessed 2011-12-17*

[18] Lange, P. (2010) "Mauritius - Telecoms, Mobile, Broadband and Forecasts", *http://www.budde.com.au/Research/Mauritius-Telecoms-Mobile-Broadband-and-Forecasts.html Last Accessed 2011-12-17*

Acronyms

BCCH	Broadcasting Control Channel (Control Info)
BSS	Base station subsystem
GSM	Groupe Spécial Mobile
KPI	Key Performance Indicator
TCH	Traffic Channels
TRX	Transmitter/Receiver (Transceiver)
PDCH	Packet Data Channel (Data Traffic)
SDCCH	Slow Dedicated Control Channel (Signalling)

Frequency and Time Domain Packet Scheduling Based on Channel Prediction with Imperfect CQI in LTE

Yongxin Wang[1], Kumbesan Sandrasegaran[2], Xinning Zhu[3], Jingjing Fei[4] Xiaoying Kong[5] and Cheng-Chung Lin [6]

[1] FEIT, University of Technology, Sydney, Australia
Yongxin.Wang-1@student.uts.edu
[2] FEIT, University of Technology, Sydney, Australia
Kumbesan.Sandrasegaran@uts.edu.au
[3]Beijing University of Post and Telecommunications, Beijing, China
zhuxn@bupt.edu.cn
[4] CSE, University of New South Wales, Sydney, Australia
jingjingf@cse.unsw.edu.au
[5] FEIT, University of Technology, Sydney, Australia
Xiaoying.Kong@uts.edu.au
[6] FEIT, University of Technology, Sydney, Australia
Cheng-Chung.Lin@eng.uts.edu.au

Abstract

Channel-dependent scheduling of transmission of data packets in a wireless system is based on measurement and feedback of the channel quality. To alleviate the performance degradation due to simultaneous multiple imperfect channel quality information (CQI), a simple and efficient packet scheduling (PS) algorithm is developed in downlink LTE system for real time traffic. A frequency domain channel predictor based on Kalman filter is first developed to restore the true CQI from erroneous channel quality feedback. Then, a time domain grouping technique employing the joint of Proportional Fair (PF) and Modified Largest Weighted Delay First (M-LWDF) algorithms is used. It was proved this proposal achieves better performance in terms of system throughput and packet loss ratio by simulation results.

Keywords

LTE, packet scheduling, channel estimation, Kalman filter, imperfect CQI

1. Introduction

To provide a long-term development of the third generation (3G) services and better quality of wireless communication in terms of high performance and capacity, simplicity and wide range of terminals, "Super 3G"[1]. standard that expands upon the 3rd generation partnership project (3GPP) specification was proposed in 2004, and it was called long-term evolution (LTE) within 3GPP. Orthogonal frequency division multiple access (OFDMA) technology was chosen in the downlink LTE instead of other access technologies (e.g. Frequency Division Multiple Access - FDMA, Time Division Multiple Access - TDMA, and Code Division Multiple Access - CDMA) due to its robustness to inter-symbol interference (ISI) and its immunity to frequency-selective fading nature of the mobile channels[2].The use of the OFDMA is best suited for simultaneous support for multiple high data rate real time applications.

In downlink LTE, packet scheduling is the term used when a packet scheduler select a user, based on a priority such as good channel quality and packet delay information, to transmit its packets on the radio resources. The LTE standard defines a resource allocation structure in both time and frequency domains. The minimum downlink LTE transmission unit that can be allocated to a user is called a physical resource block (PRB) which includes two resource blocks (RB). In the frequency domain (FD), a PRB consists of 12 consecutive sub-carries of 180 kHz total bandwidth and is of 1 ms duration in time domain (TD). This feature encourages the use of both time and frequency domain packet scheduling algorithms.

There are many well-known packet scheduling algorithms designed in time domain, such as PF algorithm[3] and M-LWDF algorithm[4]. The PF scheduler provides a good trade-off between throughput maximization and fairness guarantee by trying to serve a user whose channel quality achieves its peak maximum to average ratio in any given scheduling interval. However, because of a lack of consideration of the buffer information of each user, such as the delay of the head of line (HoL, the HoL packet of a user is the packet that has resided the longest in its buffer at the enhanced Node B (eNB)), the PF scheduling algorithm is not expected to support real time services. The M-LWDF algorithm was developed to support real time traffic in code division multiple access-high data rate (CDMA-HDR) system. This algorithm is supposed to have a reasonably overall performance due to the consideration of packet delay information, the instantaneous data rate and average throughput. The adaptation of these two packet scheduling algorithms into frequency domain for downlink LTE can be found in[5].

Various frequency domain packet scheduling algorithms that are able to allocate resources to multiple users in one transmission time interval (TTI) have been proposed. These algorithms calculate the channel quality of each user on every PRB, and select a user with the highest priority on that PRB. Recent FDPS study in [6] has shown that FDPS can provide around 40% improvement in both average system capacity and cell throughput over time domain only scheduling. In[7-10], algorithms that are based on resource allocation and resource assignment techniques were developed. The resource allocation technique calculates the number of RBs that is to be allocated to each user in each TTI. If the PRBs are insufficient to satisfy all the users' allocation requirement, the priority of each user will be calculated to decide who can be selected to obtain the competitive resource. Thereafter, the resource assignment technique allocates the PRBs to the user who has the highest priority for that PRB based on a scheduling algorithm. These resource allocation-assignment scheduling algorithms are quite complex due to the fact that it involves many steps on the RB basis, resulting in a large number of iterations. .

To balance the computational complexity and scheduling efficiency, joint time frequency schedulers are investigated in[11-13]. These schedulers operate in two steps, where the first step performs a pre-selection based on a TD scheduling algorithm to limit the number of multiplexed users that go into the frequency scheduler. The second step consists of a resource allocation to users from the TD scheduler according to a frequency scheduling algorithm. The computational complexity can be reduced by the employment of the TD scheduler which either does not consider each user's channel quality or only take the channel quality across the whole bandwidth into consideration. Furthermore, the algorithm can benefit from the limitation of the number of input users to the FD scheduler. In fact, these algorithms are still complex due to the simultaneous employment of two packet schedulers.

However, all the above mentioned packet scheduling algorithms assume that the measured CQI of each user are ideally fed back to the eNB without any error, which is impossible in a practical wireless channel. In the downlink LTE system, feedback and processing delays cause a mismatch between the current channel state and the CQI received by the scheduler[14], along with interference, multi-path fading, and shadowing. These might cause severe performance degradation especially for a channel-dependent packet scheduler that highly relies on an accurate CQI report. In [15], performance of adaptive OFDM systems due to outdated CQI was improved by channel prediction while [16] studied the effect of multiple estimations to mitigate the degradation due to CQI delay as well as OFDM channel estimation error. The problem caused by imperfect CQI in OFDMA downlink system has been studied in[17,18]. In [17], optimal resource allocation algorithms employing a dual optimization framework were developed, in which the complexity for discrete rate scenario was shown by simulation results to be much less than continuous rate scenario. In [18], a delay-sensitive cross-layer design framework with heterogeneous delay requirements was proposed so as to guarantee a fixed target outage probability in optimal power and subcarrier allocation under heterogeneous users' delay constraints

Generally, imperfect channel conditions can lead to three types of inaccurate CQI: erroneous CQI, outdated CQI and unavailable CQI when the delay of the outdated CQI is longer than the scheduling interval. Outdated CQI is the most common situation due to the fact that feedback and processing delay lead to a mismatch between the current channel state and the received CQI. The effect of outdated CQI has been analysed in [14, 19, 20]. In [14], several CQI estimators with CQI delay were studied and numerical results showed that the proposed prediction method based on Stochastic Approximation had the best performance and a lower computational complexity. In [19], the Probability Distribution Function (PDF) of the CQI corresponding to an predicted CQI was proved to improve the Multi carrier Proportional Fair (MPF) algorithm in terms of system throughput in downlink OFDMA systems. In [20], outdated and unavailable CQI scenarios were simulated by sporadic traffic. Results showed that the proposed resource allocation scheme based on an optimal HARQ and adaptive modulation and coding policy improves scheduling efficiency for LTE-advanced networks with delay constraint. In [3], simulation results presented that the improvement method can alleviate the detrimental effect caused by CQI delay.

Several multi-user systems predictors were investigated in [21]. Wiener filter for iterative channel estimation in OFDMA systems, channel estimation performance without memory (Block-Least Squares Estimation, Block-LSE), and channel estimation with memory employing Kalman filter are discussed in [22], [23], and [24], respectively.

Among the existing packet scheduling algorithms in LTE downlink system, only limited number of them have an overall acceptable performance under simultaneous multiple imperfect channel states. However, due to the large-scale computational problem, the practical implementation of the algorithms is difficult. The aim of this paper is to propose an efficient and simple packet scheduling algorithm that provides good system performance when there are simultaneous multiple CQI problems. This scheduling algorithm involves a FD predictor based on Kalman filter and a simple TD grouping technique which allows alternate use of two different FD schedulers. Simulation results show that the packet scheduling is optimized with the combination techniques, which results in a performance improvement.

The rest of this paper is organized as follows: The downlink LTE system model is provided in Section 2. The FD channel quality predictor based on Kalman filter is described in Section 3 followed by introduction of the TD grouping technique in Section 4. Section 5 gives an overview of the proposed algorithm. Simulation results are presented in Section 6. Finally, Section 7 concludes the paper.

2. Downlink LTE System Model

The LTE deploys a simple architecture that only consists of eNBs which perform packet scheduling along with other radio resource management (RRM) functions. In this paper, 7 hexagonal cells of 5 MHz bandwidth with 25 PRBs and 2 GHz carrier frequency is modelled. Each PRB is of 1 ms duration and contains 14 OFDMA symbols and a total of 168 resource elements (REs) with the usage of a normal cyclic prefix (CP). Table 1 gives the main downlink LTE system parameters.

It is assumed in this paper, each eNB is located at the centre of each cell with 100 meter radius. Users are uniformly distributed within each cell and inter-cell interference remains constant. Each user moves within the cell at a constant speed in a constant direction which is initialized as a random value and is wrapped-around whenever it reaches the cell boundary [25]. Users only have real time applications.

Users compute their instantaneous channel quality (e.g. signal-to-noise-ratio, SNR) on each RB which is mapped into a CQI, and report it to the serving eNB at each TTI. The reported CQI value is then used to make a scheduling decision and determine the data rate of a user for packet transmission. In this paper, it is assumed that all the CQI reports have 3 ms delay [26] and become unavailable at regular intervals (i.e. the CQIs are set to 0 every 10 ms). Equation (1) gives the mathematical expression of the instantaneous SINR experienced by user i on RB j at time t:

$$r_{i,j}(t) = Efficiency_{i,j}(t) * \frac{RE_{data}}{TTI} \quad (1)$$

where $Efficiency_{i,j}(t)$ is the efficiency (in bits/RE) of PRB j of user i at time t. RE_{data} is the total number of Res specified for downlink data transmission.

Table 1. Main downlink LTE system parameters

System Parameters	***Values***
Cellular layout	7 hexagonal cells
Bandwidth	5 MHz
Carrier frequency	2 GHz
Mode of operation	FDD
Number of RBs	25
Number of sub-carriers per RB	12
Total number of Sub-carriers	300
Sub-carrier spacing	15 kHz
Scheduling interval (TTI)	1 ms
Number of OFDMA symbols per TTI	14 (Normal CP)
Total number of REs	168
Total eNB transmit power	43.01 dBm
Erroneous CQI	Delay and unavailable
User speed	120km/h

3. FD CHANNEL QUALITY PREDICTOR BASED ON KALMAN FILTER

In 1960, Kalman filter was proposed as an efficient recursive solution of the least-squares method in [24]. The Kalman filter is the best possible (optimal) estimator for a wide variety of problems, due to its simple and clear mathematical model with the robustness and efficiency in the resolution of various problems.

3.1. Prediction Process

As an optimal autoregressive prediction method, Kalman filter has been applied to channel condition prediction and it enables to forecast current channel state from the state in the last time slot. Specially, the classical Kalman filter method includes two steps: estimation and correction. In the estimation step, the current CQI value is estimated from the CQI in last TTI while the estimated current CQI conditioned on last TTI is improved according to the observed CQI value that is received from each user's CQI report in the correction step. The equations are given below:
Estimation step:

$$X(t|t-1) = \Phi X(t-1|t-1) \tag{2}$$

$$P(t|t-1) = \Phi P(t-1|t-1)\Phi' + Q \tag{3}$$

where $X(t|t-1)$ is the estimate of $X(t)$ based on the time t-1. Φ is transition matrix that connects two consecutive states, $X(t)$ and $X(t$-$1)$, of the system, $P(t|t-1)$ is the covariance of $X(t|t-1)$, and $P(t-1|t-1)$ is the covariance of $X(t-1|t-1)$. The random variable Q denotes noise error matrix. The aim of calculating $P(t|t-1)$ is to estimate the accuracy of $X(t|t-1)$. If $P(t|t-1) \to 0$, which means the estimate $X(t|t-1)$ is so close to the realistic CQI value that only limited correction need to be made, otherwise, if $P(t|t-1) \to 1$, which means the estimate $X(t|t-1)$ is considered to be unreliable, and it will reduce to the observation.

Correction step:

$$Z(t) = HX(t) + R \tag{4}$$

$$K(t) = P(t|t-1)H^T[HP(t|t-1)H^T + R]^{-1} \tag{5}$$

$$\tilde{X}(t) = \tilde{X}(t|t-1) + K(t)[Z(t) - H\tilde{X}(t|t-1)] \tag{6}$$

$$P(t) = [I - K(t)H]P(t|t-1) \tag{7}$$

Equation (4) calculates the observed value $Z(t)$ from the realistic channel, where H is the channel gain, R is observed error matrix. In our situation, Z(t) can be obtained directly from the CQI value reported by the users.

Equation (5) gives the Kalman gain $K(t)$, where the superscript T represents transpose. Kalman gain can be regarded as a weight factor that "decides" the amount of improvement to the estimate $X(t|t-1)$

From the Equation (2) - (5), we can get the optimal estimation of current state $\tilde{X}(t)$ from Equation (6), which is an update from $X(t|t-1)$.

In addition, we need to update the covariance to use in the next packet scheduling interval, and this process is given by Equation (7), where I is the identity matrix.

The procedure is shown in Figure. 1.

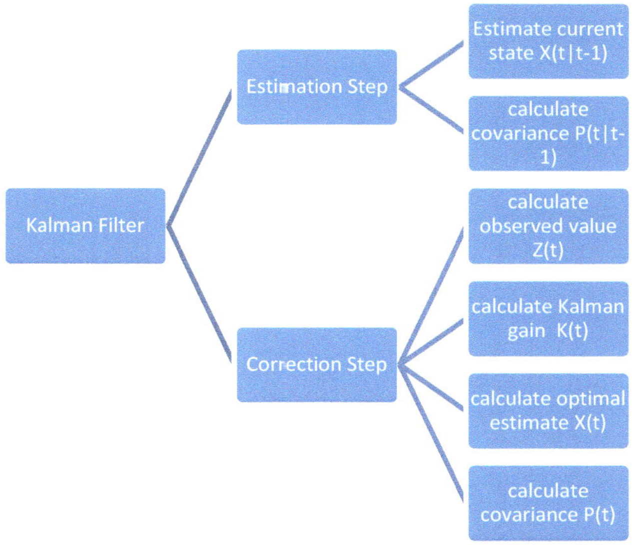

Figure 1: the procedure of Kalman filter

3.2. Initialization of Kalman Filter

Although few papers and books talk about initialization of Kalman filter, it is rather important to mention that if either the matrix Q or R are disarranged once, they keep disarranged and will not be updated and the Kalman process might give bad predictions [27].

At first, the expressions of state matrix *X(t)* and observed matrix *Z(t)* can be derived from the Taylor series:

$$x(t) = x(0) + \dot{x}(0)t + \frac{1}{2}\ddot{x}(0)t^2 + \cdots \qquad (8)$$

where $\dot{x}(0)$ indicates the first derivative of **x** at initial time *t=0*, $\ddot{x}(0)$ indicates the second derivative of **x** at initial time *t=0* etc. The Taylor series is generally reduced to:

$$x(t) = x(0) + \dot{x}(0)t + \frac{1}{2}\ddot{x}(0)t^2 \qquad (9)$$

In channel condition prediction, x(0) is the Signal-to-Interference-plus-Noise-Ratio (SINR) of user *i* on RB *j* at scheduling interval 0, denoted by γ_{ij}, we can consider $\dot{x}(0)$ as the change rate of SINR, represented as v_i and $\ddot{x}(0)$ is the gradient of v_i, denoted by b_i. γ'_{ij}, v'_i, b'_i are the observed γ_{ij}, v_i, b_i, respectively. On this basis, the state equation and observation equation can be respectively established as follows, where *T* denotes the transpose.

$$X(t) = (\gamma_{ij}, v_i, b_i)^T$$
$$Z(t) = (\gamma'_{ij}, v'_i, b'_i)^T$$

Then, the transition matrix can be determined as:

$$\Phi = \begin{vmatrix} 1 & t & t^2/2 \\ 0 & 1 & t \\ 0 & 0 & 1 \end{vmatrix}$$

Thereafter, since the observed matrix is directly obtained from the state matrix, the channel gain matrix can be determined as the identity matrix:

$$H = \begin{vmatrix} 1 & 0 & 0 \\ 0 & 1 & 0 \\ 0 & 0 & 1 \end{vmatrix}$$

After the determination of state matrix, the observed matrix, the transition matrix and the channel gain matrix, covariance matrix of the measurement error R can be initialized Let $X_{e,x}(t), X_{e,y}(t), X_{e,z}(t)$ denote the random variable that describes the measurement error. The error measurement covariance is

$$R = E[(X_{e,x}(t), X_{e,y}(t), X_{e,z}(t))^T (X_{e,x}(t), X_{e,y}(t), X_{e,z}(t))] \quad (10)$$

The elements of the principal diagonal are the covariance of $X_{e,x}(t), X_{e,y}(t), X_{e,z}(t)$ respectively. Other elements are the covariance of each two variables, since each variable is completely independent. The matrix is:

$$R = \begin{vmatrix} \sigma_{e,x}^2 & 0 & 0 \\ 0 & \sigma_{e,y}^2 & 0 \\ 0 & 0 & \sigma_{e,z}^2 \end{vmatrix}$$

Finally, the state matrix $X(t-1|t-1)$ can be simply set to zero. However, it is obvious that it is not zero, hence, we get an error in the first estimation and we can set the covariance matrix of the estimation error $P(t-1|t-1)$ as:

$$P_0 = \begin{vmatrix} {\gamma_{ij}}^2 & 0 & 0 \\ 0 & {v_i}^2 & 0 \\ 0 & 0 & {b_i}^2 \end{vmatrix}$$

4. TD Grouping Technique

Papers [13, 28] proved that packet scheduler can benefit from the use of two scheduling algorithms. However, the simultaneous use of two different scheduling algorithms can lead to an increase in scheduling processing time and computational complexity. A scheduling technique that employs FD- PF and FD-M-LWDF alternately in adjacent TTIs is proposed

FD-PF packet scheduling algorithm selects a user that maximizes $\mu_{i,j}(t)$ on PRB j at TTI i:

$$\mu_{i,j}(t) = \frac{r_{i,j}(t)}{R_i(t)} \quad (11)$$

$$R_i(t+1) = \left(1 - \frac{1}{t_c}\right) R_i(t) + \frac{1}{t_c} * rtot_i(t+1) \quad (12)$$

$$rtot_i(t+1) = \sum_{j=1}^{RB_{max}} I_{i,j}(t+1) * r_{i,j}(t+1)$$

$$I_{i,j}(t+1) = \begin{cases} 1 & \text{if packets of user } i \text{ are scheduled on PRB } j \text{ at TTI } t+1 \\ 0 & \text{if packets of user } i \text{ are not scheduled on PRB } j \text{ at TTI } t+1 \end{cases} \quad (13)$$

where $\mu_{i,j}(t)$ is the priority of user i on PRB j at scheduling interval t, $r_{i,j}(t)$ is the instantaneous data rate of user i on PRB j at scheduling interval t, $R_i(t)$ is the average throughput of user i at scheduling interval t, $I_i(t+1)$ is the indicator function of the event that packets of user i are selected for transmission at scheduling interval $t+1$ and t_c is a time constant. The design objective of FD-PF algorithm is to maximize the long term throughput of the user whose current achievable data rate is better compared to the average throughput.

FD- M-LWDF packet scheduling algorithm selects a user that maximizes $\mu_{i,j}(t)$ in Equation (14) to receive its packet in each scheduling interval.

$$\mu_{i,j}(t) = \alpha_i * W_i(t) * \frac{r_{i,j}(t)}{R_i(t)} \tag{14}$$

$$\alpha_i = -\frac{(log\delta_i)}{T_i} \tag{15}$$

where $\mu_{i,j}(t)$ is the priority of user i on PRB j at scheduling interval t and α_i is the QoS requirement of user i, $W_i(t)$ is the delay of the HoL packet of user i at scheduling interval t. Moreover, $r_{i,j}(t)$ represents the instantaneous data rate of user i on PRB j at scheduling interval t and $R_i(t)$ is the average throughput of user i at scheduling interval t. In addition, δ_i indicates the service-dependent PLR threshold of user i and T_i is the service-dependent buffer delay threshold of user i. FD-M-LWDF algorithm was proposed to support QoS of multiple real time data users sharing a wireless channel. This algorithm is based on an overall consideration of packet delay information, the instantaneous data rate and average throughput and supposed to have a reasonably good overall performance.

The TD grouping technique divides the whole scheduling TTI into two groups: odd numbered TTI group and even numbered TTI group. FD-PF and FD-M-LWDF algorithm are employed in the two groups alternately. Each packet scheduling interval only employs one single scheduling algorithm which reduces the computational complexity compared with other joint TD-FD packet scheduling algorithms that prioritize users twice. Due to the fact that this algorithm uses two different scheduling algorithms in long-term, it is supposed to have reasonable performance improvement.

5. The Joint FD Prediction and TD Grouping Packet Scheduling Algorithm

The main principle of the proposed algorithm is: firstly, at each TTI, the optimal estimated CQI value of each user is predicted by Kalman filter on each PRB, moreover the highest supported data rate of each user on each PRB can be estimated, and then the scheduling algorithm employed at this TTI is decided by the TD grouping technique which makes the scheduling decision based on the estimated CQIs.

Based on the above analysis, the proposed packet scheduling algorithm is described in Figure 2.

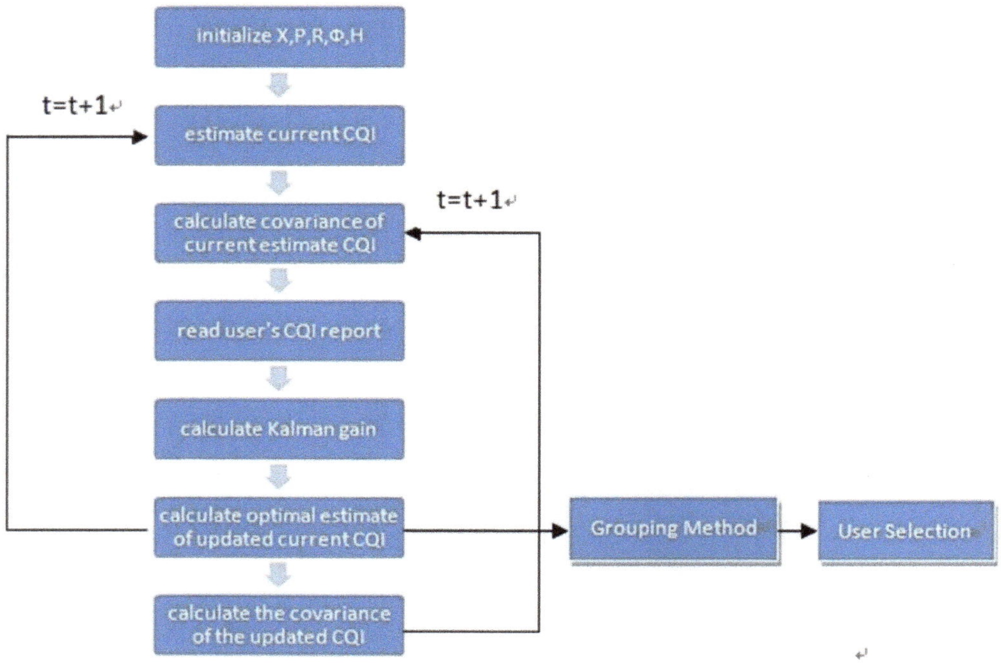

Figure 2: Flow chart of the proposed algorithm

6. Simulation Results

The metrics such as system throughput and service packet loss ratio (PLR) are used in this paper to evaluate packet scheduling performance. System throughput is defined as the total size of successfully received packets (in bits) at all users in the downlink divided by the simulation time. The system throughput has following mathematical expression:

$$system\ throughput = \frac{1}{T}\sum_{i=1}^{N}\sum_{t=1}^{T} prx_i(t) \tag{16}$$

where $prx_i(t)$ is the total size of correctly received packets (in bits) of user i at time t, T is the total simulation time and N is the total number of users.

Service PLR is an important quality of service (QoS) metric, to satisfy QoS requirements, it has to be kept below a threshold throughout a session. For real time applications, the threshold is 10^{-6} (i.e. QoS constraints of the GBR (guaranteed bit rate) services). The service PLR is defined as the ratio of the total size of discarded packets of a service to the total size of all packets of a service that have resided in the eNB buffer. The mathematical expression for the service PLR is given below:

$$PLR = \frac{\sum_{i=1}^{N}\sum_{t=1}^{T} pdiscard_i(t)}{\sum_{i=1}^{N}\sum_{t=1}^{T} psize_i(t)} \tag{17}$$

where PLR is the packet loss rate and $pdiscard_i(t)$ represent the total size of discarded packets (in bits) of user i of a service at time t. $psize_i(t)$ is the total size of all packets (in bits) that have arrived into the eNB buffer of user i of a service at time t, N is the total number of users and T is the total simulation time.

The simulation environment settings described in Table 1 is simulated. The FD-PF and FD-M-LWDF schedulers mainly serve as reference schedulers. Figure 3 and Figure 4 show the PLR of the FD-PF, FD-M-LWDF and the proposed algorithm with increasing number of users in the system (system capacity) under perfect CQI and imperfect CQI, respectively. It was observed in Figure 3 that the FD-PF algorithm has the worst PLR followed by the proposed algorithm and the FD-M-LWDF algorithm. To satisfy the QoS of GBR services, the maximum number of users that FD-PF algorithm can support is less than 30. Both of the proposed algorithm and the FD-M-LWDF algorithm can support 30 users. The use of FD-M-LWDF scheduler in the TD grouping based algorithm has significantly reduced the PLR compared with the FD-PF only scheduling and achieved the same system capacity with FD-M-LWDF algorithm, it is due to the fact that they consider packet delay information when making scheduling decision. Table 2 illustrates the proposed algorithm has 116.94% and 6.31% improvement in service PLR over FD-PF and FD-M-LWDF algorithm at 70 users, respectively.

Figure 4 shows that the proposed algorithm outperforms the other two algorithms and has the lowest PLR when the reported CQIs have 3 ms delay and become unavailable at a regular interval (10 ms). This is due to the use of the Kalman filter based CQI estimator. Packet scheduling decisions are made based on the estimated CQI instead of the erroneous CQI. It can be seen in the figure that both FD-PF and FD-M-LWDF scheduler have a substantial increase in PLR, which leads to a degradation in system capacities that fall below 30 users. This illustrates that the FD-PF and FD-M-LWDF scheduler are not tolerant of the imperfect CQI.

The system throughput of the three algorithms with increasing system capacity is shown in Figure 5 and Figure 6. A perfect CQI scenario is depicted in Figure 5 while Figure 6 demonstrates a practical system model with imperfect CQI. It can be seen from Figure 5 that the differences of the three algorithms are not obvious until the system capacity is above 50 users. FD-PF has the worst throughput performance while the proposed algorithm and the FD-M-LWDF algorithm have similar system throughput.

Similar to Figure 4, Figure 6 shows the proposed algorithm outperforms the two reference algorithm when a practical channel condition with imperfect CQI is considered. An improvement of 8.56% and 4.42% in system throughput over FD-PF and FD-M-LWDF algorithm with 70 users is illustrated in Table 3, respectively.

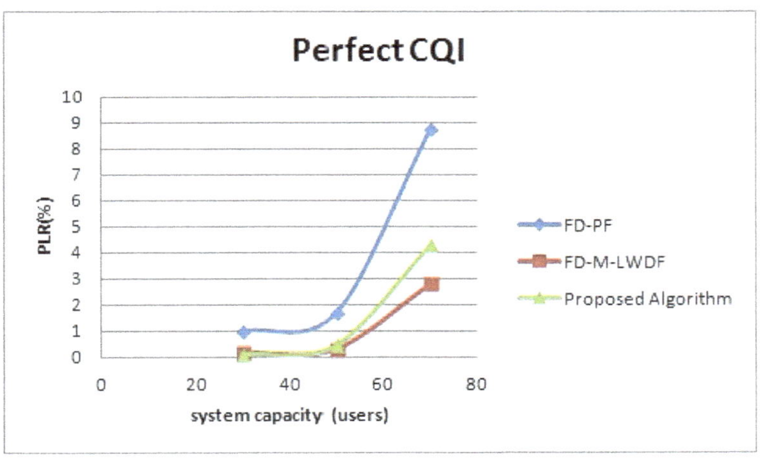

Figure 3 PLR comparison with perfect CQI

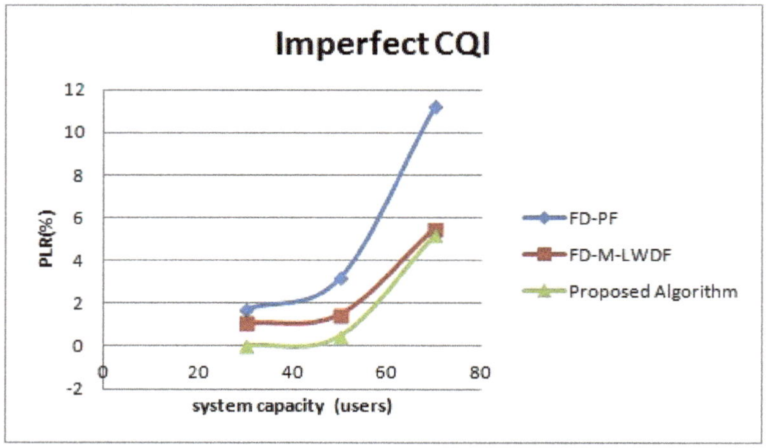

Figure 4 PLR comparison with imperfect CQI

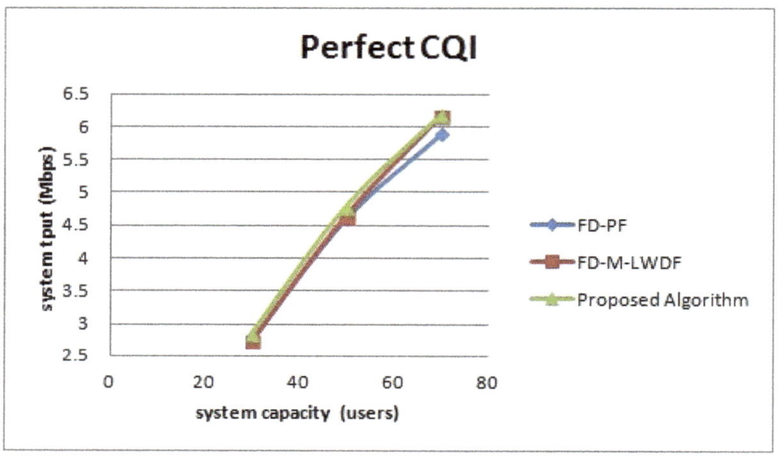

Figure 5 System throughput comparison with perfect CQI

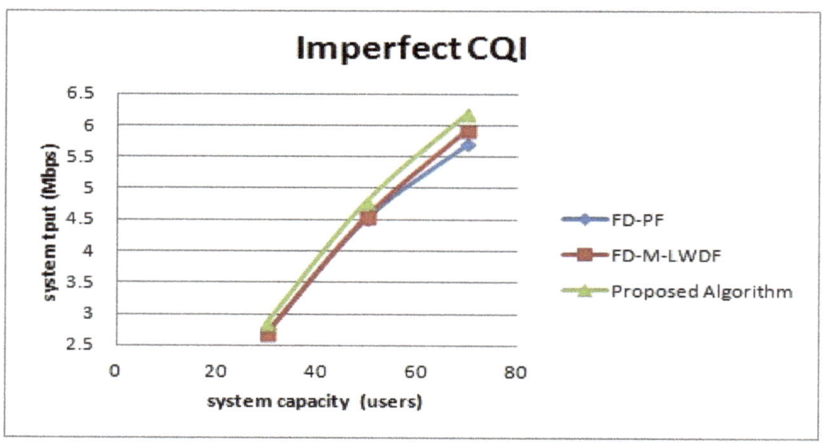

Figure 6 System throughput comparison with imperfect CQI

Table 2 PLR at 70 users

	PLR	**Performance improvement (%)**
FD-PF algorithm	11.202	116.94
FD-M-LWDF algorithm	5.4898	6.31
Proposed algorithm	5.1637	-

Table 3 System throughput at 70 users

	System throughput (Mbps)	**Performance improvement (%)**
FD-PF algorithm	5.6921	8.56
FD-M-LWDF algorithm	5.9174	4.42
Proposed algorithm	6.1792	-

7. CONCLUSIONS

This paper investigates the performance of three packet scheduling algorithms in the ideal and practical downlink 3GPP LTE system, respectively. A novel packet scheduling algorithm based on FD prediction and TD grouping technique for real time applications in the downlink LTE system is proposed in this paper. The simulation results show that the proposed algorithm outperforms FD-PF and M-LWDF algorithm in terms of service PLR and system throughput when the CQI reports has delay and errors. This simple and efficient algorithm which is robust to simultaneous multiple channel defects are proved to satisfy QoS requirements for real time users.

REFERENCES

[1] K. Kinoshita, "Current Status of "FDMA" 3G service and DoCoMo's B3G Activities," in ICB3G – 2004, May 2004, pp. 13-21.

[2] 3GPP, Physical Layer Aspects for Evolved Universal Terrestrial Radio Access (UTRA) (Release 7), 2009.

[3] A. Jalali, R. Padovani, and R. Pankaj, "Data throughput of CDMA-HDR a high efficiency-high data rate personal communication wireless system," in Vehicular Technology Conference Proceedings, 2000. VTC 2000-Spring Tokyo. 2000 IEEE 51st, 2000, pp. 1854-1858.

[4] M. Stambaugh, "HARQ Process Boosts LTE Communications," Agilent Technologies Sep, 2008.

[5] H. A. M. Ramli, R. Basukala, K. Sandrasegaran, and R. Patachaianand, "Performance of well known packet scheduling algorithms in the downlink 3GPP LTE system," in Communications (MICC), 2009 IEEE 9th Malaysia International Conference on, 2009, pp. 815-820.

[6] A. Pokhariyal, T. E. Kolding, and P. E. Mogensen, "Performance of Downlink Frequency Domain Packet Scheduling for the UTRAN Long Term Evolution," in Personal, Indoor and Mobile Radio Communications, 2006 IEEE 17th International Symposium on, 2006, pp. 1-5.

[7] J. Gross, J. Klaue, H. Karl, and A. Wolisz, "Subcarrier allocation for variable bit rate video streams in wireless OFDM systems," in Vehicular Technology Conference, 2003. VTC 2003-Fall. 2003 IEEE 58th, 2003, pp. 2481-2485 Vol.4.

[8] N. Ruangchaijatupon and J. Yusheng, "Simple Proportional Fairness Scheduling for OFDMA Frame-Based Wireless Systems," in Wireless Communications and Networking Conference, 2008. WCNC 2008. IEEE, 2008, pp. 1593-1597.

[9] P. Jeongsik, H. Sungho, and C. Ho-Shin, "A Packet Scheduling Scheme to Support Real-Time Traffic in OFDMA Systems," in Vehicular Technology Conference, 2007. VTC2007-Spring. IEEE 65th, 2007, pp. 2766-2770.

[10] X. Liu, L. Guangyi, Y. Wang, and P. Zhang, "Downlink Packet Scheduling for Real-Time Traffic in Multi-User OFDMA System," in Vehicular Technology Conference, 2006. VTC-2006 Fall. 2006 IEEE 64th, 2006, pp. 1-5.

[11] A. Wang, L. Xiao, S. Zhou, X. Xu, and Y. Yao, "Dynamic resource management in the fourth generation wireless systems," in Communication Technology Proceedings, 2003. ICCT 2003. International Conference on, 2003, pp. 1095-1098 vol.2.

[12] A. Pokhariyal, K. I. Pedersen, G. Monghal, I. Z. Kovacs, C. Rosa, T. E. Kolding, and P. E. Mogensen, "HARQ Aware Frequency Domain Packet Scheduler with Different Degrees of Fairness for the UTRAN Long Term Evolution," in Vehicular Technology Conference, 2007. VTC2007-Spring. IEEE 65th, 2007, pp. 2761-2765.

[13] B. Kian Chung, S. Armour, and A. Doufexi, "Joint Time-Frequency Domain Proportional Fair Scheduler with HARQ for 3GPP LTE Systems," in Vehicular Technology Conference, 2008. VTC 2008-Fall. IEEE 68th, 2008, pp. 1-5.

[14] R. A. Akl, S. Valentin, G. Wunder, and S. Stanczak, "Compensating for CQI Aging By Channel Prediction: The LTE Downlink."

[15] M. R. Souryal and R. L. Pickholtz, "Adaptive modulation with imperfect channel information in OFDM," in Communications, 2001. ICC 2001. IEEE International Conference on, 2001, pp. 1861-1865 vol.6.

[16] W. Yafeng and Y. Hongwen, "Retransmission priority scheduling algorithm for forward link packet data service," in Communication Technology Proceedings, 2003. ICCT 2003. International Conference on, 2003, pp. 926-930.

[17] I. C. Wong and B. Evans, "Optimal resource allocation in the OFDMA downlink with imperfect channel knowledge," Communications, IEEE Transactions on, vol. 57, pp. 232-241, 2009.

[18] D. S. W. Hui and V. K. N. Lau, "Delay-sensitive cross-layer designs for OFDMA systems with outdated CSIT," in Wireless Communications and Networking Conference, 2007. WCNC 2007. IEEE, 2007, pp. 264-269.

[19] H. Ayoub and M. Assaad, "Scheduling in OFDMA Systems With Outdated Channel Knowledge," in Communications (ICC), 2010 IEEE International Conference on, 2010, pp. 1-5.

[20] T. Villa, R. Merz, R. Knopp, and U. Takyar, "Adaptive modulation and coding with hybrid-ARQ for latency-constrained networks," in European Wireless, 2012. EW. 18th European Wireless Conference, 2012, pp. 1-8.

[21] J.-H. Jeon and Y.-H. Lee, "Prediction of channel information in multi-user OFDM systems," in Circuits and Systems, 2006. MWSCAS'06. 49th IEEE International Midwest Symposium on, 2006, pp. 468-472.

[22] J. Bonnet and G. Auer, "Chunk-based channel estimation for uplink OFDM," in Vehicular Technology Conference, 2006. VTC 2006-Spring. IEEE 63rd, 2006, pp. 1555-1559.

[23] D. Aronsson, T. Svensson, and M. Sternad, "Performance evaluation of memory-less and Kalman-based channel estimation for OFDMA," in Vehicular Technology Conference, 2009. VTC Spring 2009. IEEE 69th, 2009, pp. 1-5.

[24] R. E. Kalman, "A new approach to linear filtering and prediction problems," Journal of basic Engineering, vol. 82, pp. 35-45, 1960.

[25] A. G. Orozco-Lugo, F. A. Cruz Prez, and G. Hernandez Valdez, "Investigating the boundary effect of a multimedia TDMA personal mobile communication network simulation," in Vehicular Technology Conference, 2001. VTC 2001 Fall. IEEE VTS 54th, 2001, pp. 2740-2744 vol.4.

[26] K. I. Pedersen, G. Monghal, I. Z. Kovacs, T. E. Kolding, A. Pokhariyal, F. Frederiksen, and P. Mogensen, "Frequency Domain Scheduling for OFDMA with Limited and Noisy Channel Feedback," in Vehicular Technology Conference, 2007. VTC-2007 Fall. 2007 IEEE 66th, 2007, pp. 1792-1796.

[27] M. Kohler, Using the Kalman filter to track human interactive motion: modelling and initialization of the Kalman filter for translational motion: Citeseer, 1997.

[28] G. Monghal, K. I. Pedersen, I. Z. Kovacs, and P. E. Mogensen, "QoS Oriented Time and Frequency Domain Packet Schedulers for The UTRAN Long Term Evolution," in Vehicular Technology Conference, 2008. VTC Spring 2008. IEEE, 2008, pp. 2532-2536.

Traffic Based Analysis of Efficient & Dynamic Probabilistic Broadcasting Algorithm in MANETs Routing Protocols

Deepak Dembla * and Yogesh Chaba **

*Associate Professor, Deptt. of CSE, AIET, Jaipur INDIA
dembla.deepak@gmail.com

**Associate Professor-Deptt. of CSE GJU of Science & Technology, Hisar INDIA

**yogeshchaba@yahoo.com

Abstract

A mobile adhoc network is a collection of autonomous mobile nodes that that communicate with each other over wireless links. Hosts need to specify the requirements of the neighbors for efficient routing outside their transmission range. However, since there is no stationary infrastructure such as base stations, each host has to act as a router in itself and hence, a routing protocol runs on every host and is therefore subject to the limit of the resources at each mobile host. This paper proposes a new Efficient and dynamic probabilistic broadcasting (EDPB) approach, which solves the broadcast storm problem of AODV. Global Mobile Simulator is used to run simulations. The main performance metrics considered in simulations are routing overhead and end-to-end delays. The results show that, the normalized routing load is reduced to around 35% to 40% compared with AODV-blind flooding and AODV-fixed probability model, at a very heavy traffic load, when used with AODV-EDPB. The results show that AODV-EDPB algorithm performs better than AODV-blind flooding and AODV- fixed probability in terms of packet delivery ratio, end-to-end delay etc. especially in dense networks.

Keywords

MANET, GloMoSim, AODV, flooding, broadcast.

1. INTRODUCTION

An adhoc wireless network is a collection of two or more devices which are equipped with wireless communication and networking capability which can communicate within their radio transmission range. Adhoc networks can be categorized in two ways i.e fixed and mobile adhoc networks. MANETs are self-organizing i.e. nodes in the network move randomly with dynamic topologies and are autonomous in nature. In fixed adhoc networks are also called infrastructure mobile networks, in which all the nodes in the network are connected to base stations within their radio range. Multihop routing technique is implemented in MANETs and every node in mobile ad hoc networks act as a router. All on-demand MANETs routing protocols implement a simple abd generic form of broadcasting known as flooding to find out the routes and whenever a source has data to send to a destination, but does not have the route to the same, it will initiate a route-query process. In blind flooding, technique, the source node broadcasts a Route Request Query packet to its neighbors. Each node broadcast this RREQ packet exactly once if it is found for the first time and the destination node receives the this packets along several different paths, chooses the best route as per the to the route selection policies of the particular routing protocol and notifies the source node about the route selected through a

Route-Reply packet. In broadcasting process, a source node sends a message to all other nodes in the mobile adhoc network. Many routing protocols and applications are operated based on broadcast, such as AODV, DSR routing protocols, and data dissemination to all nodes in network. Therefore, a robust and efficient broadcasting algorithm is necessary in an ad hoc network environment. Flooding technique delivers the data to every node in the network but the topological information needs to be maintained or known in advance. In networks, where node mobility is so high that a given unicast routing protocol may fail to keep up with the rate of topology changes, flooding may become the good alternative for routing data. However, in other scenarios where node mobility is tractable by a routing protocol, flooding can be a inefficient option because the total number of transmissions to deliver a single message to a destination.is N (network size), and in blind flooding, a node transmits a packet, which is received by all neighbor nodes that are within the radio transmission range. Upon receiving the broadcast packet, each node determines if it has transmitted the packet before. If not, then the packet is retransmitted. This process allows for a broadcast packet to be disseminated through the adhoc network. Blind flooding automatically stops when all nodes have received and transmitted the packet being sent at least once. As all the nodes participate in the broadcast, blind flooding suffers from broadcast storm problem, which leads to redundant rebroadcasts, contention and packet collision.

The simplest and most trivial broadcasting algorithm is pure flooding. Every node that receives the broadcast message retransmits it to all its neighbors. The problem of pure flooding is that it produces many redundant messages, which may consume scarce radio and energy resources, and cause collision that is called broadcast storm problem. Therefore, the basic principle of designing an efficient and resource conservative broadcast algorithm is trying to reduce the redundant messages, which means to inhibit some nodes from rebroadcasting and the message can still be disseminated to all nodes in the network. Broadcasting is important in MANETs for routing information discovery, for instance, protocols such as dynamic source routing (DSR), Adhoc on demand distance vector (AODV) , zone routing protocol (ZRP) and location aided routing (LAR) use broadcasting to establish routes. The techniques for efficient broadcasting can be grouped into four families' e.g. simple flooding method, probability-based methods, area-based methods and the neighbor knowledge-based methods. In simple flooding [5], source node initiates flooding by broadcasting a packet to all its neighbors and in turn, neighbor nodes rebroadcast the packet exactly once and the process continues until each node in the network has retransmitted the packet. Finally, all the nodes reachable from the source receive the packet. But, flooding causes broadcast storm problem. In probability-based methods [6], each node is assigned a probability for retransmission depending upon the topology of the network. A common range is assumed in area based technique, wherein, a node will rebroadcast if only sufficient new area can be covered with the retransmission range. Each node stores neighborhood state information and uses it to decide whether to retransmit or not, in neighbor knowledge based methods [7] The main aim behind all approaches is to minimize the number of retransmission messages and the number of nodes transmitting the message.

2. RELATED STUDY

AODV does not maintain the topological information about the whole network and constructs a route when it is required. When a source node needs a route to some destination , it broadcasts a RREQ packet to its immediate neighbors. Each neighboring node rebroadcasts the received RREQ packet only once if it has no valid route to the destination. Each intermediate node that forwards the RREQ packet creates a reverse route pointing towards the source node S. When the destination node or an intermediate node with a valid route to the destination receives this packet, it replies by sending a route reply packet. The route reply packet RREP packet is unicast towards the source node S along the reverse path set-up by the forwarded packet. Each intermediate node that participates in forwarding the RREP packet creates a forward route pointing towards the destination.

Zhang and Agrawal [12] proposed a probabilistic scheme for AODV route discovery process, where the probability to forward an RREQ packet is determine by the number of duplicate RREQ packets received at a node. Because some packets may be lost due to collisions, the number of duplicate packets at the receiver, is not a good criterion to get the local attributes of the forwarded node.

M. Bani Yassein et. al.[13], proposed an a adaptive routing algorithm based on probabilistic approach in which different medium are used for flooding based on number of average number of neighbors nodes multiple criteria i.e. high, medium and low. These values are set according to the local neighbors' information.

Qi.Zhang and Dharma [14] have implemented approach that uses the concept of gossip, but the construct minimal dominating set is not required. Instead of that, categorizes mobile hosts into four groups according to their neighborhood information. For each group, there is a specified value of probability so the nodes with more neighbors are given higher probability, while the nodes with less neighbors are given lower probability of forwarding the route request packets. Probabilistic routing approaches [10] [11] have been proposed to help control the forwarding of the routing controls packets.

Cartigny and Simplot [15] proposed an technique which combines the advantages of both probability based and distance based schemes for retransmitting the route requests by neighbor nodes which are in the border transmission range of sender nodes. The value of probability is determined by the information collected from the nods neighbors.

3. PROPOSED MODEL (EDPB SCHEME)

The probabilistic scheme is similar to ordinary flooding, except that the nodes only rebroadcast with a predetermined probability. In dense network, it is much likely that multiple nodes share similar transmission coverage. Thus, having some random nodes not to rebroadcast saves network resources without harming packet delivery effectiveness. In sparse network, there is much less shared network coverage and, therefore, not all nodes will receive all the broadcast packets with this scheme unless the probability parameter is high and when the probability is 100%, this scheme is identical to ordinary flooding scheme. The proposed algorithm is a combination of the probability based schemes and knowledge based schemes and automatically adjusts the re- broadcast probability at each mobile node according to the value of the immediate number of neighbors in its transmission range.

The value of probability changes when the host node moves to a different location in the neighborhood. The rebroadcast probability is higher in sparse networks as compared to denser networks. EDPB algorithm achieves higher saved rebroadcast as compared with the probabilistic approach where probability is fixed and in this scheme, the node rebroadcasts a message according to a high probability if the message is received for the first time, and the number of neighbors of node A is less than average number of neighbours on hearing a broadcast message m at node A, and if node A has a low degree, retransmission will be done accordingly, otherwise, if A has a high degree and its rebroadcast probability is set at low. In probabilistic scheme, a node rebroadcasts the message with a fixed probability so that every node has the same probability to rebroadcast the message, when receiving a broadcast message for the first time. In dense networks, multiple nodes share similar transmission ranges.

These probabilities control the frequency of rebroadcasts requests and thus save network resources. In sparse networks, there is much less shared coverage, thus some nodes will not receive all the broadcast packets unless the probability is high. So if the rebroadcast probability is set to a far smaller value. On the other hand, if probability is set to a far larger value, many duplicate rebroadcast packets will be generated which arises the need for automatic adjustment.

To select the value of probability p, the average neighbor number is analyzed. Let X be the area of the adhoc network, N is the total no. of mobile nodes in the network and R is the radio transmission range then, average number of nodes (n bar) can be computed from the formula n bar = (n-1)(p)(πr^2)/X. The value of fixed probability that used in AODV-FP is set at p= 0.8. It has been shown that this probability value enable fixed probabilistic flooding to achieve a good performance. The brief outline of the proposed algorithm is shown in figure1. The rebroadcast probability is set high at the hosts in sparser areas where number of neighbor nodes is less and low at the hosts in denser areas where the number of nodes is high. This scheme for density based rebroadcasting implementation analysis requires mobile hosts which periodically exchange "HELLO" messages between neighbors to and construct a one hop neighbor list in routing table at each host. A high number of neighbours imply that the host is in a dense area and a low number of neighbors imply that the host is situated in a sparser area.

```
On hearing a broadcast packet m at node A
Get the broadcast ID from the message: n
bar, average number of neighbor(threshold)
Get degree n of node X (no. of neighbors
of node X)
If packet m received for the first time then
If n<n bar then
Node X has a low degree
Set high rebroadcast probability p=p1;
Else
n>=n bar
Node A has a high degree
Set low rebroadcast probability p=p2;
Endif
Endif
Generate a random number RN over [0,1]
If RN <= p
then rebroadcast the received message;
else
 drop the message
endif
```

Figure 1. EDPB algorithm

4. SIMULATION MODEL

GloMoSim is used as the simulation tool. Performance evaluation of the protocols is done on parameters such as routing overhead and end-to-end delay in different network density, normalized routing load, etc. The simulation, different mobile nodes are moving in different network area. Each node has 250 meter radio transmission range. The node mobility model used is random-waypoint model [21]. In which each node at the beginning of the simulation remains stationary for a pause time, then chooses a random destination and starts moving towards it with a speed selected from a uniform distribution range between 0 to maximum speed. The speed varies from 5 to 20 m/sec. The simulation parameters used, are as per Table 1.

Table1. Simulation Parameters

Parameter	**Value**
Pause time	0,10,20,40,600 sec
Number of nodes	25,50,75,100
Traffic load	5,10,15,20,25 connections
Mobility model	Random waypoint model

No. of trials	30
Data Traffic	CBR
Transmission range	250 meters
Bandwidth	2Mbps
Simulation time	1100 secs
Packet size	512 bytes
Topology size	800x800 m2
Node speed	5,10,15,20, m/sec

5. PERFORMANCE METRICS AND RESULTS ANALYSIS

AODV is already implemented in GloMoSim simulator and analysis is done and the results are obtained for AODV in which traditional blind flooding scheme of route discovery operation is followed and implemented for route rebroadcast request packets. Now, the aim is to reduce the flooding of route request packets during the rout discovery process, and as a result it reduces the broadcast storm problem and net effect is that overall network is improved by the reduced average end-to-end delay and also routing overhead.

Traffic Load Analysis: The performance of three protocols in terms of routing overhead vs. offered traffic load is shown below in figure 2. It is found that at a very heavy traffic load, the normalized routing load is reduced to 35% to 40% as compared with AODV blind flooding and AODV fixed probability scheme, at a very heavy traffic load as depicted in below mentioned figure no.3, when used with AODV-EDPB. The delays for different traffic loads is also shown in below mentined figure no.3. The packet transmission load has a direct impact on latency and if the number of packets is high, than the number of collisions is high, and in turn leads to more retransmissions. As a result, packet experiences more latency and because of higher number of duplicate route request packets of rebroadcasts, AODV-EDPB experiences lower latencies than AODV-Blind Flooding and AODV-Fixed Probability.

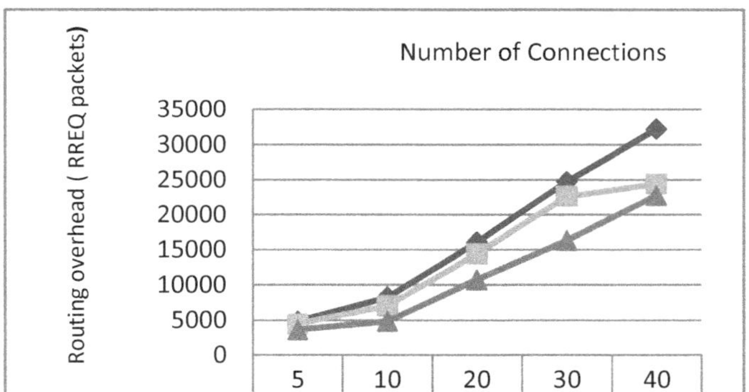

	5	10	20	30	40
AODV-BF	4850	8275	16100	24775	32225
AODV-FP	4510	7050	14450	22650	24450
AODV-EDPB	3620	4850	10750	16324	22750

Figure 2: Routing Overhead vs. Traffic load (no. of Connections)

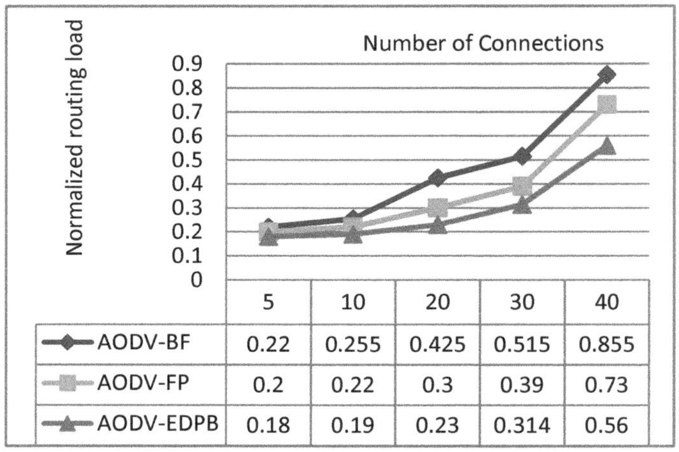

	5	10	20	30	40
AODV-BF	0.22	0.255	0.425	0.515	0.855
AODV-FP	0.2	0.22	0.3	0.39	0.73
AODV-EDPB	0.18	0.19	0.23	0.314	0.56

Figure 3. Routing load vs. Traffic load (no. of connections)

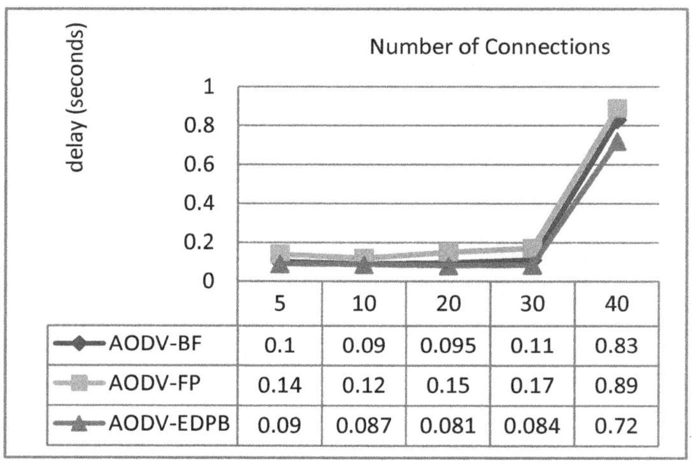

	5	10	20	30	40
AODV-BF	0.1	0.09	0.095	0.11	0.83
AODV-FP	0.14	0.12	0.15	0.17	0.89
AODV-EDPB	0.09	0.087	0.081	0.084	0.72

Figure 4. Delay vs. Traffic load (No. of Connections)

Network pause time load analysis: The results in figure 5 show the routing overhead generated by three protocols when the pause time is increased from 10 to 100 sec.

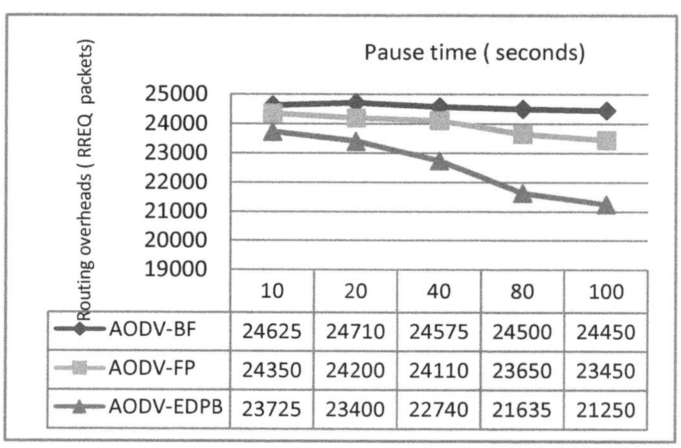

	10	20	40	80	100
AODV-BF	24625	24710	24575	24500	24450
AODV-FP	24350	24200	24110	23650	23450
AODV-EDPB	23725	23400	22740	21635	21250

Figure 5. Routing overhead vs. pause time.

The normalized routing load generated by all the three protocols with different network pause times and the number of constant bit rate is set at 30 as shown in figure 6. The results reveal that when the network pause time is increased, the mobility of nodes is decreased which leads to decrease in normalized routing load. It shows that AODV-EDPB has better performance over the other two protocols. The delays incurred by all the three protocols for different network pause time as shown in figure 7. The longer the average pause time is, the less is the node movement within the network, and this means that the nodes look like fixed rather than mobile, so the number of generated RREQ packets will be low at network with high pause time.

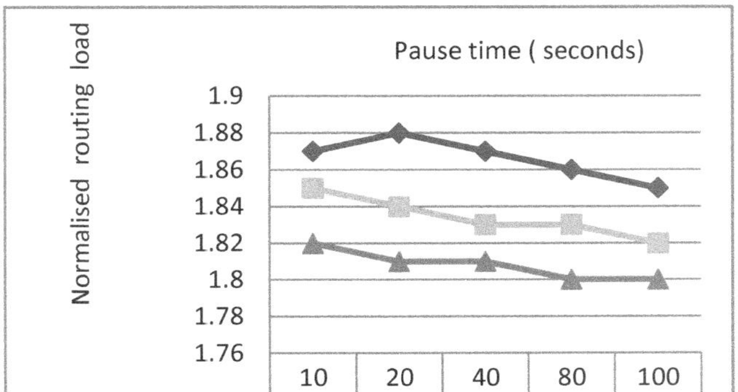

	10	20	40	80	100
AODV-BF	1.87	1.88	1.87	1.86	1.85
AODV-FP	1.85	1.84	1.83	1.83	1.82
AODV-EDPB	1.82	1.81	1.81	1.8	1.8

Figure 6. Routing load vs. pause time.

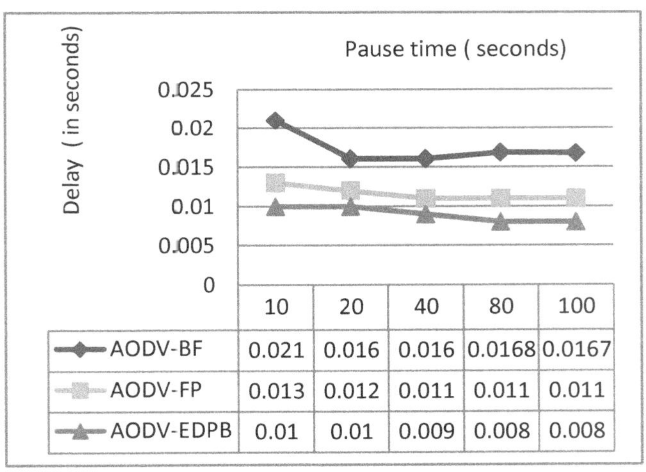

	10	20	40	80	100
AODV-BF	0.021	0.016	0.016	0.0168	0.0167
AODV-FP	0.013	0.012	0.011	0.011	0.011
AODV-EDPB	0.01	0.01	0.009	0.008	0.008

Figure 7. Delay vs. pause time.

6. CONCLUSION

The simulation results show that new proposed algorithm i.e. EDPB algorithm has definitely superior performance over traditional AODV Blind flooding and AODV fixed probability versions of routing protocols and generates much lower routing overhead and end-to-end delay, as a consequence, the packet collisions and contention in the network is reduced. The proposed algorithm is a combination of the probabilistic and knowledge based approaches and automatically adjusts the rebroadcast probability at each mobile node according to the knowledge of the number of neighbours. The value of probability changes when the nodes move to a different neighborhood nodes. In a sparse regions, the rebroadcast probability is high as compared to probability in dense regions where no. of local neighbors is high. Compared with the probabilistic approach where p is fixed, EDPB algorithm achieves higher saved rebroadcast. The new proposed algorithm determines the rebroadcast probability by taking in to account the network density. The results also show that although the traffic load is increased, the normalized routing load is still low. As a continuation of this research in the future, a plan to combine the EDPB with different approaches is formulated this suggests solving the broadcast storm problem, and will analyze the effect of this improvement on the performance of DSR and other on demand routing protocols.

REFERENCES

1. K. Viswanath, and K Obraczka. "Modeling the performance of flooding in wireless multi-hop Adhoc n/w". In Comp Comm 29, 949-956, 2006.
2. C. Perkins, E. Beldig-Royer and S. Das. "Adhoc on Demand Distance Vector (AODV) Routing". Request for Comments 3561,July2003,http://www.ietf.org/rfc/rfc3561.txt.,RFC, 2007.
3. B. Williams and T. Camp, "Comparison of Broadcasting Techniques for Mobile Adhoc Networks", Proc. of the 3rd ACM International Symposium on Mobile Adhoc N/W and Comp, pp. 194 – 205 (2002).
4. K. Viswanath, and K Obraczka. "Modeling the performance of flooding in wireless multi-hop Adhoc networks". Comp Comm, 949-956, 2006.
5. G. Lim, K. Shin, S. Lee, H. Yoon and J. S. Ma, " Link Stability and Route Lifetime in Adhoc Wireless Networks", Proc. of International Conference on Parallel Processing Workshops, pp. 116 – 123 (2002).
6. Mohammed, M. Ould-Khaoua, L. M. Mackenzie, and J. Abdulai, "An Adjusted Counter-Based Broadcast Scheme for Mobile Adhoc Networks," in Proc. of the Tenth International Conference on Computer Modeling and Simulation, 2008, pp. 441 - 446.
7. J. Barritt, B. Malakooti, Z. Guo. Intelligent Multiple- Criteria Broadcasting in Mobile Ad-hoc Networks. In Performance and Management of Wireless and Mobile Networks, LCN, pp 761-768, 2006
8. Perkins, E. Belding-Royer, and S. Das, "Adhoc On- Demand Distance Vector (AODV) Routing," IETF Mobile Adhoc Networking Working Group INTERNET DRAFT, RFC 3561, July 2003, http://www.ietf.org/rfc/rfc3561.txt, RFC, 2007.
9. Johnson, Y. Hu, and D. Maltz, "The Dynamic Source Routing Protocol (DSR) for Mobile Adhoc Networks," IETF MANET Working Group INTERNET DRAFT, http://www.ietf.org/rfc/rfc4728.txt.
10. Sasson Y, Cavin D, Schiper A. "Probabilistic broadcast for flooding in wireless mobile Adhoc networks." In Proc. of IEEE wireless communication and networking conference (WCNC), March 2003.
11. Dimitrios Liarokapis, Ali Shahrabi, Andreas Komninos. "DibA:An Adaptive Broadcasting Scheme in Mobile Adhoc Networks". In Communication N/W and Services Research, CNSR, 224-231, 2009.
12. Q. Zhang and D. P. Agrawal, "Dynamic probabilistic broadcast in MANETs," J of Parallel and Dist Comp, vol. 65, pp. 220-233, 2005.
13. Bani-Yassein M, Ould-Khaoua M, Mackenzei LM, Papanastasiou S. "Performance analysis of adjusted probabilistic broadcasting in mobile Adhoc networks". Int. J. Wireless Inform N/W, pp:127–40,2006.

14. Qi Zhang and Dharma P. Agrawal, "Analysis of Leveled probabilistic Routing in Mobile Adhoc Networks", OBR Center for distributed and Mobile Computing, ECECS, Uniy Cincinnati incinnati, OH 45221-0030.

15. J. Cartigny, D. Simplot. "Border node retransmission based probabilistic broadcast protocols in ad-hoc networks", Telecommunication Systems, vol. 22, no 1–4, pp. 204, 2003.

16. Y.-C. Tseng, S.-Y. Ni, E.-Y. Shih, "Adaptive approaches to relieving broadcast vol.storm in a wireless multihop mobile Adhoc network" ,IEEE Transactions on Computers, 52, no 5,2003.

17. M.B.yassin, M.Ould Khaoua, L.M.Mackenzie et. al "Improving Route Discovery in on demand routing protocol using Local Topology Information in MANETs" ACM 2006.

18. M. D. Colagrosso, "Intelligent broadcasting in mobile Adhoc networks: Three classes of adaptive protocols," EURASIP Journal on Wireless Communication and Networking, vol. 2007, pp. 16, 2007.

19. Perkins C.E. "Ad Hoc networking", Adison Wesley, ISBN – 0-201-30976-9, 2001.

20. C-K. Toh, Ad Hoc Mobile Wireless Networks: Protocols and Systems (Prentice-Hall, New York, 2002).

21. C. Bettstetter, H. Hartenstein and P. Costa: Stochastic properties of the random-waypoint mobility model, Wireless Networks, Vol.10, no. 5, pp. 555-567 (2004).

17

AMPLITUDE AND PHASE MODULATION FOR ULTRASONIC WIRELESS COMMUNICATION

Nan Gao [1] and Shigeru Shimamoto [2]

[1 2]Graduate School of Global Information and Telecommunication Studies
Tokyo, Japan

ABSTRACT

Short range wireless communications have been used more and more frequently in our life. But the electromagnetic fields waves also have some disadvantages. One of these disadvantages is health problems. Many studies shows the electromagnetic field waves using for communication may damage our health. And in most hospitals, they also have bans on the use of mobile phones and wide area networks because of Electromagnetic Interference. So this paper studied the use of ultrasound for wireless communication in air, instead of using electromagnetic field wave. In order to find an advisable modulation method for ultrasound wireless communication, Amplitude modulation method and Phased modulation method has been test.

KEYWORDS

wireless communication; electromagnetic pollution; ultrasound, modulation

1. INTRODUCTION

In recent years, short range wireless communications have been used more and more frequently in our life. Just like RF systems[1] which using various kinds of technologies, including many popular commercial protocols. Other wireless communications like cell phones or wireless local area networks are also used very often today. Almost all of these communication systems link to each other by electromagnetic field waves, otherwise referred to electromagnetic radiation. But the electromagnetic fields waves also have some disadvantages. One of these disadvantages is health problems. Now several research work have already been taken out to study the adverse health effect of electromagnetic field waves. The earliest scientific study in this area was following the work of epidemiologist Nancy Wertheimer in 1979[2]. Her research showed there was a high incidence of leukemia among the children who lived near high tension power lines, where electromagnetic fields were stronger. After that, more research has broadened in the biophysical field of electromagnetic field waves and health. In May 2011, the WHO's International Agency for Research on Cancer published a review of the evidence on health risks of electromagnetic fields wave. This review concluded limited evidence that cell phone users might be at increased risk of glioma and acoustic neuroma, and there was inadequate evidence of any other health risks posed by electromagnetic fields wave [3].

Also in most hospitals, they have bans on the use of mobile phones and wide area networks for different reason. Because these wireless communications may be dangerous. In hospitals, there are many electronic devices which keep watch on patient's heartbeats and other things when they are getting surgery. But if we use mobile phone near these devices, the electromagnetic waves may cause them stop working. This will cause bad accidents when the doctors and nurses are trying to save lives. So any communications system within the hospital should not generate significant EMI and have no EMC problems

So we consider using ultrasound wave for wireless communication instead of using electromagnetic field wave here. The ultrasound wave has already been used for communication in the past time. For example, through-wall communication[4] and underwater communication[5]. Now with the development of technology, it is possible to send ultrasound wave through air within a certain distance. This means that ultrasound wave can be used for short-range wireless communication in air. Some studies of ultrasound communication system have also been published, just like a wireless keyboard using ultrasound[6]. Ultrasound wave can provide a better environment for our health, or a safer wireless communication environment in hospital. Using ultrasound will also offer other advantages over position and security. Because the propagation speed of acoustic wave is much slower than electromagnetic field wave and it is effectively blocked by most barriers.

2. Apparatus and Experiment

Today wireless communication network is envisioned to provide a wide range in our life. More and more applications of wireless network have been developed. Just like wireless sensor network using in monitoring[7,8], or mobile ad hoc network[9]. At the same time, wireless communication network is also used in medical area. For example, wireless body area network to capture continuous data from patients in real time [10,11]. Or wireless sensor network used in healthcare applications for vital sign collection from medical devices[12,13]. Even a wireless sensor network for air pollution monitoring[14]. Some other technology just like ZigBee is also used wildly[15]. However, all of these wireless communication systems may cost EMS problems. So in most hospitals, they have bans on the use of common wireless communication systems, just like cell phone or wireless location area network. Because EMS problems may be dangerous in hospital. As we know, there are many electronic devices which keep watch on patient's heartbeats and other things when they are getting surgery. But if we use mobile phone near these devices, the electromagnetic waves may cause them stop working. This will cause bad accidents when the doctors and nurses are trying to save lives.

Any communications system within the hospital should not generate significant EMI and have no EMC problems. Some new wireless systems have been proposed, just like optical wireless network[16]. Here we consider using ultrasound wave for wireless communication system in hospital.

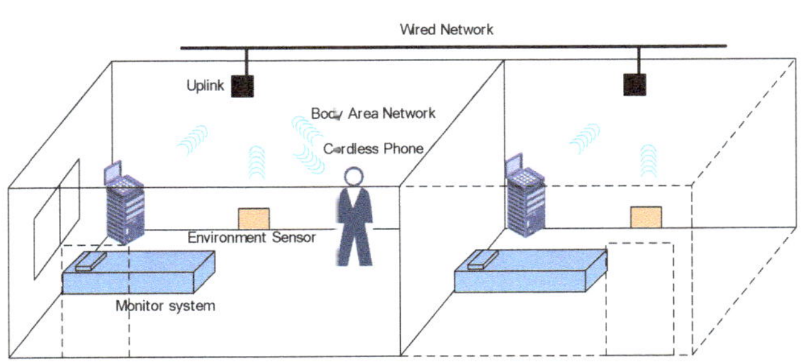

Figure 1. Ultrasound wave communication system

Figure 1 is a diagram of ultrasound wireless communication system. We set one main ultrasound transceiver in every room. This transceiver is used to receive ultrasound signals sending by other devices and communicate with other room through wired network. In this system, all the wireless communications use ultrasound wave. Therefore, a modulation method using for ultrasound communication is needed. And the main objective of this paper is finding a suitable modulation method for this communication system.

2.1. Basic Testing

In order to test different modulation method, a transport model for ultrasound wireless communication is studied. Figure 2 shows the typical experimental arrangement. In this experimental, the source was driven by a signal generator. This signal generator can supply the required digital signal for experimental. A linear power amplifier with a gain of 32dB was used to boost the output of the signal generator. The received signal was fed into signal analyzer for offline signal processing.

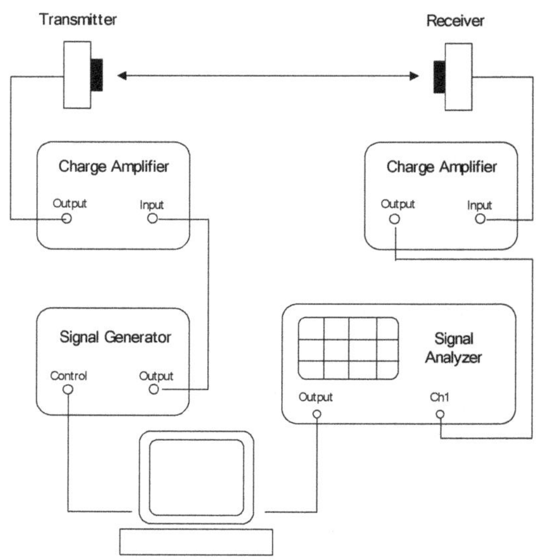

Figure 2. Schematic diagram of the experimental arrangement

In this experiment, the first important parameter is operating frequency of the ultrasound transducer. Figure 3 shows the test result of operating frequency. In this test, two transducer was set face to face, and the distance between them was 1 meter. It can be seen that optimum frequency is 40kHz, while 60kHz is also can be used.

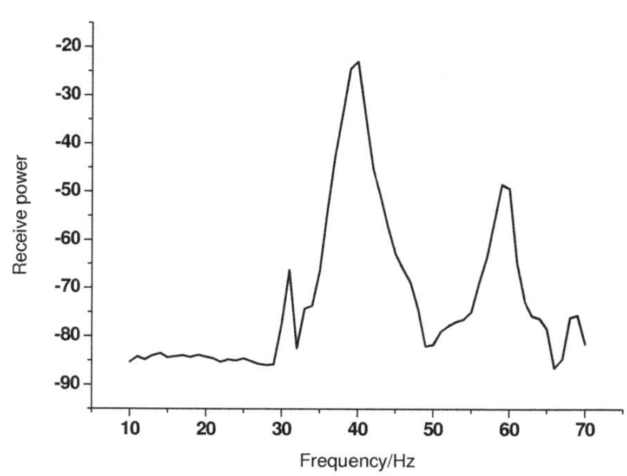

Figure 3. Experiment result of frequency response

As known to all, transfer rate may become higher when the 60kHz operating frequency is used for transmission. But high frequency will also cause high air absorption. Air absorption is one of the major factors which will significantly affect ultrasound attenuation. This absorption will be affect by many things, for example air temperature, air compression, relative humidity and so on. Weather condition will also affect it, just like wind and rain. In general the air absorption can be defined by the following equation, when air temperature is 20 degree centigrade:

$$A_e = 7.4\frac{f^2 r}{\phi}\times 10^{-8} \tag{1}$$

Where f is frequency of ultrasound wave, and ϕ is relative humidity.

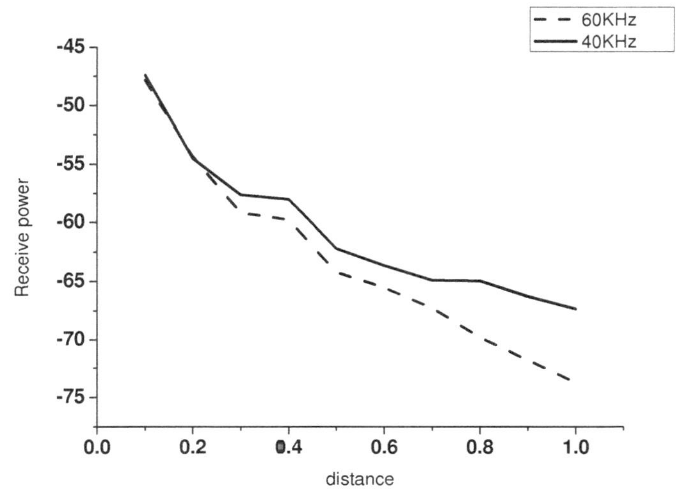

Figure 4. Transmission attenuation for different frequency

Figure 4 shows the different ultrasound attenuation between two frequency. Obviously, the attenuation is much higher when the operating frequency is 60kHz. Because the transmission distance is very importance for one indoor communication system, so this paper choose 40kHz as the operating frequency.

Because the final goal of this paper is building an ultrasound wireless communication system, so the next part will study the modulation method using for ultrasound wave. Modulation is the addition of information to a signal. Mainly, modulation method can be divided into three types. They are Amplitude Modulation, Frequency Modulation and Phase modulation. But as showing in figure 3, the bandwidth of ultrasound transducer is very narrow, and the attenuation will change with frequency. So the Frequency Modulation method is not suit for ultrasound wireless communication in air.

2.2. Amplitude Modulation for Ultrasound Communication

Amplitude modulation(AM) was the earliest modulation method used to transmit voice by radio. It is a modulation technique used in electronic communication, most commonly for transmitting information via a radio carrier wave. AM works by varying the strength of the carrier in proportion to the waveform being sent. Because it is very easy to vary the amplitude of ultrasound, Am is the first modulation method which is studied in this paper.

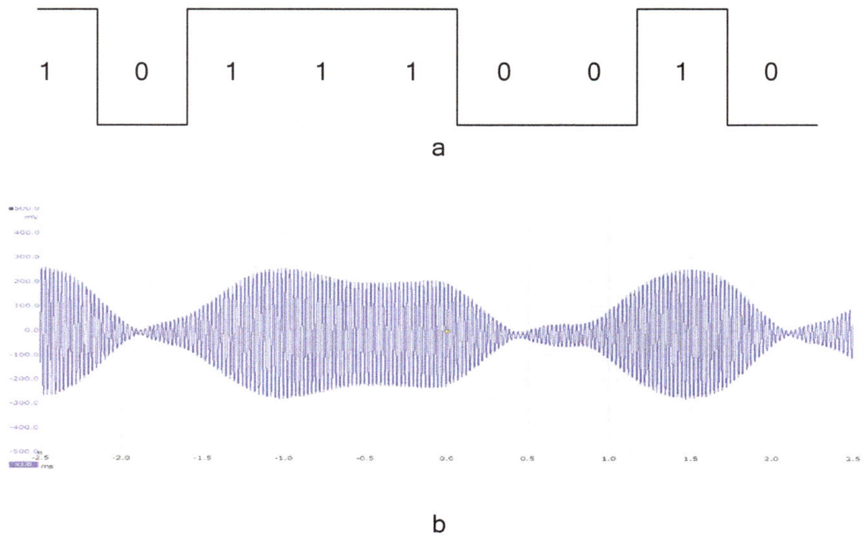

Figure 5. Transmission experiment using amplitude modulation

The experiment result is showing in figure 5. Figure 5(a) is the digital date used for modulation and figure 5(b) is the received ultrasound signal. The experiment system just as showing in figure 1 while the distance between two transmitter is 1meters. On the receive side, a PC oscilloscope has been used to measure the received signal. As showing in this result, AM modulated signal can be sent and received by a pair of ultrasound transmitters. But there are also many disadvantage of AM modulation.

One of the disadvantage is amplitude of received signal is not steady. Both of transmission distance and environment will significantly affected the received amplitude. Other disadvantage of all amplitude modulation techniques is that the receiver amplifies and detects noise and electromagnetic interference in equal proportion to the signal. Another disadvantage of AM is that it is inefficient in power usage; at least two-thirds of the power is concentrated in the carrier signal. The carrier signal contains none of the original information being transmitted. So the amplitude modulation method is not very suit for ultrasound wireless communication. In the next experiment, this paper will test phase modulation method.

2.3. Phase Modulation for Ultrasound Communication

The same to the amplitude modulation, phase modulation also uses a finite number of distinct signals to represent data. When the data is binary digits, phase shift keying(PSK) can be used for digital modulation. Usually, each phase in the signal encodes an equal number of bits.

One of the most widely used phase modulation method is Binary Phase shift keying(BPSK). It is particularly well suited to the growing area of data communications. It use two phases which are separated by 180 degree and so can also be termed 2-PSK. This modulation method in the most robust of all phase shift keying method. BPSK has the highest level of noise and distortion. So the demodulator can reach an incorrect decision. However, the transmit speed is only able to 1 bit per symbol and not suitable for high speed applications.

The general form for BPSK can be described by the following equation:

$$s_n(t) = \sqrt{\frac{2E_b}{T_b}} \cos\left(2\pi f_c t + (1-n)\pi\right), n = 0,1 \tag{2}$$

where E_b is the energy per bit, T_b is the bit duration, and f_c is the frequency of carry wave.

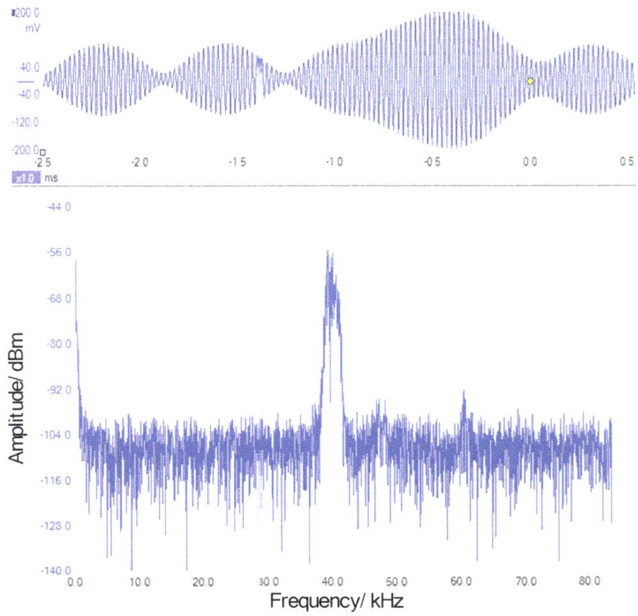

Figure 6. Transmission experiment using BPSK modulation

The first experiment is about BPSK method. The experiment system is the same to previous one. Figure 6 shows the received signal and spectrum. From this result , it can be seen that phase modulated 40kHz signal has been successfully received by the receive side. Because BPSK modulation signal is different from amplitude modulation, we choose to measure constellation diagram and error vector magnitude(EVM) of this signal.

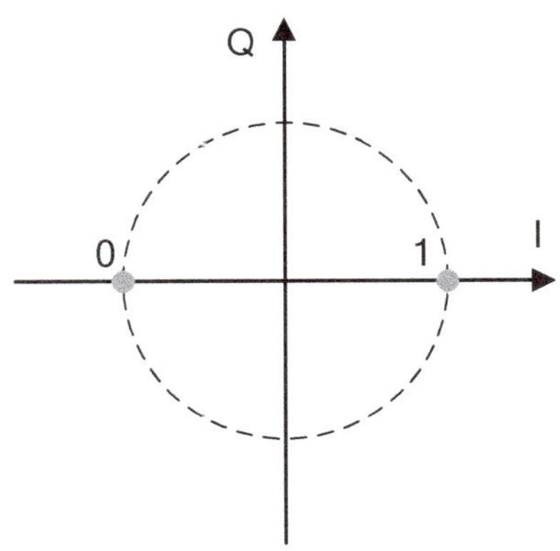

Figure 7. Example of constellation diagram for BPSK

Constellation diagram is a representation of a signal modulated by a digital modulation scheme such as quadrature amplitude modulation or phase shift keying. It displays the signal as a two-dimensional scatter diagram in the complex plane at symbol sampling instants. Measured constellation diagrams can be used to recognize the type of interference and distortion in a signal. The example of constellation diagram for BPSK is showing in figure 7.

While the error vector magnitude is a measure used to quantify the performance of a digital radio transmitter or receiver. Under the ideal conditions, one signal sent by a transmitter and received by a receiver would have all constellation points at the ideal locations. However, in actual cases the constellation points will deviate from the ideal locations. Then error vector magnitude is a measure of how far the points are from the ideal area. Many things can affect the error vector magnitude, just like noise, spurious signals, distortion, noise and phase noise. Therefore, error vector magnitude can provide a comprehensive measure of the quality of a communication system.

The error vector magnitude can be described by the following equation:

$$EVM\left(\%\right)=\sqrt{\frac{P_{error}}{P_{reference}}}*100\% \tag{3}$$

Where P_{error} is the RMS power of error vector, and $P_{reference}$ is defined as the reference constellation average power. Generally, the error vector magnitude should be lower than 20%, otherwise the digital data can not be sent correctly.

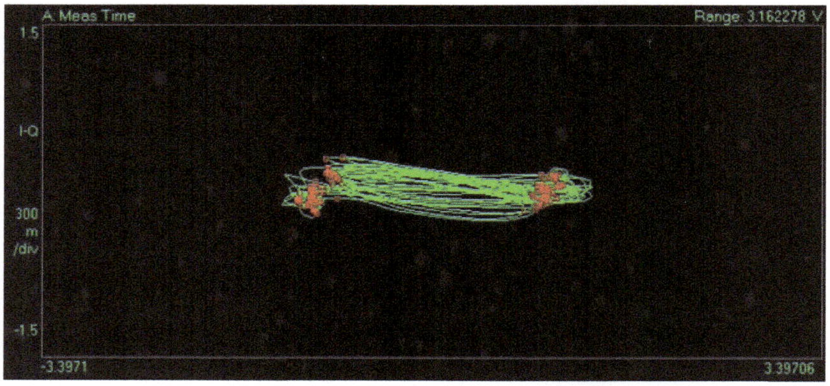

Figure 8. Constellation diagram for BPSK

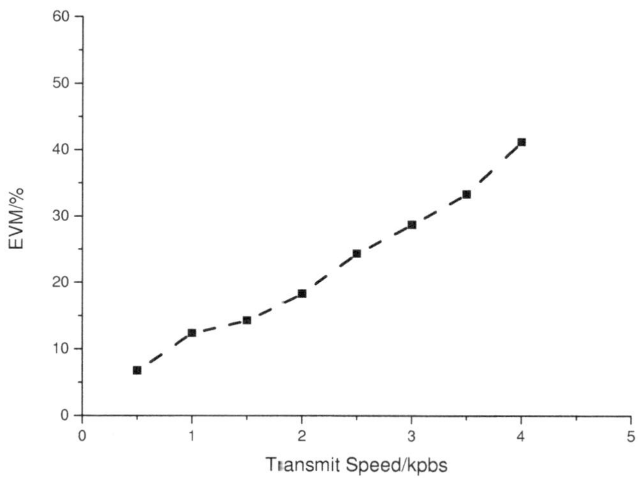

Figure 9. Transmit speed of BPSK

Figure 8 shows the experiment result of constellation diagram of BPSK signal. In this result, almost all constellation points are shown on the real axis, at 0 degree and 180 degree. It proves that phase modulation method can be used for ultrasound communications. In this experiment, the transmit speed is set as 1 kpbs. So in figure 9, error vector magnitude for different transmit speed has been measured. The error vector magnitude, as conventionally defined for signal carrier modulation, should lower than 20 percent. So the highest transmit speed is about 3 kpbs when we use BPSK method.

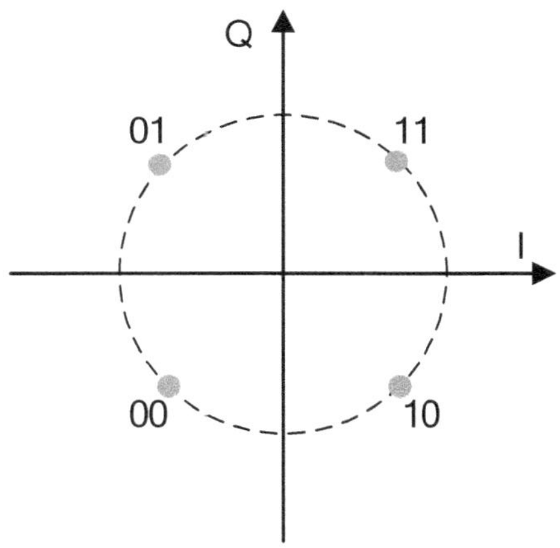

Figure 10. Example of constellation diagram for QPSK

In order to increase transmit speed, another phase modulation method has been tested. This modulation method is Quadrature Phase Shift Keying(QPSK). Sometimes this is also known as 4PSK. This method uses four points on the constellation diagram, equispaced around a circle. The example of constellation diagram for QPSK is showing in Figure 10. Because using four phases, this method is able to encode two bits per symbol. Generally, the main advantage of QPSK is that it can reach a data rate twice of BPSK and preserving similar levels of bit error rate.

Also, the general form for BPSK can be described by the following equation:

$$s_n(t) = \sqrt{\frac{2E_s}{T_s}} \cos\left(2\pi f_c t + (2n-1)\frac{\pi}{4}\right), n = 1,2,3,4 \tag{4}$$

Different from BPSK, in this equation E_s means energy per symbol and T_s means symbol duration. Other parameters in this equation is the same to the BPSK equation.

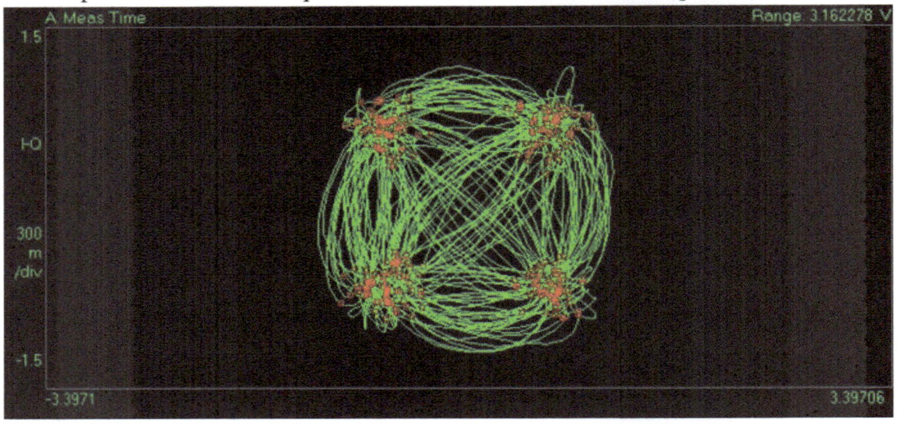

Figure 11. Constellation diagram for QPSK

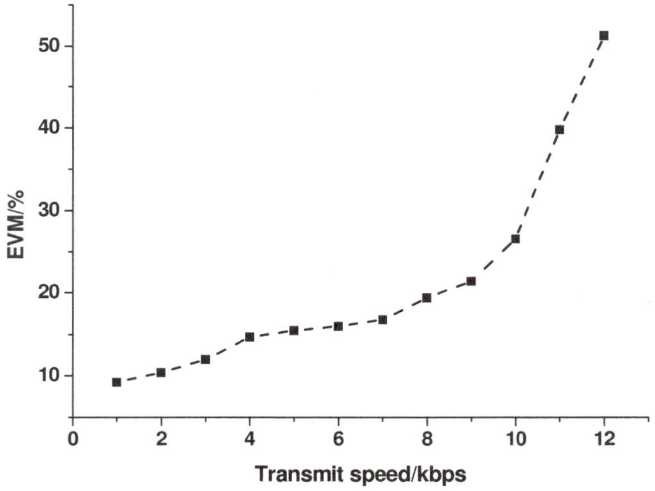

Figure 12. Transmit speed of QPSK

Figure 11 shows the constellation diagram of QPSK signal, and figure 12 shows the error vector magnitude for different transmit speed. This experiment proves that QPSK method can also been used for ultrasound wireless communication, the same to BPSK method. The highest transmit speed for QPSK can reach 8 kbps, twice higher than BPSK.

Actually, phase shift keying not only contains BPSK and QPSK. Any number of different phases may be used for a PSK constellation, just like 8-PSK or 16-PSK. Generally, these modulation methods are known as higher order PSK Higher order PSK can offer higher transmit speed, however the error rate may become too high. And the requirement of device will also increase. This paper has tested 8PSK method, in order to make sure whether higher order PSK can be used for ultrasound wireless communication.

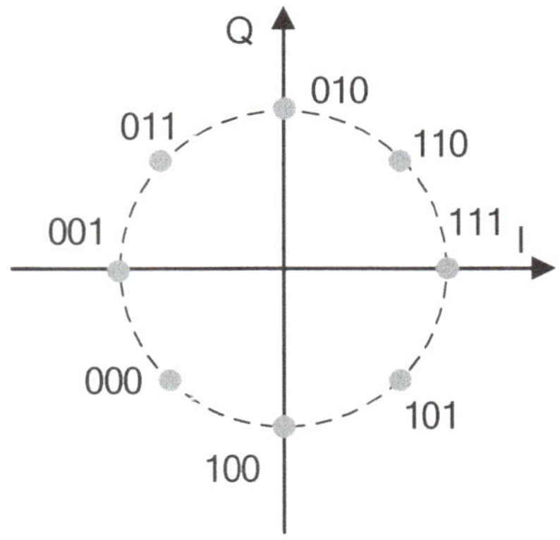

Figure 13. Example of constellation diagram for 8PSK

The constellation diagram for 8PSK is showing in figure 13. This method uses four phases, and is able to encode three bits per symbol. Also, the general form for BPSK can be described by the following equation:

$$s_n(t) = \sqrt{\frac{2E_s}{T_s}} \cos\left(2\pi f_c t + \frac{(n-1)\pi}{4}\right), n = 1, 2, \cdots, 8 \quad (5)$$

The parameters in this equation is the same to the QPSK equation.

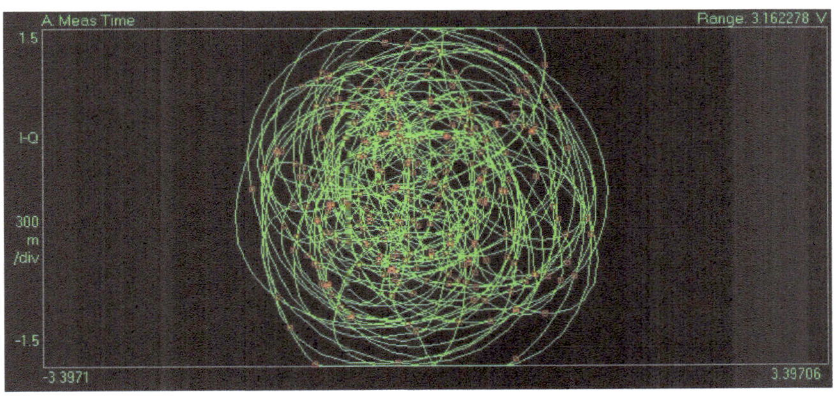

Figure 14. Constellation diagram for 8PSK

The experiment result is showing in figure 14. Obviously, there was no signal arrived at the receive side. In the further experiment we test many other cases, including different distance, different transmit speed and different power. However the 8PSK signal still can not transmit successfully. This experiment proved higher order PSK was not suited for this ultrasound communication system. The QPSK method should be the best choice.

3. CONCLUSIONS

The modulation method using for ultrasound communication system has been studied in this paper. Initial experiments identified the frequency response of this ultrasonic transmitter. This was used to decide the modulation method. Then three modulation methods have been tested. According to the experiment result, QPSK modulation method would be a good choice for ultrasound communication system in air. Both of EVM and transmit speed have been obtained in the experiment. These results have indicated that QPSK can be used to propagate ultrasonic signals in air. Other research about ultrasound communication, including simulation method and ultrasound phased array, has been published in Wireless Communications and Networking Conference 2014.

REFERENCES

[1] A. Bensky, *Short-Range Wireless Communication:*"Fundamentals of RF System Design and Application", *2nd ed. Burlington, MA:Newnes*, 2004, pp. 115-150.

[2] Keen, C.M., "Possible biological effects of EM fields associated with electric power systems - History and status of the issue", *EMF Services.*

[3] Baan R, Grosse Y, Lauby Secretan B, El Ghissassi F,"Carcinogenicity of radiofrequency electromagnetic fields", *The Lancet Oncology* 12(7): 624-6.

[4] G. J. Saulnier, H. A. Scarton, and A. J. Gavens, "Through-Wall Communication of Low-Rate Digital Data Using Ultrasound", *IEEE Ultrasonics Symposium*, 1051-0017,2006.

[5] Milica Stojanovic, "Recent Advances in High-speed Underwater Acoustic Communications", *IEEE Journal of Oceanic Engineering,* vol. 21, No. 2, Appil 1996

[6] C. Li, D. A. Hutchins, and R. J. Green,"Short-range ultrasonic digital communications in air", *IEEE Trans. Ultrason. Ferroelectr. Freq. Control*, vol. 55, no. 4, pp. 908-918, Apr. 2008

[7] Ying-Wen Bai, Li-Sih Shen and Zong-Han Li, "Design and Implementation of an Embedded Home Surveillance System by Use of Multiple Ultrasonic Sensors", *IEEE Transactions on Consumer Electronics*, Vol.56, 2010, pp. 119-124

[8] D.D.Chandhary, S.P.Nayse, L.M.Waghmare, "Application of Wireless Sensor Networks for Greenhouse Parameter Control in Precision Agriculture" *International Journal of Wireless & Mobile Networks* Vol.3, No.1 February 2011, pp. 140-149

[9] Saurbh Goyal, Indu Singh, "An Improved Inverted table Approach to Detect Selfish Node in Mobile Ad Hoc Network", *International Journal of Applied Engineering Research,* Vol. 7 No.11, 2012

[10] Ranjana R.Chittal, Kamod Kumar,"Medical Applications of Wireless Networks", *Second International Conference on System and Networks Communications,* 2007

[11] Hyosum Hwang and Nak Myeong Kim, "An Enhaced Frame Transmission Method for Medical Devices with Ultra Low Power Operation", *IEEE Transactions on Consumer Electronics,* Vol.58, No.1 February 2012, pp.154-160

[12] Minsoo Lee and Xiaoyue, "Han Complex Window Qurey Support for Monitoring Streaming Data in Wireless Body Area Network", *IEEE Transactions on Consumer Electronics,* Vol.57, No.4, November 2011, pp.1710-1718

[13] Mr.Ankit R.Bhavsar, Dr. Harshal A. Arolkar, "Wireless Sensor Networks: A Possible Solution for Animal Health Issues in Rural Area of Gujarat" *Specialized, Refereed and Indexed Journal in International Scientific and Corporate Databases,* Vol. 2 Issue 2, July 2012.

[14] Kavi K. Khedo, Rajiv Perseedoss and Avinash Mungur. "A Wireless Sensor Network Air Pollution Monitoring System", *International Journal of Wireless &Mobile Network,* Vol.2, No.2 May 2010, pp.31-45

[15] Tareq Alhmiedat, Amer O. Abu Salem, Anas Abu Taleb,"An Improved Decentralized Approach for Tracking Multiple Mobile Targets Through ZigBee WSNs", *International Journal of Wireless & Mobile Networks,* Vol.5 No.3, June 2013, pp.61-76

[16] Jiang Liu, Wasinee Noonpakdee, Shigeru Shimamoto, "Design and Performance Evaluation of OFDM-Based Wireless Services Employing Radio over Optical Wireless Link", *International Journal of Wireless & Mobile Networks,* Vol.3, No.5 October 2011, pp.173-184.

18

PERFORMANCE EVALUATION OF LEAST DISRUPTIVE TOPOLOGY REPAIR ALGORITHM (LEDIR) USING NS-2

Niharika Gorre, Prudhvi Raj Kyatham and Mohamed El-Sharkawy

Purdue School of Engineering and Technology, Indianapolis, Indiana, USA

ABSTRACT

In wireless sensor networks, sensor nodes fail either when some critical event occurs at the node or when the battery of the nodes is completely drained. If the remaining nodes in the sensor network are not aware of the node failure, the network might under undergo significant broadcast delay and path loss. An alarm packet has to be broadcast throughout the network when any critical event occurs and the transmission path must be recovered to achieve better quality of service in the sensor network. This paper utilizes least disruptive topology repair (LeDir) algorithm to minimize the power consumption and to handle such critical events. LeDir algorithm is used to recover the transmission path and to ensure that the quality of service issues in the network is met. The variations in the throughput, roundtrip time, broadcast delays and packet delivery ratio of the network are observed by performing multiple NS-2 simulations. It is observed that the throughput of the network could be quickly regained when the transmission path is recovered by node replacement.

KEYWORDS

LeDir, NS-2, AODV.

1. INTRODUCTION

Wireless sensor networks contain a huge number of low power sensor nodes that are placed heterogeneously in the deployable area. Each node acts a transceiver and has processing capabilities. They detect sudden changes in the environment such as temperature, pressure, humidity and report the information to the end users immediately. They are employed in applications such as environmental monitoring, gas monitoring in coal mines, in military applications to broadcast any urgent information to the end users. They prevent further damage to the area of interest by giving status signals to the end users. Wireless sensor networks are limited by the number of nodes used and their battery power, energy and computational capacity.

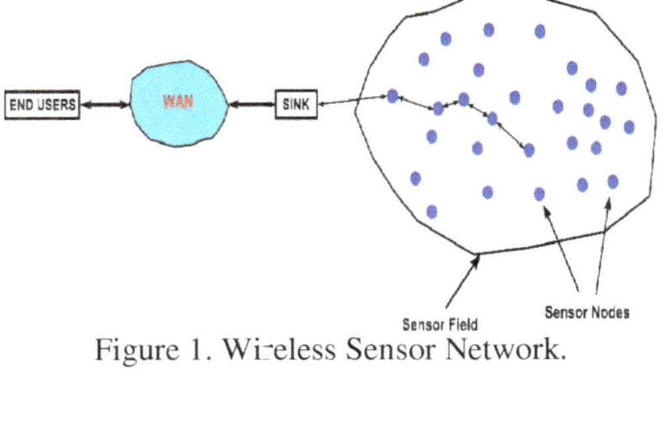

Figure 1. Wireless Sensor Network.

The source node sends the information to the destination node by choosing the shortest path based on the protocols and routing tables used in the network. The difference between wireless sensor networks and ad-hoc networks is that the former uses point to point communication whereas the latter uses broadcast communication. When a node in the network fails, all the other nodes in the network need to be aware of the node failure. The central node is made to broadcast an alarm packet that contains information about the node failure to the remaining nodes in the network [1, 2].

A single node failure might cause the entire network to become disjoint. As a solution to this problem we reposition the failure node and replace it with a working node already present in the network. Especially in applications like search and rescue operations, the timely coordination among the nodes is required. In this paper, network restoration algorithms-DARA (Distributed Actor Recovery Algorithm) and Least Disruptive topology repair algorithm (LeDir) are studied. LeDir maintains constant path even after the node failure. LeDir is implemented in NS2 simulator and various parameters are calculated. The paper considers the problem of the connectivity restoration [3, 5]. Throughput, End-to-End delay of the Packets, packet RTT and Packet Delivery Ratio are calculated from the trace files generated during the simulation. Data trace file contains information about the network. GNUPLOT tool is used to generate graphs and analyze the results.

1.1. System Model

The used system contains a central node and groups of other sensor nodes. The central node is placed such that it is the transmission range of all the other nodes in the network. This acts as the main hub for the all the other nodes in the network that is all the data transmission happen through this node. The remaining nodes are grouped together in the form of blocks such that the center node is positioned in the middle of the node blocks (node 9) as shown in Figure 2. The central node maintains a registry that has information about the nodes participating in the network such as routing tables, battery levels, etc.

Figure 2 shows the model of the used network. Node 9 is the central node and all the remaining nodes are arranged in the form of blocks. Nodes 0,1,2,3 form the first block, nodes 4,5,6,7,8 form the second block, nodes 10, 11,12,13,14 form the third block. We consider two data transmissions in the network scenario. During the first transmission node 0 sends data packet to node 7 with node 3 and node 9 being the intermediate nodes. Node 8 sends the data packet to node 12 with node 9 and node 11 as the intermediate nodes. The data transmission occurs successfully until node failure occurs at node 3.

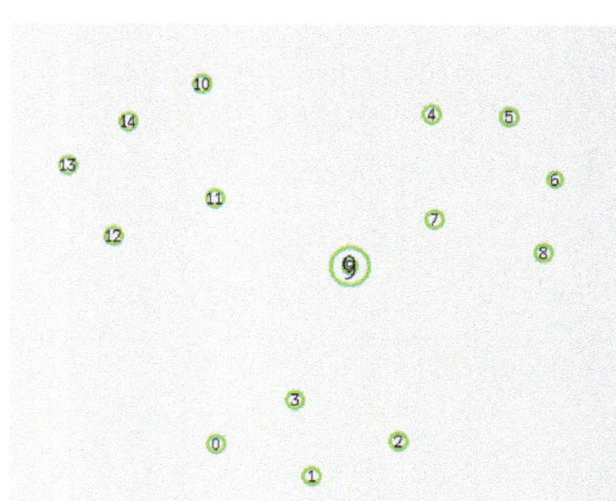

Figure 2. Network Model.

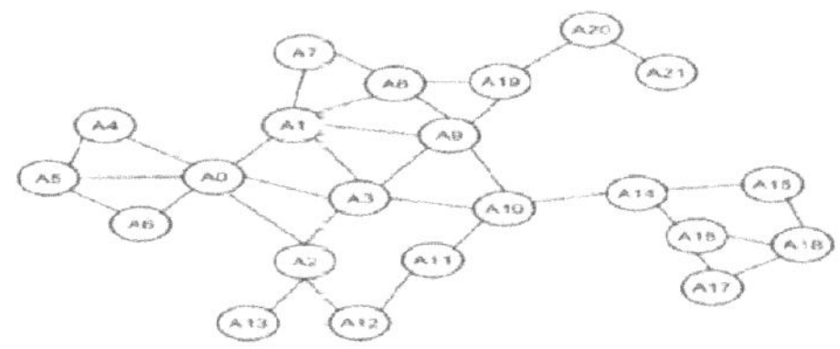

Figure 3. Network Example.

1.2. Distributed Actor Recovery Algorithm (DARA):

When a node fails, the best node used for replacement is selected from its one hop neighbors. The node having lowest degree and least distant from the failure node is selected. Any child nodes disconnected during this process are recovered by recursive relocation procedure. Thus shortest path between the nodes is not maintained [11].

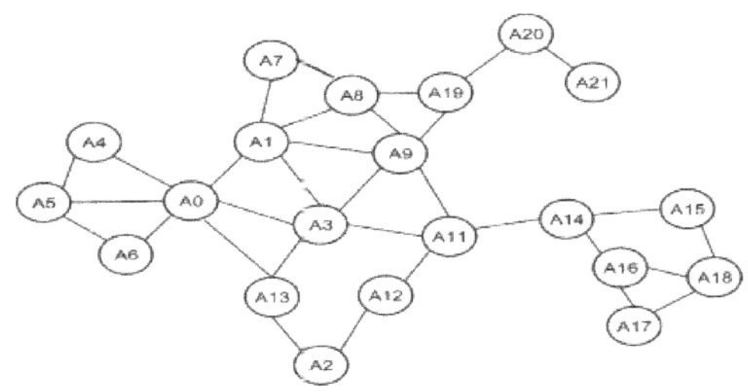

Figure 4. Network after the Implementation of DARA.

Let us assume that node A10 fails. Nearest neighbor with the lowest degree is selected. Thus A11 moves to the location of A10. The process is repeated by moving A12 to the position of A11, A2 to the location where A12 was earlier. Finally, A2 is replaced with A13. The final network after the implementation of DARA is shown in Figure 4. The shortest path length is increased by one-hop after implementing DARA. A0 and A3 that were one-hop neighbors of A2 earlier are moved far with the involvement of node A13. This algorithm may not be applicable for applications sensitive to delay. LeDir algorithm is a path restoration algorithm that is used to sustain the path lengths before the failure.

1.3. Least-Disruptive Topology Repair (LeDir) Algorithm:

LeDir involves cascaded movement only for the lead node. This algorithm helps the restoration of connection without extending the length of the shortest path among nodes. The LeDir algorithm allows the movement of nodes in blocks in contrast to other path restoration algorithms in which the nodes move in a serial fashion.

The algorithm performs the following steps:

- **Detecting the node failure:** The sensor nodes transmit messages to their one-hop distant neighbors periodically. When a node does not transmit the message it is understood it node has failed. The one-hop neighbors of the failed node find if the failed node is critical to the network. A node is considered to be important if it is on the cut-vertex. Let us assume that the node A10 has failed. It is important to recover as it is on the cut-vertex.

- **Identifying the smallest block:** The smallest block is the block with the least number of nodes. The reachable set of nodes for each direct neighbor of the failed node is identified and the set containing less number of nodes is selected. Thus the recovery overhead is reduced. Here, the block containing nodes 14, 15,16,17,18 is the smallest block.
- **Replacing the failure node:** The faulty node is replaced with the one-hop neighbor of the smallest block. The nearest neighbor of node 10 that belongs to the smallest block is node 14. Thus node 14 moves to the position of node 10.
- **Movement of child nodes:** The nodes that are at two hops distance from the faulty node are called children and those at a distance of three hops are called grandchildren. Once the child knows that its parent is moving to a new location, it tells its one hop neighbors that is grandchildren and follows the path of its parent. If a child has two parent nodes that are moving, it relocates to a position to maintain its connection with both the parent nodes.

In the example, node 14 informs its neighbor's node 15 and node 16 and moves to the position of node 10. The children of node 14 that is node 15 and node16 follow the path of node 14 to maintain the transmission link with node 14. This is highly advantageous as the routing tables of the children need not be updated. Before following the path of node 14, nodes 15 and 16 inform their children node 17 and node 18 about their movement. Since node 18 has three parent nodes, node 15, node 16 and node 17, it moves to a new location such that it is connected to these three nodes. Thus the node 17 does not relocate. Figure 5 shows how the LeDir algorithm is implemented [7].

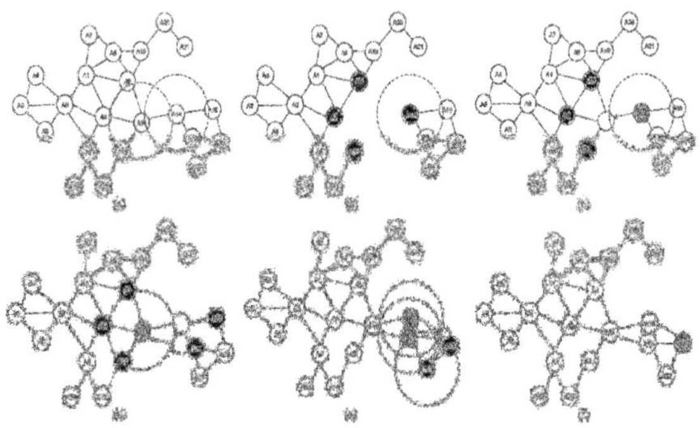

Figure 5. Proposed Algorithm.

Table 1. Comparison of DARA and LeDir [3].

Property	DARA	LeDir
Maximum no. of nodes required	N-3	0.5(N-1)
Maximum no. of messages to be sent	5N-3	1.5(N-1)
Maximum distance travelled by the node	r	r
Maximum distance travelled by other nodes engaged in the path recovery	rN	0.5(rN)

From table 1, DARA involves N-3 nodes whereas LeDir involves N-1/2 nodes for path restoration, where N is the number of nodes. In our network model, 15 sensor nodes are used. DARA requires 12 nodes whereas LeDir needs 7 nodes for path recovery. LeDir needs to send 21 messages and DARA needs to send 72 messages. The less number of messages provides less overhead. Thus, LeDir performs better than DARA. While the distance travelled by the peak node is the same in both the algorithms, the maximum distance travelled by the children is more in DARA i.e. in our system the child nodes need to travel a distance of 15r for DARA and 7.5r for LeDir, where r is the transmission range. Thus, LeDir if found to outperform in many ways and tries to recover the network with less recovery overhead while maintaining the shortest path lengths. This paper implements LeDir in NS-2 for the system model considered. In the used network model, it is assumed that the node 3 is the faulty node. The one hop neighbors of node 3, node 0, node2 and node 9, identify the failure when node 3 stops sending periodical status messages. LeDir algorithm identifies the smallest blocks as blocks containing node 0 and node 2. Node 2 is assumed to be least distant from node 3. Thus node 2 relocated to the position of node 3 and node 3 is moved to the previous location of node 2. Thus, the data transmission path is recovered and the network starts functioning normally. The quality of service parameters could be met.

2. IMPLEMENTATION

2.1. NS-2 Simulator:

NS-2 version 2.35 is used to simulate the network. NS-2 is an object oriented, discrete event simulator. NS-2 supports the use of TCP protocol is primarily useful for simulating wireless networks. NS-2 can be programmed using the Tool Command Language (TCL) and uses C++ in the backend of the system for interpretation. The TCL script is used to invoke the event scheduler, setup the network topology, set the start and stop of data transmission, call functions in the inbuilt NS-2 library [8, 14].

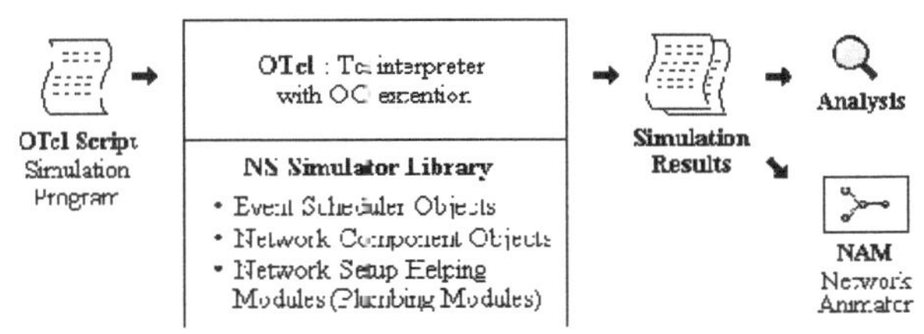

Figure 6. Network simulator.

When the TCL file is run in NS-2 the simulation results could be either in the form of a trace file or a Network Animator (NAM). The results are analyzed to estimate the performance of the communication network in the real world. For the simulator, a wireless sensor network containing 15 sensor nodes is considered. They are assumed to have full energy (100%) at the beginning of the simulation. Nodes are arranged in the form of blocks with each block having nodes placed at different levels. In the topology, node 9 acts as the central node for the all the other blocks. All the transmission occurs through the central node.

Software requirements:

Programming language: TCL, C++, PERL, GAWK
Simulator: NS-2 2.35
User interface: NAM, Text Editor
Operating System: Ubuntu 12.00

Hardware Requirements:

Minimum RAM requirement: Min 1 GB
Minimum Hard Disk space needed: 20GB
Minimum Processor Speed: 1.8 GHz

Other parameters used in the Network:

Channel type: Channel/Wireless Channel
Radio-propagation model: Two-way Ground
Network Interface type: Physical/Wireless Interface;
MAC type : Mac/802.11
Interface queue type: Drop Tail/Priority Queue
Link layer type: LL
Antenna model: Omnidirectional Antenna
Queue length: 50
Number of nodes: 15
Routing Protocol: AODV
Topography: 3000 X 3000
Initial Energy: 100 Joules
Transmission Protocol: TCP/New Reno
Application: CBR (Cluster based routing) application

AODV protocol is used for the routing the packets and is used to find the shortest path in shorter duration of time [12]. Transmission protocol (TCP) along with new Reno Algorithm is used. TCP helps in the avoidance of congestion in the network. The new Reno algorithm uses congestion window techniques to control the congestion in the network by reducing the flow of data packets. The application is set to have constant bitrate to maintain constant speed of flow of data.

The network has two TCP connections- node 0 to node 7 and node 8 to node 12. Node 3 is made to fail after certain period of time. LeDir algorithm finds the best node to replace the faulty node and the connection is recovered. We used the agent trace information is used to calculate the throughput and mac trace information is used to calculate packet latency. Agent trace information is obtained while performing simulations for the packet delivery ratio and end-to-end delay of the network. Particle extraction report language (Perl) is used for calculating the throughput and GAWK is used for the remaining calculations. The graphs are plotted using the GNUPLOT tool. It is a 3-D graph generating tool used for the research purposes.

2.2. Nam:

Nam is an animation tool used to view the trace file data produced during the network simulations in the form of animations. Layout of the topology, data transmissions and other changes occurring in the network with time are shown graphically. Nam contains various controls such as play, pause, fast forward etc. which are useful to observe the simulations occurring at the required instant of time. The step size of time could be varied to change the speed of simulation [11].

Nodes are represented using circles. Different colors are used for nodes to identify the nodes belonging to a particular block. The inner circles represent the energy level of a particular block. They are differentiated as-green color for 100% energy, yellow for less energy, red for no energy. The central node contains information about all the nodes participating in the sensor network. All the nodes update their status periodically to the central node.

The data transmission begins at 5 seconds. The network is made to perform two data transmissions-one from node 0 to node 7 with node 3 and node 9 being the intermediate nodes, the other from node 8 to node 12 with node 7 and node 11 being the intermediate nodes. When the data packet reaches the destination, it sends an acknowledgement about the packet reception to the source. If any node in the path fails, the transfer of packets comes to a halt due to disconnection in the network. There is no transmission of data between node 0 and node 7 until the path is recovered.

The LeDir algorithm identifies node 2 as the nearest node in the smallest block. Node 2 is moved to the location of node 3 to take the responsibilities of node 3. This is illustrated in Figure 7. The data flows from node 0 to node2 and then to node 7 through the central node. The functionality of the sensor network is regained and hence the quality of service parameters of the network could be achieved by using the LeDir algorithm. As the data transmission take place continuously, the battery levels of the nodes participating in the data flow keeps decreasing from time to time. The decrease in battery level is indicated by inner yellow circles. Node 9 gets exhausted first since it participates in both the transmissions.

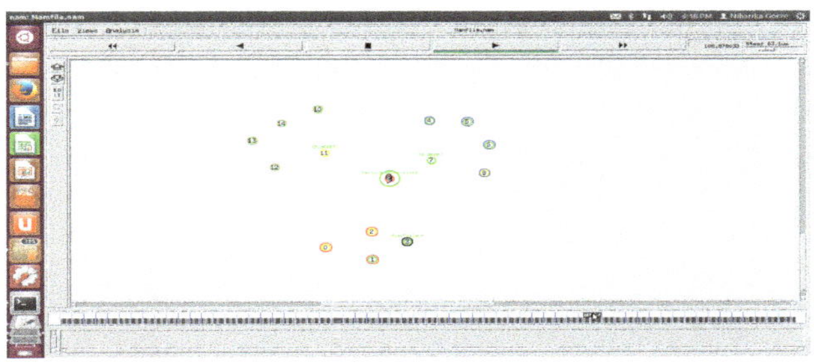

Figure 7. Network Animation (Nam).

3. RESULTS AND DISCUSSION

3.1. Throughput:

Network throughput is the rate of successful delivery of data packets over a communication channel. The data might be transmitted either using a physical channel or a wireless channel. It is measured in bits per second (bps) or data packets per second. It is the most important quality of service measure of a communication network [6].

The trace file generated while simulating the main source code is given as input to the source code that generates throughput. The throughput code is written using the Perl language. The following equation is used to calculate the throughput based on the information in the trace file.

$$\text{Throughput} = \text{total number of bits received/ time duration}$$

Perl script utilizes the agent trace information. Agent trace has information about the data packets only. The information about the routing and MAC packets is not considered in the calculation of throughput.

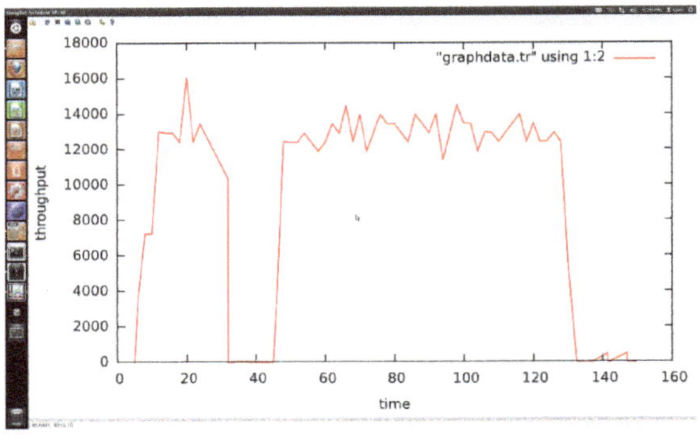

Figure 8. Throughput.

Figure 8 shows the graph of Throughput for the proposed network. The x-axis indicates time and y-axis indicates throughput of the network. The data transmission starts at 5 seconds. The throughput of the network changes from 5 seconds to 25 seconds due to the traffic and transmission delays in the network. When the node 3 fails at 26th second, it is observed that the throughput drops to zero. This is because the network becomes disjoint as there is connecting path between node 0 and node 7.

Later LeDir algorithm is applied and node 3 is replaced with node 2. The repositioning of the nodes is done at the 32nd second. Now, node 2 starts to take the responsibilities of node 3. The total simulation time is 120 seconds. At the 45th second the data transmission starts again and thus the throughput is found to increase from 45 seconds to 120 seconds. The throughput gradually falls to zero again after few seconds. The maximum throughput of 16000 bytes is observed at time of twenty seconds.

3.2. End-to-End Delay:

End to end delay is the average time taken by a packet that is transmitted from the source to reach the destination. End-to-end delay includes all the delays in the network along with the propagation delay i.e. delays caused in the network for finding the shortest path and queues are also considered. The network is found to perform better when the end-to-end delay is low. The following equation is used in the calculation of end-to-end delay in the network:

$$\text{End to end Delay} = \text{arrival time of packet} - \text{send time of the packet}$$

The source code for calculating the end-to-end delay is written using the Gawk scripting language. We calculate packet delay using their packet ID. Packet ID contains information about the sending time and receiving time of the packet. The data trace file is given as an input to the gawk code to produce a trace file containing the delay information. The packet delays at different time intervals are contained in the trace file. A graph of time versus end-to-end delay is plotted using the GNU plot. The graph is shown in Figure 9. When the data transmission starts at 5th second the end-to-end delay is found to be the lowest, 0.04 seconds. As the traffic in the network varies the delay is found to vary till node 3 fails where the end-to-end delay falls to zero. After the node recovery at 45th second the delay is found to have the highest value of 0.11 seconds.

When the network has recovered from failure all the nodes take time updating the routing tables. Since the end-to-end delay considers all the delays in the network along with the transmission delay, the peak of end-to-end delay is found immediately after the node recovery.

3.3 Average Round trip time:

Average Round-trip time is the time duration between transmission of the packet to destination and reception of the acknowledgement signal. Round trip time depends on the number of nodes between the source and destination, traffic in the network, the number of other requests that are handled by the intermediate nodes. The average round trip time could vary from a few milliseconds to seconds. The lower the round-trip time the better is the throughput in the network.

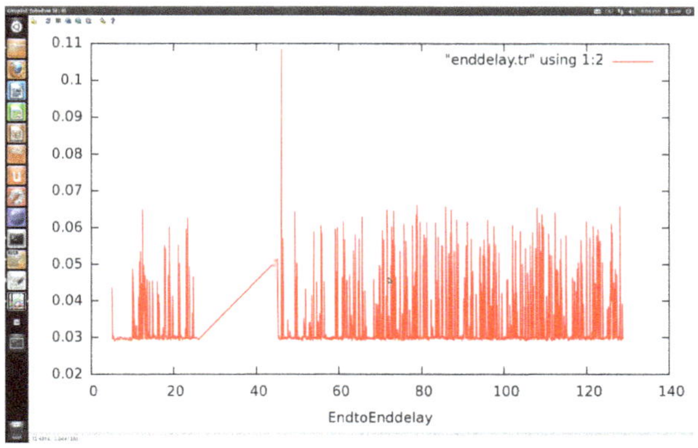

Figure 9. End-to-End Delay of Network.

As the transmission of packets starts at 5th seconds, the round trip time is the lowest of 0.3 seconds. This is because the traffic and other delays are limited at the beginning of the simulation. As time progresses, the round trip increases and varies based on number of other concurrent transmissions. When node 3 fails at 24th second, the round trip time falls to the lowest value. After the recovery of the data path between node 0 and node 7, the simulation begins again at the 45th second round trip time starts increases again. The variations in the round trip time from 45 seconds to the end of the simulation at 120 seconds are shown in Figure 10. When the round trip time is the highest, 0.08 seconds, the network performance degrades.

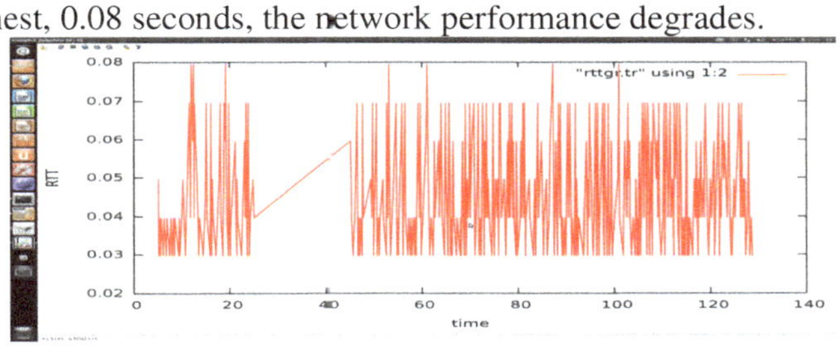

Figure 10. Graph of Time vs Round Trip Time.

3.4. Packet Delivery Ratio:

Packet deliver ratio is the ratio of successful packet delivery in the network. It is defined as the ratio of packets that successfully reached the destination to the packets produced by the source nodes. Higher packet delivery indicates better performance of protocols used in the network. It could be defined by using the equation below:

$$\text{Packet Delivery Ratio} = N1/N2$$

Where N1 is the number of data packets received at the destination node and N2 is the number of data packets generated at the source node. Gawk is used for the calculation of packet delivery ratio. Figure 11 shows the packet count at the source and the destination. 1238 packets are received at the destination while 1253 packets are originally transmitted from the source. The packet delivery ratio of the network is 98.88%. This shows that the network has better performance. Thus, LeDir algorithm could be applied in wireless sensor networks to maintain a constant performance of the network even in the presence of node failures and other critical events in the network.

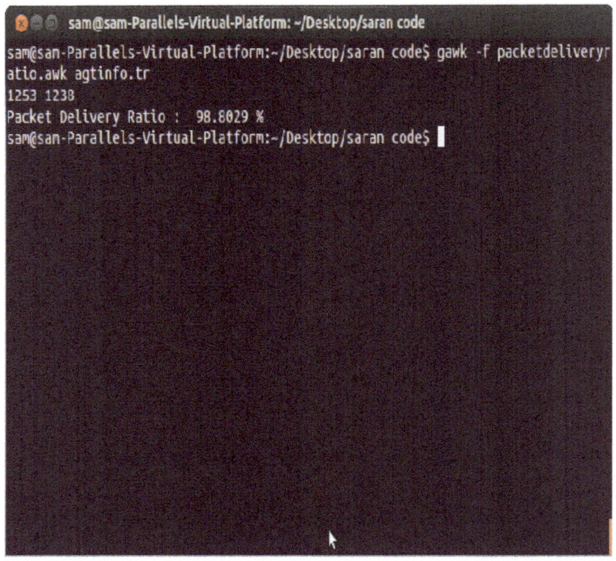

Figure 11. Packet Delivery Ratio.

4. CONCLUSION

Two different path restoration algorithms for wireless sensor networks-DARA and LeDir are studied. LeDir algorithm could be used to restore the wireless sensor network connectivity without extending the short path of nodes in the network and hence is more beneficial compared to DARA. It helps in improving the quality of the sensor network all times. LeDir is implemented on NS-2 to study the throughput and other parameters of the network. The algorithm could be used to recover only one transmission path at a time. In case of multiple node failures the procedure cannot be applied. But the probability that many nodes fail at some time is less and occurs when the deployment area is subject to a very hazardous event.

REFERENCES:

[1] P. Guo, T. Jiang, Q. Zhang, and K. Zhang, "Sleep Scheduling for Critical Event Monitoring in Wireless Sensor Networks," IEEE Transactions On Parallel And Distributed Systems, Vol. 23, No. 2, February 2012.

[2]. F. Akyildiz and I. H. Kasimoglu, "Wireless Sensor and Actor Networks: Research Challenges," Ad Hoc Netw. J., Vol. 2, No. 4, PP. 351–367, Oct. 2004.

[3] A. Abbasi, M. Younis and U. Baroudi, "Recovering from a Node Failure in Wireless Sensor-actor Networks with Minimal Topology Changes", IEEE Transactions on Vehicular Technology, Vol. 62, No. 1, January 2013.

[4] T. Zheng, S. Radhakrishnan and V. Sarangan, "PMAC: An Adaptive Energy-Efficient MAC Protocol for Wireless Sensor Networks," Proc. 19th IEEE Int'l Parallel and Distributed Processing Symposium, PP. 224-231, Apr. 2005.

[5] G. Wang et al., "Sensor Relocation in Mobile Sensor Networks," Proc. of 24th Annual Joint Conf. (INFOCOM), Miami, FL, Mar. 2005, PP. 2302–2312.

[6] P. Rohal, R. Dahiya and P. Dahiya," Study and Analysis of Throughput, Delay and Packet Delivery Ratio in MANET for Topology Based Routing Protocols (AODV, DSR And DSDV)," International Journal for Advance Research in Engineering and Technology, Vol. 1, Issue II, Mar. 2013.

[7] R. J. Egusha, B. Mathivanan and E. Konguvel, "A Distributed LeDiR Algorithm for WSA Networks," IRACST, 2250-3498, Vol.4, No.3, June 2014.

[8] "Introduction to Network Simulator NS-2" by T Issariyakul and E Hossain.

[9] T. He, P. Vicaire, T. Yan, L. Luo, L. Gu, G. Zhou, R. Stoleru, Q. Cao, J. Stankovic and T. Abdelzaher, "Real-Time Analysis of Tracking Performance in Wireless Sensor Networks," IEEE Real-Time Applications Symposium, May 2006

[10] M. Ismail and M. Y. Sanavullah "Security Topology in Wireless Sensor Networks with Routing Optimization", WCSN 2008.

[11] A. Abbasi, M. Younis and K. Akkaya, "Movement-assisted Connectivity Restoration in Wireless Sensor and Actor Networks," IEEE Trans. Parallel Distrib. Syst., Vol. 20, no. 9, PP. 1366–1379, Sep. 2009.

[12] http://nsnam.isi.edu/nsnam/index.php/Nsnam.

[13] http://en.wikipedia.org/wiki/Ad_hoc_On- Demand_Distance_Vector_Routing.

[14] http://www.isi.edu/nsnam/ns/tutorial.

DESIGN AND ANALYSIS OF HIGH GAIN DIODE PRE-DISTORTION LINEARIZER FOR TWTA

Anuraag Misra[1], Archan Sarkar[2] and Bhaswar Dutta Gupta[2]

[1]Accelerator Physics Group, Variable Energy Cyclotron Centre, Kolkata, India
[2]Department of Applied Electronics and Instrumentation Engineering, Netaji Subhash Engineering College, Kolkata, India

ABSTRACT

This paper presents the design and analysis of a high gain, broadband Schottky and PIN diode based RF pre-distortion linearizer for TWTA. The circuit is using ABCD matrix approach. The simulation is performed using Agilent ADS software. We have proposed a new linearizer circuit which can achieve a high gain compared to existing linearizer designs.

KEYWORDS

PIN diode, Schottky diode, pre-distortion linearizer, gain

1. INTRODUCTION

Traveling wave tube amplifier (TWTA) is used to amplify RF signals in the microwave range. The major advantage of TWTA over some other microwave amplifiers is that it has substantial gain over a broad range of frequencies compared to klystron tubes. TWTA are commonly used in satellite transponders as amplifiers and also very much popular in high frequency ion sources for particle accelerators. TWTA suffer from a major drawback that at high frequencies the response becomes very non linear. Linearizers are electronic circuits which correct the non-linear behaviour of amplifiers to boost maximum output power and efficiency. One way to implement linearizers is to create a circuit with inverted behavior to that of the amplifier. These circuits counteract the non-linearity of the amplifier and minimize the distortion of the signal. This creates an increase in linear operating range of the amplifier. Linearized amplifiers have a quite higher efficiency with enhanced signal quality. There are various concepts to linearize an amplifier, which includes pre-distortion and post-distortion and also feedback linearization. Pre-distortion linearizer is the most popular type of linearizer which creates inverse amplitude and phase non linearity to that of TWTA, which improves the non-linearity of the communication system which operates the TWTAs.

Amplifiers when operated in saturation, due to decreasing amplification and changing phase, non-linearity occurs. This behavior is generally termed as gain or phase compression. These changes can be compensated by pre-distortion linearizers. The corresponding behavior is generally called gain expansion or phase expansion. Pre-distortion linearizer functions in the small signal area and boost the DC power consumption of the system slightly. Linearizers are favorably used in high power amplifiers which use electron tubes or solid state amplifiers. These systems are used in satellite communication or High Definition (HD) television which require high signal quality.
An analog linearizer employing an amplifier with series feedback connection and a high source inductance is small in size and very simple. Although this technique aids low DC power

consumption has the limitations that it can be used only in power amplifiers whose input power is above 20dBm. A linearizer made of a diode in series connection with a capacitor in parallel reduces the shortcomings of the aforesaid linearizer. This linearizer although simple and small in configuration needs an extra isolation mechanism to isolate it from the power amplifier which needs to be linearized. Furthermore this linearizer has a very narrow degree of control on the attained gain and phase characteristics distortions and hence is very limited in its applications. The advantage of the diode based linearizer is that it is flexible and gives us a high degree of control in attaining the gain and phase characteristics but still needs isolation mechanism between it and the power amplifier. There are methods to overcome the isolation issue by using hybrid couplers. Thus diode based linearizers has a distinct edge over analog linearizers. All these linearizers have certain benefits. However they lack the ability to linearize power amplifiers which exhibits dual-inflection points in their gain characteristics.

This paper presents a diode based pre-distortion linearizer with an aim to improve the overall gain of the linearizer. The PIN diode provides an added level of freedom in achieving the required amplitude level and depth of inflection in the characteristic. The quarter wave transmission line helps in providing the dual inflection points in the Output power vs. Input power characteristic.

Highly efficient and linear power amplifiers are in great demands with the advent of wireless standards. The transistor devices which are employed in power amplifiers are operated in saturation mode to attain high efficiency, but this also results in amplitude and phase distortions in the output.

2. DESIGN OF LINEARIZER

A quadrature hybrid coupler is used where port 1 is used for input signal, in port 2 and port 3 a circuit configuration L which is shown in Fig. 2 is connected, and port 4 is used as output.

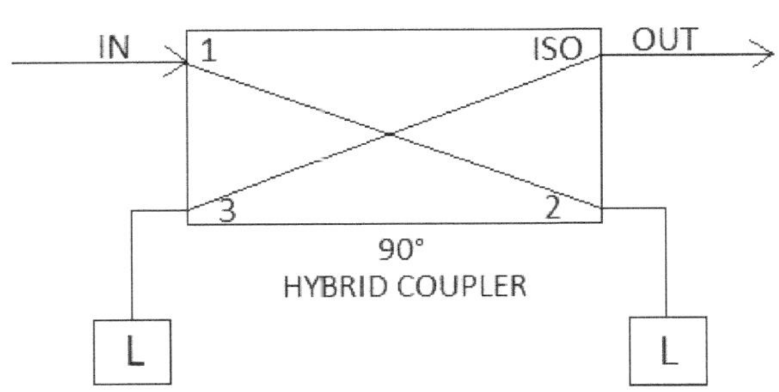

Fig. 1: Block diagram of the quadrature hybrid coupler

The circuit L consists of a quarter wave transmission line with a network of diodes on both sides of it. The biasing of the diodes are done with voltage source V_O through resistors R_{b1}, R_{b3} and R_{b3}.

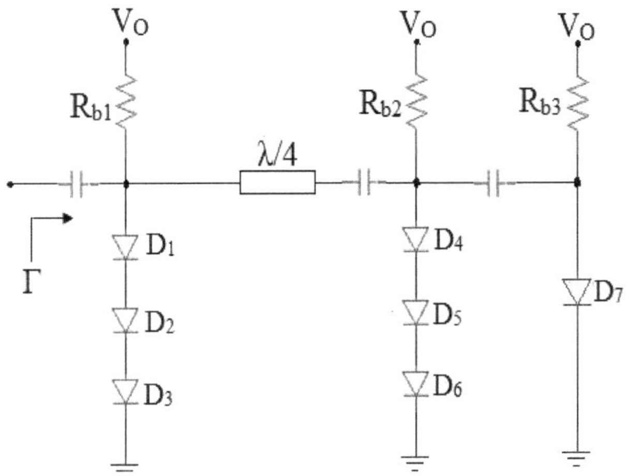

Fig. 2: Schematic Diagram of the circuit configuration L

Here D_1, D_2, D_3, D_4, D_5 and D_6 are Schottky diodes and D_7 is a PIN diode.
R_{b1}, R_{b2} and R_{b3} are biasing resistances.
Γ is the input voltage reflection coefficient S_{11}.

3. EQUIVALENT CIRCUIT OF LINEARIZER

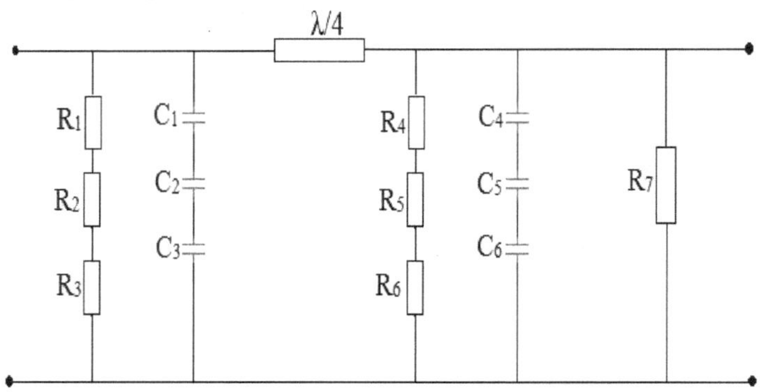

Fig. 3: Equivalent circuit of the Linearizer

R_1, R_2, R_3, R_4, R_5 and R_6 denote the dynamic resistances of the Schottky diodes.
C_1, C_2, C_3, C_4, C_5 and C_6 denote the junction capacitances of the Schottky diodes.

$$R_a = R_1 + R_2 + R_3 \tag{1}$$

$$C_a = 1/C_1 + 1/C_2 + 1/C_3 \tag{2}$$

$$\therefore Y_a = G_a + j\omega C_a \tag{3}$$

Here Y_a is the admittance of R_a and G_a is the conductance of R_a

.

$$R_b = R_4 + R_5 + R_6 \tag{4}$$

$$C_b = \frac{1}{C_4} + \frac{1}{C_5} + \frac{1}{C_6} \tag{5}$$

$$\therefore Y_b = G_b + j\omega C_b \tag{6}$$

Here Y_b is the admittance of R_b and G_b is the conductance of R_b.

$ABCD$ matrix of diodes D_1, D_2 and D_3:

$$\begin{bmatrix} A & B \\ C & D \end{bmatrix} = \begin{bmatrix} 1 & 0 \\ Y_a & 1 \end{bmatrix} \tag{7}$$

$ABCD$ matrix of diodes D_4, D_5 and D_6:

$$\begin{bmatrix} A & B \\ C & D \end{bmatrix} = \begin{bmatrix} 1 & 0 \\ Y_b & 1 \end{bmatrix} \tag{8}$$

$ABCD$ matrix of transmission line:

$$\begin{bmatrix} A & B \\ C & D \end{bmatrix} = \begin{bmatrix} \cos\beta\ell & jZ_0 \sin\beta\ell \\ j\dfrac{\sin\beta\ell}{Z_0} & \cos\beta\ell \end{bmatrix} \tag{9}$$

For quarter wave transmission line,

$$\beta\ell = \frac{\pi}{2} \tag{10}$$

Therefore equation (9) reduces to:

$$\begin{bmatrix} A & B \\ C & D \end{bmatrix} = \begin{bmatrix} 0 & jZ_0 \\ j\dfrac{1}{Z_0} & 0 \end{bmatrix} \tag{11}$$

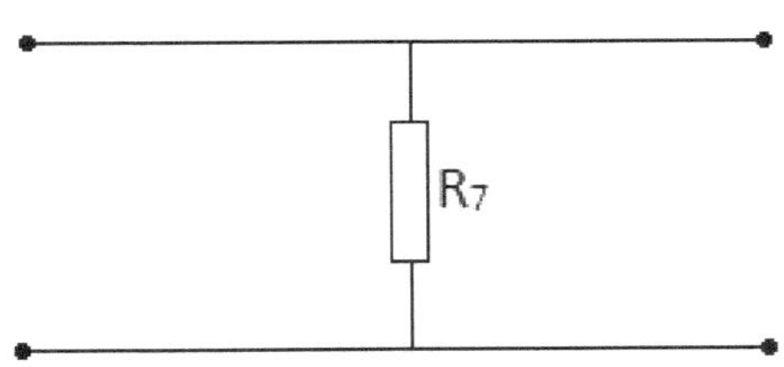

Fig. 4: Equivalent circuit of the PIN diode

$ABCD$ matrix of PIN diode D_7:

$$\begin{bmatrix} A & B \\ C & D \end{bmatrix} = \begin{bmatrix} 1 & 0 \\ G_7 & 1 \end{bmatrix} \tag{12}$$

Here G_7 is the conductance of the PIN diode R_7.

Total *ABCD* matrix:

$$\begin{bmatrix} A & B \\ C & D \end{bmatrix} = \begin{bmatrix} 1 & 0 \\ Y_a & 1 \end{bmatrix} \begin{bmatrix} 0 & jZ_0 \\ j\frac{1}{Z_0} & 0 \end{bmatrix} \begin{bmatrix} 1 & 0 \\ Y_b & 1 \end{bmatrix} \begin{bmatrix} 1 & 0 \\ G_7 & 1 \end{bmatrix} \tag{13}$$

$$\Rightarrow \begin{bmatrix} A & B \\ C & D \end{bmatrix} = \begin{bmatrix} jY_2Z_0 + jG_3Z_0 & jZ_0 \\ j\frac{1}{Z_0} + jY_1Y_2Z_0 + jY_1G_3Z_0 & jY_1Z_0 \end{bmatrix} \tag{14}$$

To find the input voltage reflection coefficient, we have to find S_{11}. The relation between *ABCD* parameters and S_{11} is:

$$S_{11} = \frac{AZ_0 + B - CZ_0 - DZ_0}{AZ_0 + B + CZ_0 + DZ_0} \tag{15}$$

$$\therefore |S_{11}| = |\Gamma| = \frac{-Z_0^2(Y_aY_b + Y_aG_7) + Z_0(-Y_a + Y_b + G_7)}{Z_0^2(Y_aY_b + Y_aG_7) + Z_0(Y_a + Y_b + G_7) + 2} \tag{16}$$

Equation (16) shows the magnitude of the input voltage reflection coefficient of the circuit.
To find the forward voltage gain, we have to find S_{21}. The relation between forward voltage gain S_{21} and *ABCD* parameters is:

$$S_{21} = \frac{2Z_0}{AZ_0 + B + CZ_0^2 + DZ_0} \tag{17}$$

$$\therefore |S_{21}| = \frac{2}{Z_0^2(Y_aY_b + Y_aG_7) + Z_0(Y_a + Y_b + G_7) + 2} \tag{18}$$

$$\angle S_{21} = -90^\circ \tag{19}$$

Equation (18) shows the magnitude of the forward voltage gain of the linearizer. Equation (19) shows the phase of the forward voltage gain of the linearizer.

3.1. SIMULATIONS

We have simulated the circuit shown in Fig. 2 using Agilent ADS software. The PIN and Schottky diodes used for simulation were from Agilent. The models were HMPP-3890 and HSMS-2820 respectively. We modelled a diode to match the specifications of HSMS-2820. The biasing resistances were chosen as 1kΩ and voltage sources as 10V. The characteristic impedance of the quarter wave transmission line was 45Ω and since *S*-parameters were being measured two 'Port Impedance Termination for *S*-parameters' were positioned on both ends of the circuit.
Here are some screenshots of ADS 2009 where we have simulated the circuit using one , three and five number of Schottky diodes in series connection.

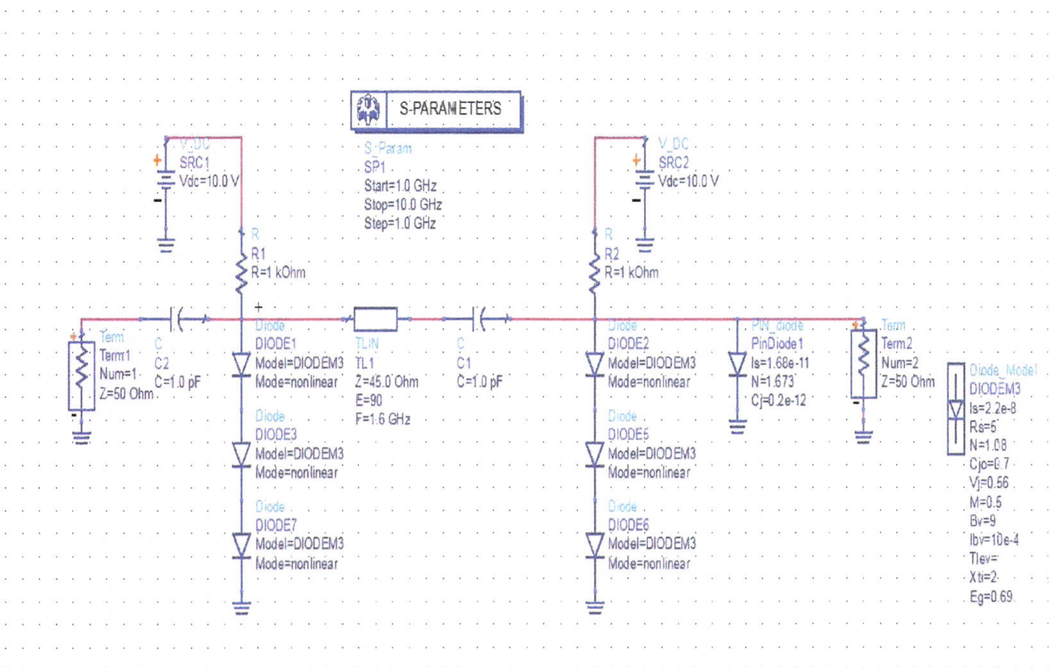

Fig. 5: ADS simulation screenshot for the simulation of three diodes in series

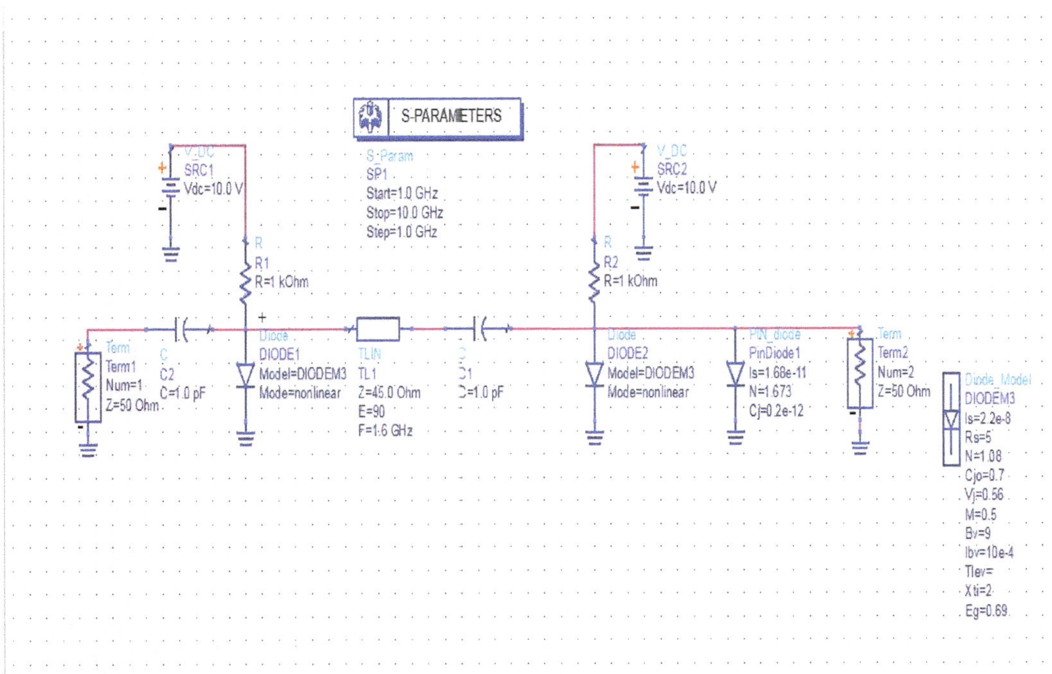

Fig. 6: ADS simulation screenshot for the simulation of one diode

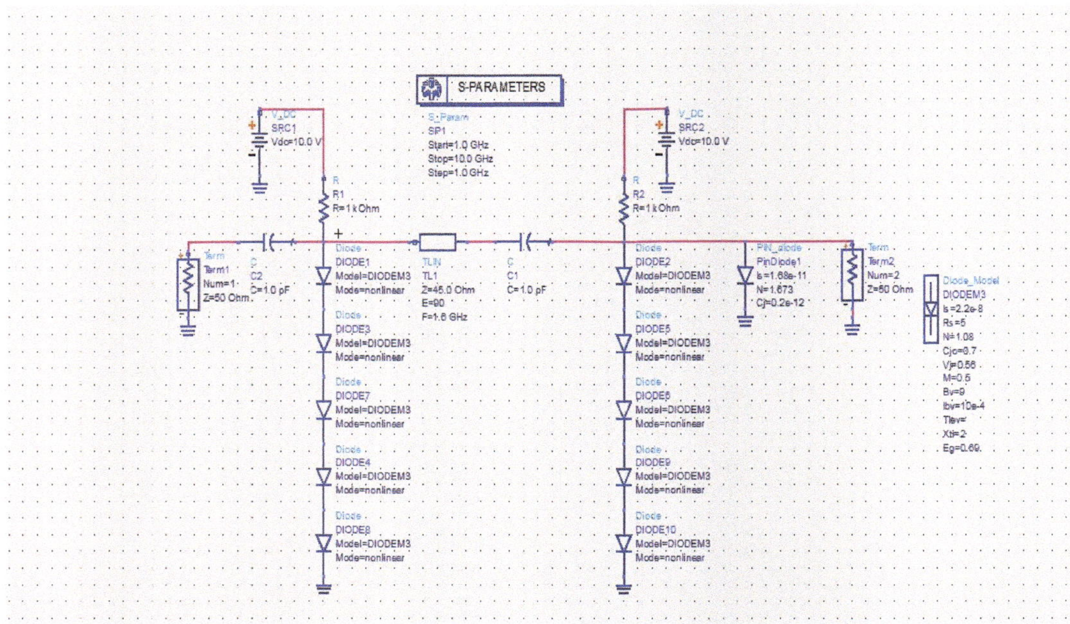

Fig. 7: ADS simulation screenshot for the simulation of five diodes in series

The corresponding graphs of the simulated circuits are shown below and it can be seen that the gain of the circuits is increasing as the number of Schottky diodes in series connection are being increased.

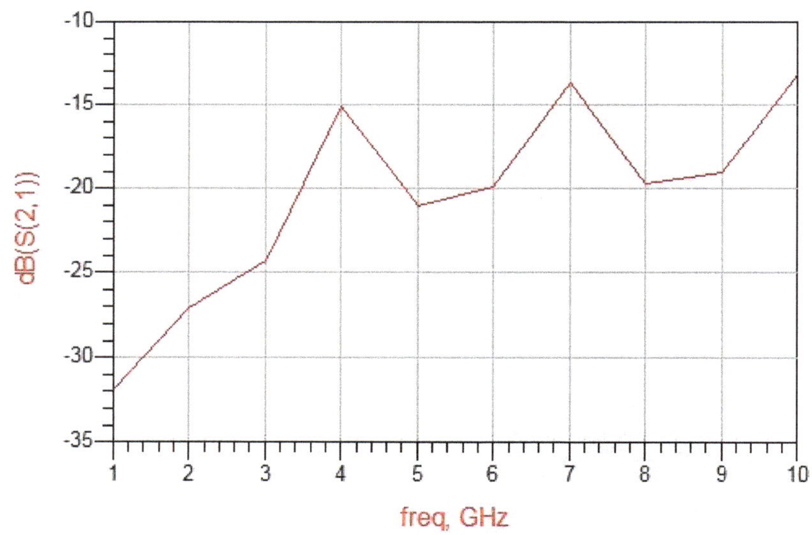

Fig. 8: | S_{21} | at 1.6 GHz of circuit shown in Fig. 2

Here, while using 3 Schottky diodes in series, at frequency 4GHz the corresponding gain is approximately -15dB.

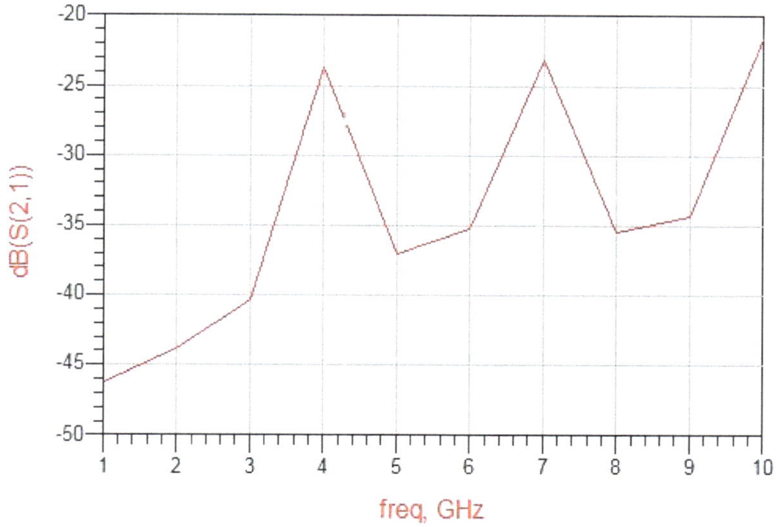

Fig. 9: $|S_{21}|$ at 1.6 GHz of circuit using one Schottky diode

Here, while using one Schottky diode, at frequency 4GHz the corresponding gain is approximately -24dB.

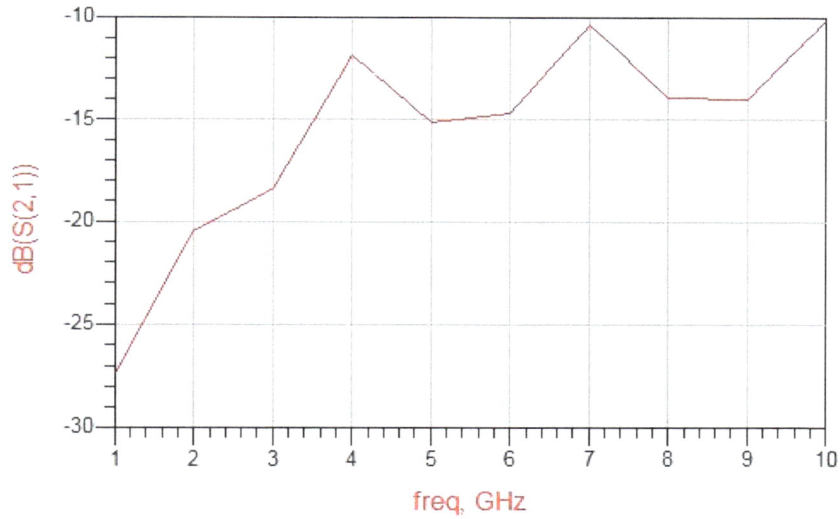

Fig. 10: $|S_{21}|$ at 1.6 GHz of circuit using 5 Schottky diodes

Here, while using five Schottky diodes in series, at frequency 4GHz the corresponding gain is approximately -12dB.

6. CONCLUSIONS

This paper presented the increase in overall gain of Schottky diode and PIN diode based pre-distortion linearizers operating at microwave frequencies. The results of the simulation demonstrate that the gain of the linearizer increases as we increase the number of diodes in series connection. Therefore, a high gain linearizer can be implemented using our proposed topology

5. ACKNOWLEDGEMENT

The authors would like to thank the training coordinator Dr. P.Y.Nabhiraj of VECC and HOD of the Department of Applied Electronics and Instrumentation Engineering of Netaji Subhash Engineering College for allowing us to explore this interesting area.

REFERENCES

[1] M. S. Hashmi, Z. S. Rogojan, and F. M. Ghannouchi, "A flexible dual-inflection point RF pre-distortion linearizer for microwave power amplifiers", *Progress In Electromagnetics Research C*, Vol. 13, 1-18, 2010

[2] M. S. Hashmi, Z. S. Rogojan, S. R. Nazifi and F. M. Ghannouchi, "A broadband dual-inflection point RF pre-distortion linearizer using backward reflection topology", *Progress In Electromagnetics Research C*, Vol. 13, 121-134, 2010

[3] S. C. Bera, R. V. Singh and V. K. Garg, "Temperature behavior and compensation of Schottky barrier diode", *International Journal of Electronics,* Vol. 95, No. 5, May 2008, 457-465

[4] S. C. Bera, Virender Kumar, Surinder Singh and Dipak Kumar, "Temperature behavior and compensation of diode based pre-distortion linearizer", *IEEE MICROWAVE AND WIRELESS COMPONENTS LETTERS, VOL. 23, NO. 4, APRIL 2013*

[5] Yamauchi, K., K. Mori, M. Nakayama, Y. Mitsui, and T. Takagi, "A microwave miniaturized linearizer using a parallel diode with a bias feed resistance," *IEEE Trans. on Microwave Theory Tech.*, Vol. 45, No. 12, 2431 - 2435, December 1997.

[6] F. M. Ghannouchi and J. S. Cardinal, "A new adaptive double envelope feedback linearizer for mobile radio power amplifiers", *IEEE MTT-S Int. Microwave Symp. Digest*, 573 - 576, San Diego, May 1994.

[7] K. Yamauchi, K. Mori, M. Nakayama, Y. Mitsui, T. Takagi, "A microwave miniaturized linearizer using a parallel diode", *IEEE MTT-S Int. Digest*, Denver, CO, pp. 1199 - 1202, June 1997.

[8] Katz A, Gray R, and Dorval R. "Wide/multiband linearization of TWTAs using predistortion", *IEEE Transactions on Electron Devices*, 2009, 56(5): 959 - 964.

[9] S. C. Bera, P. S. Bhardhwaj, R. V. Singh, and V. K. Garg, "A diode linearizer for microwave power amplifiers", *Microw. J., vol. 46*, no. 11, pp. 102–113, Nov. 2003.

[10] Jacob Millman and Christos C. Halkias, "Integrated Electronics", McGRAW – HILL Publications

Permissions

All chapters in this book were first published in IJWMN, by AIRCC Publishing Corporation; hereby published with permission under the Creative Commons Attribution License or equivalent. Every chapter published in this book has been scrutinized by our experts. Their significance has been extensively debated. The topics covered herein carry significant findings which will fuel the growth of the discipline. They may even be implemented as practical applications or may be referred to as a beginning point for another development.

The contributors of this book come from diverse backgrounds, making this book a truly international effort. This book will bring forth new frontiers with its revolutionizing research information and detailed analysis of the nascent developments around the world.

We would like to thank all the contributing authors for lending their expertise to make the book truly unique. They have played a crucial role in the development of this book. Without their invaluable contributions this book wouldn't have been possible. They have made vital efforts to compile up to date information on the varied aspects of this subject to make this book a valuable addition to the collection of many professionals and students.

This book was conceptualized with the vision of imparting up-to-date information and advanced data in this field. To ensure the same, a matchless editorial board was set up. Every individual on the board went through rigorous rounds of assessment to prove their worth. After which they invested a large part of their time researching and compiling the most relevant data for our readers.

The editorial board has been involved in producing this book since its inception. They have spent rigorous hours researching and exploring the diverse topics which have resulted in the successful publishing of this book. They have passed on their knowledge of decades through this book. To expedite this challenging task, the publisher supported the team at every step. A small team of assistant editors was also appointed to further simplify the editing procedure and attain best results for the readers.

Apart from the editorial board, the designing team has also invested a significant amount of their time in understanding the subject and creating the most relevant covers. They scrutinized every image to scout for the most suitable representation of the subject and create an appropriate cover for the book.

The publishing team has been an ardent support to the editorial, designing and production team. Their endless efforts to recruit the best for this project, has resulted in the accomplishment of this book. They are a veteran in the field of academics and their pool of knowledge is as vast as their experience in printing. Their expertise and guidance has proved useful at every step. Their uncompromising quality standards have made this book an exceptional effort. Their encouragement from time to time has been an inspiration for everyone.

The publisher and the editorial board hope that this book will prove to be a valuable piece of knowledge for researchers, students, practitioners and scholars across the globe.

List of Contributors

Moez Hizem
6'Tel Research Unit, Higher School of Communications of Tunis, Sup'Com, Tunisia

Ridha Bouallegue
Sup'Com, Tunisia

B. S. Premananda
Department of Telecommunication Engineering, R.V. College of Engineering, Bangalore, India

Dr. B. V. Uma
Department of Electronics and Communication Engineering, R.V. College of Engineering, Bangalore, India

Hussein Al-Bahadili
Faculty of Information Technology, Petra University, Jordan

Ghassan Issa
Faculty of Information Technology, Petra University, Jordan

Alia Sabri
Faculty of Information Technology, Applied Science University, Jordan

Amir Esmailpour
Department of Electrical and Computer Engineering, University of New Haven, West Haven, USA

Veljko Knezevic
Department of Computer Networks, Ryerson University, Toronto, Canada

Maria Gracias
Department of Computer Networks, Ryerson University, Toronto, Canada

Gelareh Kokabian
Department of Electrical and Computer Engineering, University of New Haven, West Haven, USA

A.Shastri
Prof & Vice-Chancellor, Banasthali University, Banasthali, India

R. Dadhich
Associate Prof. & Head Department of MCA, Govt. College of Engineering, Ajmer, India

Ramesh C. Poonia
Research Scholar, Apaji Institute of Mathematics & Applied Computer Technology, Banasthali University, Banasthali, India

Rishipal Singh
Department of CSE, Guru Jambheshwar University of Science and Technology, Hisar, India
School of Computer and Systems Sciences, Jawaharlal Nehru University, Delhi, India

D.K.Lobiyal
School of Computer and Systems Sciences, Jawaharlal Nehru University, Delhi, India

M.N. SreeRangaRaju
Department of Telecommunication Engg Bangalore Institute of Technology Bangalore, India

Dr. Jitendranath Mungara
CMR Institute of Tech & Mgt Bangalore, India

Lalli
Department of Computer Science, Bharathidasan University, Trichy

Palanisamy
Department of Computer Science & Engineering, Alagappa University, Karaikudi

N. Gomathi
JNTU, Hyderabad

Dr. P. Seethalakshmi
Anna University, Triuchirappalli

Dr. A.Govardhan
JNTUH College of Engg., Hyderabad

L. Nithyanandan
Department of Electronics and Communication Engineering, Pondicherry Engineering College, Puducherry, India

J. Susila Sandhiya
Department of Electronics and Communication Engineering, Pondicherry Engineering College, Puducherry, India

D.D.Seth
School of Electronics, KIIT University, Bhubaneswar, Odisha, India

Srikanta Patnaik
Interscience Institute of Technology & Management, Bhubaneswar, Odisha, India

Srikanta Pal
Department of Electronics and Communication Engineering, Birla Institute of Technology, Mesra, Ranchi, India

Hasan Al-Refai
Department of Computer Information Systems, Philadelphia University, Amman Jordan

Ali Alawneh
Department of Management Information Systems, Philadelphia University, Amman Jordan

Khaldoun batiha
Dean of IT, Philadelphia University, Amman Jordan

Oussama Ghorbel
National Engineers School of Sfax, Sfax University, Tunisia. CES Research Unit

Walid Ayedi
National Engineers School of Sfax, Sfax University, Tunisia. CES Research Unit

Mohamed Wasim Jmal
National Engineers School of Sfax, Sfax University, Tunisia. CES Research Unit

Mohamed Abid
National Engineers School of Sfax, Sfax University, Tunisia. CES Research Unit

Oliver Mitch Maguitte
School of Innovative Technologies and Engineering, University of Technology, Mauritius

Mohammad Sameer Sunhaloo
School of Innovative Technologies and Engineering, University of Technology, Mauritius

Ben Oodit
School of Innovative Technologies and Engineering, University of Technology, Mauritius

Vinaye Armoogum
School of Innovative Technologies and Engineering, University of Technology, Mauritius

Yongxin Wang
FEIT, University of Technology, Sydney, Australia

Kumbesan Sandrasegaran
FEIT, University of Technology, Sydney, Australia

Xinning Zhu
Beijing University of Post and Telecommunications, Beijing, China

Jingjing Fei
CSE, University of New South Wales, Sydney, Australia

Xiaoying Kong
FEIT, University of Technology, Sydney, Australia

Cheng-Chung Lin
FEIT, University of Technology, Sydney, Australia

Deepak Dembla
Associate Professor, Deptt. of CSE, AIET, Jaipur INDIA

Yogesh Chaba
Associate Professor-Deptt. of CSE GJU of Science & Technology, Hisar INDIA

Nan Gao
Graduate School of Global Information and Telecommunication Studies Tokyo, Japan

Shigeru Shimamoto
Graduate School of Global Information and Telecommunication Studies Tokyo, Japan

Niharika Gorre
Purdue School of Engineering and Technology, Indianapolis, Indiana, USA

Prudhvi Raj Kyatham
Purdue School of Engineering and Technology, Indianapolis, Indiana, USA

Mohamed El-Sharkawy
Purdue School of Engineering and Technology, Indianapolis, Indiana, USA

Anuraag Misra
Accelerator Physics Group, Variable Energy Cyclotron Centre, Kolkata, India

Archan Sarkar
Department of Applied Electronics and Instrumentation Engineering, Netaji Subhash Engineering College, Kolkata, India

Bhaswar Dutta Gupta
Department of Applied Electronics and Instrumentation Engineering, Netaji Subhash Engineering College, Kolkata, India